DICTIONARY OF CONCEPTS IN PHYSICAL GEOGRAPHY

Recent Titles in
Reference Sources for the Social Sciences and Humanities
Series Editor: Raymond G. McInnis

Research Guide for Psychology
Raymond G. McInnis

Dictionary of Concepts in Human Geography
Robert P. Larkin and Gary L. Peters

Dictionary of Concepts in History
Harry Ritter

A Research Guide to the Health Sciences:
Medical, Nutritional, and Environmental
Kathleen J. Haselbauer

DICTIONARY OF CONCEPTS IN PHYSICAL GEOGRAPHY

Thomas P. Huber,
Robert P. Larkin,
and Gary L. Peters

GB
10
.H82
1988
WEST

Reference Sources for the Social Sciences and Humanities, Number 5

Greenwood Press
New York • Westport, Connecticut • London

Library of Congress Cataloging-in-Publication Data

Huber, Thomas Patrick.
  Dictionary of concepts in physical geography.

  (Reference sources for the social sciences and humanities, ISSN 0730-3335 ; no. 5)
  Includes bibliographies and index.
  1. Physical geography—Dictionaries. I. Larkin, Robert P. II. Peters, Gary L. III. Title. IV. Series.
  GB10.H82   1988   910′.02′0321   87-29582
  ISBN 0-313-25369-2

British Library Cataloguing in Publication Data is available.

Copyright © 1988 by Thomas P. Huber, Robert P. Larkin, and Gary L. Peters

All rights reserved. No portion of this book may be reproduced, by any process or technique, without the express written consent of the publisher.

Library of Congress Catalog Card Number: 87-29582
ISBN: 0-313-25369-2
ISSN: 0730-3335

First published in 1988

Greenwood Press, Inc.
88 Post Road West, Westport, Connecticut 06881

Printed in the United States of America

The paper used in this book complies with the Permanent Paper Standard issued by the National Information Standards Organization (Z39.48-1984).

10 9 8 7 6 5 4 3 2 1

Contents

Series Foreword vii

Preface ix

The Dictionary 1

Appendix: Outline of Concepts 279

Index 285

Series Foreword

In all disciplines, scholars seek to understand and explain the subject matter in their area of specialization. The object of their activity is to produce a body of knowledge about specific fields of inquiry. As they achieve an understanding of their subject, scholars publish the results of their interpretations (that is, their research findings) in the form of explanations.

Explanation, then, can be said to organize and communicate understanding. When reduced to agreed-upon theoretical principles, the explanations that emerge from this process of organizing understanding are called concepts.

Concepts serve many functions. They help us identify topics we think about, help classify these topics into related sets, relate them to specific times and places, and provide us with definitions. Without concepts, someone has said, "man could hardly be said to think."

Like knowledge itself, the meanings of concepts are fluid. From the moment an authority introduces a concept into a discipline's vocabulary, where it is given a specific meaning, that concept has the potential to acquire a variety of meanings. As new understandings develop in the discipline, inevitably the meanings of concepts are revised.

Although this pattern in the formation of the meaning of concepts is widely recognized, few dictionaries—certainly none in a consistent manner—trace the path a concept takes as it becomes embedded in a research topic's literature.

Dictionaries in this series uniformly present brief, substantive discussions of the etymological development and contemporary use of the significant concepts in a discipline or subdiscipline. Another feature that distinguishes these dictionaries from others in the field is their emphasis upon bibliographic information.

Volumes contain about 100 entries. Consistently, entries comprise four parts. In the first part, brief statements give the current meaning of a concept. Next, discursive paragraphs trace a concept's historical origins and connotative development. In part three, sources mentioned in part two are cited, and where appropriate, additional notes briefly highlight other aspects of individual references. Finally, in part four, sources of additional information (that is, extensive reviews, encyclopedia articles, and so forth) are indicated.

Thus with these volumes, whatever the level of their need, students can explore the range of meanings of a discipline's concept.

For some, it is the most fundamental need. What is the current meaning of Concept X? Of Concept Y? For others with more intensive needs, entries are departure points for more detailed investigation.

These concept dictionaries, then, fill a long-standing need. They make more accessible the extensive, often scattered literature necessary to knowing a discipline. To have helped in their development and production is very rewarding.

Raymond G. McInnis

Preface

This volume provides definitions and historical perspective for specialized and technical concepts used in current research and writing in the field of physical geography. It is of particular value to serious students and professionals in geography and related fields. The book will take research students far into the discipline of geography and will place individual concepts within the proper historical research context.

For each concept included in the book, we provide definitions, arranged generally from the earliest to the most recent. We then describe the historical growth of each term, using the original research works for references when possible. This historical perspective will help researchers understand how a concept has evolved over time, how it is currently used, and what the general direction of research might be in the future. Each entry includes a list of the prominent, mostly English-language, references concerning a concept; the entry ends with a helpful list of other sources of information. These sources include but are not limited to texts, literature reviews, important articles, and bibliographies. Often, concepts are connected in some way or are used in conjunction with other terms. If a concept is used in the description of another concept, it is set in SMALL CAPITAL LETTERS for ease in cross-referencing. The appendix indicates how the concepts fit into a framework grouping the entries into appropriate broad categories.

Researchers in geography, professionals in related fields, and librarians will find this reference volume very useful. Although we could not include all of the thousands of concepts in physical geography, we have made a significant effort to choose those that are deemed most important by our colleagues. We have received considerable help and advice, but the responsibility for errors or shortcomings in the book lies strictly with the authors.

We would like to express our appreciation to Carole Huber for her many insightful comments and criticisms and to Elaine Schantz for typing the manuscript.

THE DICTIONARY

A

ACTUAL EVAPOTRANSPIRATION. See EVAPOTRANSPIRATION.

ADIABATIC HEATING AND COOLING. Temperature changes in an air mass that result from expansion or contraction of that air mass as it rises or descends, respectively.

The term *adiabatic* in physics refers to a situation in which heat neither enters nor leaves a substance. It was introduced into the physics literature during the latter half of the nineteenth century and was mentioned by Siemens (1882), although its application to atmospheric processes came somewhat later.

With respect to the atmosphere, the concern is with the temperature change that occurs in a parcel of air that is either rising or descending. Without any additional heat being gained or lost from outside, the internal temperature of the parcel of air will change as a result of the change in the air pressure itself. As the parcel of air rises, for example, it expands, and as it expands, it cools. Conversely, if a parcel of air descends, the air pressure within the parcel increases, because the air is being compressed. This compression then increases the internal temperature of the air mass. As R. G. Barry and R. J. Chorley noted, "When a parcel of air moves vertically the changes that take place often follow an adiabatic pattern because air is fundamentally a poor thermal conductor, and the air parcel as a whole tends to retain its own thermal identity which distinguishes it from the surrounding air masses" (1970: 64).

Herbert Spiegel and Arnold Gruber helped clarify what is going on with respect to temperature changes and adiabatic processes as follows:

> As the parcel rises, it encounters lower pressures, which causes it to expand. The air parcel, in order to expand, pushes on its surroundings and, therefore, does work on its environment. Since work is a form of energy and there are no external

sources of this energy for the air parcel to use, it uses its own heat energy. By using its own energy to do this work, the temperature of the air parcel drops as it expands. (1983: 46)

Adiabatic changes in air masses as they rise or descend are measured by two types of LAPSE RATE. One is the dry adiabatic lapse rate and the other is the wet adiabatic lapse rate.

References

Barry, R. G., and Chorley, R. J. 1970. *Atmosphere, Weather, and Climate*. New York: Holt, Rinehart and Winston. Remains an excellent source of information for atmospheric processes.

Siemens. 1882. *Nature* 25. Cited in Murray, James A. H., et al., eds. (1933), *The Oxford English Dictionary*. Oxford: Clarendon Press. Complete citation not given.

Spiegel, Herbert J., and Gruber, Arnold. 1983. *From Weather Vanes to Satellites: An Introduction to Meteorology*. New York: John Wiley and Sons. One of the best places to start when seeking information on meteorology, this book is straightforward and readable.

Sources of Additional Information

Adiabatic processes and adiabatic lapse rates are discussed in most texts in physical geography, meteorology, and climatology, as well as in more esoteric discussions of atmospheric physics. Typical discussions can be found in John Oliver, *Physical Geography: Principles and Applications* (North Scituate, Mass.: Duxbury Press, 1979); James S. Gardner, *Physical Geography* (New York: Harper's College Press, 1977); and Arthur N. Strahler and Alan H. Strahler, *Elements of Physical Geography*, 3d ed. (New York: John Wiley and Sons, 1983). Aside from meteorological texts mentioned already, a typical discussion of adiabatic processes can also be found in Joe R. Eagleman, *Meteorology: The Atmosphere in Action* (New York: D. Van Nostrand Co., 1983). Those interested in pursuing adiabatic processes within the more technical realm of cloud and atmospheric physics might consult R. A. Craig, *The Upper Atmosphere: Meteorology and Physics* (New York: Academic Press, 1965).

AIR MASS. An extensive body of air occurring within the lower atmosphere and over a horizontal plane that exhibits relatively uniform properties of temperature, moisture, and lapse rate.

The concept of an air mass has been and continues to be useful for describing weather phenomena. P. R. Crowe related the regional air mass to other regional concepts and put it into perspective:

> Like all other regional entities, air masses are concepts of reality, valid only within the criteria employed to delimit them. They can be illustrated only by selected samples; they cannot be portrayed in full. Indeed, as compared with other regional concepts, geomorphological, botanical, economic or social, they offer two special difficulties: (a) they are much more subject to change and transformation over comparatively brief periods of time . . . ; (b) the most exciting events occur at their margins where they are thrown into relationships, the one with the other. (1971: 257–258)

The use of the term FRONTAL SYSTEM implies that air masses of differing composition will come into contact with each other. The idea of a clash of these different air masses dates from the work of H. W. Dove (1862) in his *Law of Storms*. J. Bjerknes and H. Solberg (1922), however, clearly established the terms *tropical* and *polar* when describing fronts that separate air masses moving from these respective source regions and the ambient air masses already in place. This work and subsequent work by Bjerknes (e.g., 1930) make clear the link between air masses and fronts.

The earliest general investigation into the properties of air masses is credited to A. R. Goldie (1923). The formal beginnings of air-mass classification and characterization came from T. Bergeron (1928, 1930), whose work stimulated research activity, especially within the next half-decade (Moese and Schinze, 1929; Schinze, 1932a, 1932b; Willett, 1933). This early work was all critical to the general acceptance of the air-mass theory. Bergeron devised two procedures for classifying air masses. The first uses the most conservative properties of the air, which reflect the source region and subsequent modification. The second method uses the most variable characteristics of the air mass and is the procedure most accepted in more recent work (Barry and Perry, 1973: 178–179). This classification framework is well expressed in R. G. Barry and A. H. Perry:

> This classification has the following bases: A primary division relating to the geographical source area—arctic (A), polar (P), tropical (T). A secondary division designating the maritime (m) or continental (c) nature of the source region. A tertiary subdivision indicating the relative temperature and stability conditions. k denotes air moving toward a warmer surface and w toward a colder surface.
>
> The major types are:
>
> *Continental Maritime*
>
> | | | |
> |---|---|---|
> | Arctic | cA | mA |
> | Polar | cP | mP |
> | Tropical | cT | mT |
>
> mA and mP represent modifications of cA and cP air, respectively, due to passage over ocean areas. cA and cP air masses originate in polar anticyclones . . . while cT and mT originate in the subtropical anticyclone cells. (1973: 179–180)

The term *polar* may have misleading connotations. It is well established in the literature, but A. A. Borisov (1965) and R. W. James (1970) preferred *temperate* and *subpolar*, respectively.

The above classification system is not the only one in use. Several others have been prepared. For example, A. N. Strahler (1951) introduced a classification system based on global models. His general system uses the Bergeron nomenclature for air masses and differentiates three main groups of climates: those dominated by equatorial and tropical air masses, those dominated by polar air masses, and those dominated by the interaction of the first two (i.e., mid-latitude). In 1952 J. E. Belasco (1952) distinguished 23 types of "air" over Britain. P. R. Crowe (1971: 339–340) generalized these types into eight categories based on source-region location and other air-mass specifics. Another method of air-mass classification is empirical classification based solely on a long series of recorded observations. Baur (see Bluthgen, 1964) created such a system for Europe. What one finds in such an exercise is almost unending variety and a difficult classification scheme to use.

In addition to classifying procedures and types, research has also concerned air-mass homogeneity (James, 1969) and modification (e.g., Burke, 1945; Craddock, 1951; Manobe, 1957; Gillooly and Lutjeharms, 1984). Much recent research has concentrated on relating worldwide weather patterns to tropical air masses in general and the Southern Oscillation in particular (e.g., Horel and Wallace, 1981; Zebiak, 1982).

References

Barry, R.G., and Perry, A. H. 1973. *Synoptic Climatology: Methods and Applications.* London: Methuen & Co. An excellent, comprehensive work on climate-related topics.

Belasco, J. E. 1952. "Characteristics of Air Masses over the British Isles." *Geophysical Memorandum* 87, M.O. 530b. A good example of localized air-mass studies.

Bergeron, T. 1928. "Uber die Dreidimensional Varknupfende Wetteranalyse." *Geofys. Publik.* (Oslo) 30:1–111.

———. 1930. "Richtlinien einer Dynamischen Klimatologie." *Met. Zeit.* 47: 246–262. These works by Bergeron are the original, classic air-mass studies.

Bjerknes, J., and Solberg, H. 1922. "Life Cycle of Cyclones and the Polar Front Theory of Atmospheric Circulation." *Geofys. Publik.* (Oslo) 3: 1–18. The original frontal theory work.

Bjerknes, J. 1930. "Practical Examples of Polar Front Analysis over the British Isles (1925–6)." *Geophysical Memo.* 50, M.O. 307j.

Bluthgen, J. 1964. *Allgemeine Klimogeographie.* Berlin: De Gruyter.

Borisov, A. A. 1965. *Climates of the USSR.* 2d ed. (trans. R. A. Ledwood; ed. C. A. Halstead). Edinburgh: Oliver and Boyd.

Burke, C. J. 1945. "Modification of Continental Polar Air over a Water Surface." *Journal of Meteorology* 2: 94–112.

Craddock, J. M. 1951. "The Warming of Arctic Air Masses over the Eastern North Atlantic." *Quarterly Journal of the Royal Meteorological Society* 77: 351–364.

Crowe, P. R. 1971. *Concepts in Climatology.* New York: St. Martin's Press. An excellent reference for most aspects of climatology with particular emphasis on the historical development of the science.

Dove, H. W. 1862. *The Laws of Storms Considered in Connexion with the Ordinary Movements of the Atmosphere.* 2d ed. London: Longmans, Green and Co. The conflict between air masses was first hypothesized in this work.

Gillooly, J. F., and Lutjeharms, J.R.E. 1984. "The Ocean and Climate: Large-Scale Ocean-Atmosphere Interactions in the Southern Hemisphere." *South African Journal of Science* 80: 36–40.

Goldie, A. R. 1923. "Circumstances Determining the Distribution of Temperature in the Upper Air under Conditions of High and Low Barometric Pressure." *Quarterly Journal of the Royal Meteorological Society* 49: 6–16. An early work on the properties of air masses.

Horel, J. D., and Wallace, J. M. 1981. "Planetary-Scale Atmospheric Phenomena Associated with the Southern Oscillation." *Monthly Weather Review* 109: 813–829.
James, R. W. 1969. "Elementary Air Mass Analysis." *Meteorologische Rundshaur* 23: 75–79.
———. 1970. "Air Mass Climatology." *Meteorologische Rundshaur*, 23: 65–70.
Manobe, S. 1957. "On the Modification of Air Mass over the Japan Sea When the Outburst of Cold Air Predominates." *Journal of the Meteorological Society of Japan* 2: 311–325.
Moese, O., and Schinze, G. 1929. "Zur Analyze von Neubildungen." *Ann. Hydrogr. Marit. Met.* 57: 76–81.
Schinze, G. 1932a. "Die Erkennung der Tropospharische Luftmassen aus ihren Einzelfeldern." *Met. Zeit.* 49: 169–179.
———. 1932b. "Tropospharische Luftmassen und Vertikoler Temperaturgradient." *Beitr. Physik. für Atmos.* 19: 79–90.
Strahler, A. N. 1951. *Physical Geography.* 1st ed. New York: John Wiley and Sons.
Willett, H. C. 1933. "American Air Mass Properties." *Pap. Phys. Oceanog. Met.* 2: 1–116.
Zebiak, S. E. 1982. "A Simple Atmospheric Model of Relevance to El Nino." *Journal of the Atmospheric Sciences* 39: 2017–2027.

Sources of Additional Information

Themes related to the global/ocean system affecting air-mass characteristics are included in K. S. Bryan, S. Manabe, and R. C. Pacanowski, "A Global Ocean-Atmosphere Climate Model: II The Ocean Circulation," *Journal of Physical Oceanography* 5 (1975): 30–46; J. T. Houghton, ed., *The Global Climate* (Cambridge: Cambridge University Press, 1984); S. Manabe, K. S. Bryan, and M. J. Spelman, "A Global Ocean-Atmosphere Climate Model with Seasonal Variation for Future Studies of Climate Sensitivity," *Dynamics of Atmospheres and Oceans* 3 (1979): 393–426; and World Meteorological Organization, "Global Atmospheric Research Programme," *WMO Bulletin* 25 (1976): 32–38. For other work on climate modeling, especially as it pertains to air masses, see D. H. Lenchow, "The Air Mass Transformation Experiment (AMTEX)," *Bulletin of the American Meteorological Society* 53 (1976): 353–357, and W. D. Sellers, "A New Global Climate Model," *Journal of Applied Meteorology* 12 (1973): 241–254. For a good view of how air-mass/climate studies are changing see J. E. Thornes, "A Paradigmatic Shift in Atmospheric Studies?" *Progress in Physical Geography* 5 (1981): 429–440. General sources of information on air masses include F. A. Berry and E. Bollay, *Handbook of Meteorology* (New York: McGraw-Hill Book Co., 1973), and R. W. Fairbridge, *The Encyclopedia of Atmospheric Sciences and Astrogeology*, vol. 2 (New York: Reinhold Publishing Co., 1967).

ALLUVIAL FAN. A cone-shaped body of alluvium that radiates downslope from where the stream experiences an abrupt change in gradient that causes it to rapidly lose its carrying capacity, typically, as it flows down a mountainous area onto a flatter surface.

Although such features had been described perhaps 50 years earlier, the term *alluvial fan* was first employed by Frederick Drew (1873), who studied these features in the Himalaya Mountains. Although this work was done in a humid environment, alluvial fans would gradually be recognized as features that occur primarily in the world's drier environments.

Much of the early research on alluvial fans was done under the auspices of the U.S. Geological Survey. Studies of G. K. Gilbert (1875) and W. J. McGee (1897) are excellent examples of this early work. By the turn of the century alluvial fans were being recognized as typical features of desert landscapes, as was apparent in the work of William Davis (1905). Similar views were expressed by K. Bryan (1922) and D. Johnson (1932), and detailed work on the origins of alluvial fans, along with entrenchment on parts of fans, was presented by Rollin Eckis (1928). Erich Blissenbach (1954) provided an excellent and detailed summary of the work that had been accomplished on alluvial fans as of that date, including some of his own earlier work on interrelationships between stream characteristics and the characteristics of alluvial fan deposits.

Since 1960 a number of new research articles on alluvial fans have appeared, including a thorough analysis of clastic sediments in alluvial fans by L. K. Lustig (1965), a geographic interpretation of alluvial fans in the western United States by C. B. Beaty (1963), consideration of steady-state or equilibrium relationships in alluvial fans by R. L. Hooke (1967), a discussion of alluvial fan environments by W. B. Bull (1977), and a notable geographic interpretation of certain characteristics of alluvial fans by R. J. Wasson (1974). The definitive book on alluvial fans by Andrzj Rachocki (1981) still represents the state of the art with respect to these landforms and provides a much more detailed look at the evolution of our knowledge about those features than can be presented here.

References

Beaty, C. B. 1963. "Origin of Alluvial Fans, White Mountains, California, and Nevada." *Annals of the Association of American Geographers* 53: 516–535. One of the major geographical studies of alluvial fans.

Blissenbach, E. 1954. "Geology of Alluvial Fans in Semi-arid Regions." *Geological Society of America Bulletin* 65: 175–190. An extremely important general source of information on alluvial fans.

Bryan, K. 1922. "Erosion and Sedimentation in the Papago Country." U.S. Geological Survey Professional Paper, no. 730. Washington, D.C.: U.S. Geological Survey. One of several early classic studies of alluvial fans and depositions that were sponsored by the U.S. Geological Survey.

Bull, W. B. 1977. "The Alluvial Fan Environment." *Progress in Physical Geography* 1: 222–270. A useful summary of geographical works on alluvial fans and their environments.

Davis, William M. 1905. "The Geographical Cycle in an Arid Climate." *Journal of Geology* 13: 381–407. An early and definitive statement on the erosion cycle in arid regions.

Drew, Frederick. 1873. "Alluvial and Lacustrine Deposits and Glacial Records of the Upper Indus Basin." *Quarterly Journal of the Geological Society* (London) 29: 441–471.

Eckis, R. 1928. "Alluvial Fans of the Cucamongo District, Southern Arizona." *Journal of Geology* 36: 224–247. An important early study of alluvial fans in the United States.

Gilbert, G. K. 1875. "Report on the Geology of Portions of Nevada, Utah, California, and Arizona." *Report upon Geographical and Geological Explorations and Surveys West of the 100th Meridian* 3: 21–187. This and other of Gilbert's works remain essential foundations upon which modern geologists have built.

Hooke, R. L. 1967. "Steady-State Relationships on Arid-Region Alluvial Fans in Closed Basins." *American Journal of Science* 223: 609–629.

Johnson, D. 1932. "Rock Fans of Arid Regions." *American Journal of Science* 223: 432–439.

Lustig, L. K. 1965. "Clastic Sedimentation in Deep Springs Valley, California." U.S. Geological Survey Professional Paper, no. 352-F. Washington, D.C.: U.S. Geological Survey, pp. 131–190.

McGee, W. J. 1897. "Sheetflood Erosion." *Geological Society of America Bulletin* 8: 87–112. An important contribution to the early geological literature in the United States.

Rachocki, Andrzj. 1981. *Alluvial Fans*. New York: John Wiley and Sons. The ultimate source for those who want to know everything there is to know about alluvial fans.

Wasson, R. J. 1974. "Intersection Point Deposition of Alluvial Fans: An Australian Example." *Annals of the Association of American Geographers* 56: 83–92. Another excellent geographic study of alluvial fans.

Sources of Additional Information

Brief discussions of alluvial fans and related features can be found in the following: Arthur N. Strahler and Alan H. Strahler, *Elements of Physical Geography*, 3d ed. (New York: John Wiley and Sons, 1984); Tom L. McKnight, *Physical Geography: A Landscape Appreciation* (Englewood Cliffs, N.J.: Prentice-Hall, 1984); and Edgar W. Spencer, *Physical Geology* (Reading, Mass.: Addison-Wesley Publishing Co., 1983). A more detailed discussion can be found in H. F. Garner, *The Origin of Landscapes: A Synthesis of Geomorphology* (New York: Oxford University Press, 1974). Alluvial fans, as well as numerous other features characteristics of geomorphic processes in dry environments, are discussed in W. G. McGinnies and B. J. Goldman, eds., *Arid Lands in Perspective* (Tucson: University of Arizona Press).

ALLUVIUM. Earth materials, primarily sands, silts, and clays, that have been deposited by running water.

According to James Murray et al. (1933), the term *alluvium* appeared as early as 1665, when reference was made in *Philosophical Transactions* to "our earth, where Alluviums are made in some places, and the Sea gains upon the Land in others." Its use to describe stream-deposited materials was established by the end of the seventeenth century. In 1830 it was included by Charles Lyell in his early text on geology, and according to L. Dudley Stamp (1961), it was Lyell who first distinguished alluvium from other kinds of deposits, thus leading to its current usage in the geological literature.

Generally, it is now agreed that alluvium does not refer to sediments laid down in lakes or to marine sediments, as noted by Andrew Goudie and his associates (1985) and Rhodes Fairbridge (1968). Alluvium is studied within the broader concept of sedimentation and is characterized by occurrence in layers or strata, where sorting by grain size is commonly apparent. Streams sort their deposited materials out by size as they lose their ability to carry materials, so that the coarsest deposits are dropped first as the stream loses energy. Thus in the upper reaches of streams, sands and even gravels are found in alluvium, whereas in the lower courses such deposits shift more toward silts and clays. According to R. H. Jahns, we also find that "because larger particles are usually transported nearer the channel bed, a vertical section of alluvium from river bed to flood-plain surface will often be characterized by a decrease in particle size from coarser, cross-bedded sands at the base to finer, laminated sands and silts near the surface" (1947: 158). Furthermore, transport by rivers tends to change the shapes of particles by rounding them, so that rounded particles are common signs within alluvial deposits.

Alluvium is common at the earth's surface. Landforms such as ALLUVIAL FANS, FLOODPLAINS, and DELTAS are important landscape features. Furthermore, alluvial soils have long been of human significance. However, not all alluvial soils are good for agriculture, since fertility is determined not only by whether or not a soil is alluvial or of some other origin but by things such as texture, chemical composition, and other factors that are related more to soil-forming processes than to the geologic origin of the soil. Yet it is true that alluvial soils along many of the world's great rivers are fertile and support significant agricultural population densities. This is especially true along rivers like the Yangtze in the People's Republic of China, the Ganges in India, the Nile in the United Arab Republic, and even the Mississippi in the United States. Above current flood stages, alluvial terraces are often used for agricultural purposes also, and anyone interested in viticulture has no trouble finding associations between quality wine-grape production and alluvial terraces, from those along the Napa and Salinas rivers in California to those along the Marne in the Champagne region of France.

References

Fairbridge, Rhodes W., ed. 1968. *The Encyclopedia of Geomorphology*. Encyclopedia of Earth Sciences Series. Vol. 3. New York: Reinhold Book Corp.

Goudie, Andrew, et al. eds. 1985. *The Encyclopaedic Dictionary of Physical Geography*. Oxford: Basil Backsell. The best dictionary of its type to be found anywhere.

James, R. H. 1947. "Geologic Features of the Connecticut Valley, Massachusetts, as Related to Recent Floods." U.S. Geological Survey Water Supply Paper, no. 996. Washington, D.C.: U.S. Geological Survey. An important study of alluvium and depositional processes.

Lyell, Charles. 1830. *Principles of Geology: Being an Attempt to Explain the Former Changes of the Earth's Surface by Reference to Causes now in Operation*. Vol. 1. London: J. Murray. A landmark in the evolution of geologic thought.

Murray, James A. H., et al., eds. 1933. *The Oxford English Dictionary*. Oxford: Clarendon Press.
Stamp, L. Dudley. 1961. *A Glossary of Geographical Terms*. London: Longmans, Green and Co. A classic of its genre.

Sources of Additional Information

Some reference is made to alluvium in most texts on physical geography and geology, as well as in more detailed discussions of sedimentation and sedimentary studies. Among physical geography texts see Arthur N. Strahler and Alan H. Strahler, *Elements of Physical Geography*, 3d ed. (New York: John Wiley and Sons, 1983), or John E. Oliver, *Physical Geography: Principles and Applications* (North Scituate, Mass.: Duxbury Press, 1979). Similar introductions to the term can be found in introductory geology texts, such as Edgar W. Spencer, *Physical Geology* (Reading, Mass.: Addison-Wesley Publishing Co., 1983). Discussions of alluvium in relationship to sedimentation and sedimentary processes can be found in J.R.L. Allen, *Physical Processes of Sedimentation* (London: Allen and Unwin), and H. G. Reading, ed., *Sedimentary Environments and Facies* (Oxford: Blackwell Scientific, 1978). Discussions of alluvial soils can be found in B. T. Bunting, *The Geography of Soil* (Chicago: Aldine Publishing Co., 1965), or R. M. Basile, *A Geography of Soils* (Dubuque, Ia.: W. C. Brown, 1971).

ANTICYCLONE. See ATMOSPHERIC PRESSURE.

AQUIFER. 1. A geological formation or water-bearing rock body with sufficient porosity and permeability to yield economic supplies of groundwater. 2. A rock stratum that will hold and permit the movement of water.

The word *aquifer* is often synonymous with the term *water-bearing formation*. An aquifer, therefore, is either a group of ROCK formations, a single rock formation, or part of a rock formation that is capable of holding and transmitting water. Some earth scientists also include a requirement that the water-bearing formation be capable of yielding appreciable quantities of water to wells or springs. According to O. M. Hackett (1972: 473), "aquifers perform a twofold function: they serve as natural storage reservoirs and as distribution conduits." Aquifers have been divided into six primary types on the basis of rock types (Thomas, 1952): sandstone and conglomerate, crystalline and metamorphic rocks, sand and gravel, limestone and other soluble rocks, basalt and other volcanic rocks, and porous but poorly permeable materials.

Aquifers have been familiar to people for many years. Early aquifer studies are associated with GROUNDWATER investigations and have a long history of interest by both scientists and nonscientists. Earliest interest can be traced to biblical times, and many biblical verses are concerned with water in aquifers. According to O. E. Meinzer, "the twenty-sixth chapter of Genesis . . . reads like a water-supply paper" (1934: 6).

Archaeological and documentary evidence also points to interest in aquifers by the ancient Egyptians and Persians (Tolman, 1937) as well as the later Greeks and Romans (Ward, 1975: 3). A significant step in understanding aquifers was achieved when the Roman architect Vitruvius explained the now-accepted infiltration theory that large amounts of rain percolate through the rock strata and emerge in the form of a stream or springs (Todd, 1959: 3). This infiltration theory was further elaborated in the sixteenth century by Bernard Palissy and in the seventeenth century by Pierre Perrault and Edme Mariotte.

The nineteenth century saw a great many investigations into the dynamics of water movement in aquifers. Much of this early work was done in France. Of particular significance is the work done by H. Darcy (1856) on the movement of water through sand aquifers. In the later part of the nineteenth century many Americans became involved in aquifer research. The writings of famous engineers such as A. Hazen (1893) and F. H. King (1899) as well as the classic contributions on artesian systems by geologists such as T. C. Chamberlin (1885) and N. H. Darton (1896) were significant advances in the study of aquifers.

Since 1900 a great many scientific studies of aquifers have been undertaken. Much of the impetus for these studies came from the need for more information in order to develop new water-supply sources.

Perhaps the most significant advances in the study of aquifers and groundwater were pioneered by O. E. Meinzer of the U.S. Geological Survey during the first half of the twentieth century. According to O. M. Hackett, Meinzer

> recognized that aquifers are functional components of the hydrologic cycle [see HYDROLOGIC CYCLE] and insisted that groundwater should be studied with this fact in mind. As a consequence, he drew together the two traditional approaches to groundwater investigations—that of the geologist, concerned principally with describing the earth medium, and that of the engineer or physicist, concerned principally with fluid mechanics. (1972: 476)

Another important contribution at this time was the mathematical model developed by C. V. Theis (1935) to describe the effect of a well withdrawing groundwater from an aquifer. This was the first study using advanced quantitative methods to understand groundwater hydraulics.

Many of the earlier aquifer studies were more descriptive, and although research of that type is still being conducted, much of the recent work is directed toward the analysis of water as it passes into, through, and out of aquifer systems (Hackett, 1972: 476).

The emphasis on environmental POLLUTION problems that started in the 1960s and 1970s has continued, and much recent work has focused on aquifer pollution problems. Several studies by the U.S. Geological Survey (Hult and Schoenberg, 1984; LeBlanc, 1985; Davis and Matthews, 1983) have dealt with regional pollution issues. Aquifers in coastal regions have also undergone close scrutiny by earth scientists (Gupta, 1985; Aucott and Speiran, 1985).

The modern computer and the use of simulation models have greatly aided aquifer research (Lai, 1986); yet additional methodological research continues in order to understand more fully the workings of aquifer systems.

References

Aucott, W. R., and Speiran, G. K. 1985. "Ground Water Flow in the Coastal Plain Aquifers of South Carolina." *Groundwater*, November-December, pp. 736–745. A good explanation of aquifer locations in reference to topography and rock type.

Chamberlin, T. C. 1885. *Requisite and Qualifying Conditions of Artesian Wells*. U.S. Geological Survey Fifth Annual Report. Washington, D.C.: U.S. Geological Survey. One of the first Americans to study aquifers.

Darcy, H. 1856. *Les Fontaines, Publiques de la Villa de Dijon*. Paris. A classic paper on movement of water through aquifers.

Darton, N. H. 1896. "Artesian Well Prospects on the Atlantic Coastal Plain Region." *U.S. Geological Survey Bulletin*, no. 138. Washington, D.C.: U.S. Geological Survey.

Davis, R. W., and Matthews, Edward W. 1983. "Chloroform Contamination in Part of the Alluvial Aquifer, Southwest Louisville, Kentucky." U.S. Geological Survey Water Supply Paper, no. 2202. Washington, D.C.: U.S. Geological Survey. A good analysis of the extent of pollution in an aquifer and how underground water movement affects pollution.

Gupta, A. Das. 1985. "Approximation of Salt-Water Interface Fluctuation in an Unconfined Coastal Aquifer," *Groundwater*, November-December, pp. 783–794. A highly technical theoretical discussion.

Hackett, O. M. 1972. "Groundwater." In Fairbridge, Rhodes, ed., *The Encyclopedia of Geochemistry and Environmental Sciences*. Encyclopedia Earth Science Series. Vol. 4A. New York: Van Nostrand-Reinhold Co. A good discussion of the basic principles of groundwater and aquifers.

Hazen, A. 1893. *Some Physical Properties of Sands and Gravels*. Massachusetts State Board of Health, 24th Annual Report for 1982. Boston: Massachusetts State Board of Health.

Hult, Marc F., and Schoenberg, Michael E. 1984, "Preliminary Evaluation of Ground-Water Contamination by Coal-Tar Derivatives, St. Louis Park Area, Minnesota." U.S. Geological Survey Water Supply Paper, no. 2211. Washington, D.C.: U.S. Geological Survey. A thorough and easy-to-understand discussion of a pollution problem.

King, F. H. 1899. *Principles and Conditions of Movement of Ground Water*. U.S. Geological Survey, Nineteenth Annual Report, Part II. Washington, D.C.: U.S. Geological Survey.

Lai, Chintu. 1986. "Numerical Modeling of Unsteady Open-Channel Flow." In Ven Te Chow, ed., *Advances in Hydroscience*. Vol. 14. Orlando, Fla.: Academic Press, pp. 103–111. A new book concerned with recent quantitative trends in aquifer study.

LeBlanc, Dennis R. 1985. "Sewage Plume in a Sand and Gravel Aquifer, Cape Cod, Massachusetts." U.S. Geological Survey Water Supply Paper, no. 2218. Washington, D.C.: U.S. Geological Survey. A good discussion, with maps and diagrams, of what happens to various pollutants once they enter an aquifer.

Meinzer, O. E. 1934. "The History and Development of Ground-Water Hydrology." *Journal of the Washington Academy of Sciences* 24: 6–32. One of America's foremost groundwater and aquifer authorities discusses the history of research.

Theis, C. V. 1935. "Relation between the Lowering of the Peizometric Surface and the

Rate and Duration of Discharge of a Well Using Groundwater Storage.'' *American Geophysical Union Transactions* 16: 519–524. A classic paper using quantitative techniques.

Thomas, H. E. 1952. "Ground-Water Regions of the United States—Their Storage Facilities." U.S. Congress, House Interior and Insular Affairs Committee, Physical and Economic Foundation of Natural Resources. Vol. 3.

Todd, David Keith. 1959. *Groundwater Hydrology*. New York: John Wiley and Sons.

Tolman, C. F. 1937. *Groundwater*. New York: McGraw-Hill Book Co.

Ward, R. C. 1975. *Principles of Hydrology*. London: McGraw-Hill Book Co. (UK).

Sources of Additional Information

Short definitions and discussions of aquifers can be found in John Whitton, *Dictionary of Physical Geography* (New York: Penguin Books, 1984), p. 34; Stella Stiegeler, ed., *A Dictionary of Earth Sciences* (New York: Pica Press, 1976), p. 18; Robert W. Durrenberger, *Dictionary of the Environmental Sciences* (Palo Alto, Calif.: National Press Books, 1973), p. 16; and Michael Allaby, *A Dictionary of the Environment* (New York: Van Nostrand-Reinhold Co., 1977), p. 36. Many excellent textbooks contain more detailed discussions of aquifers: L. Manddel, *Groundwater Resources, Investigations, and Development* (New York: Academic Press, 1981); Stanley N. Davis and Roger I. DeWiest, *Hydrogeology* (New York: John Wiley and Sons, 1966); and V. Pye, R. Patrick, and J. Quarles, *Groundwater Contamination in the United States* (Philadelphia: University of Pennsylvania Press, 1983). An excellent source for the latest research results is the journal *Ground Water*, published every other month.

ASTHENOSPHERE. See LITHOSPHERE.

ATMOSPHERIC CIRCULATION. Generalized pattern of wind systems and related pressure systems that exist in the lower atmosphere.

As Richard Pfeffer (1967) has noted, although an interest in winds and their causes can be traced to Aristotle, the first real model of a large-scale circulation was developed by George Hadley, in his work on the causes of the trade winds, a work that was published in 1735. Hadley's name lives on today in meteorology because of his discovery of what we now call the Hadley Cell, an atmospheric convection cell that was hypothesized by Hadley as an explanation for the pattern of the trade winds.

Wind (see GENERAL CIRCULATION) is primarily a horizontal movement of air relative to the earth's surface in response to differences in atmospheric temperature from place to place. Winds always move from higher pressure to lower pressure or down the pressure gradient. The atmospheric circulation system is comprised of major winds systems and pressure belts. Other wind systems exist also, such as LAND AND SEA BREEZES. Furthermore, wind acts as one of the geomorphic agents of erosion; thus it has produced a specific set of related LANDFORMS and effects.

The essence of Hadley's hypothesized cell begins with heating at the equator, which causes equatorial air to rise, with that rising air gradually cooling and spreading out toward the poles and then sinking back to the surface and beginning its flow equatorward, drawn there by the low pressure that resulted from the initial rising air. Originally believed to be a cell that began with rising air at the equator and ended with descending air in the very high latitudes, Hadley's initial ideas have been revised into a more realistic three-cell model by C-G Rossby (1949). The term *Hadley Cell* now is used to refer to the cell in which warm air rises from the equator; spreads out in both directions poleward; then descends in the subtropics, forming the subtropical high-pressure belts; and finally, at the surface, results in the equatorward-moving trade winds.

Returning to the overall atmospheric circulation pattern, what has emerged is the following generalized picture of winds and pressure systems: low pressure in the equatorial zone, high pressure in the subtropics, low pressure in the vicinity of the polar front, and high pressure at the poles, setting in motion the easterly belts of trade winds, the broad middle-latitude belt of westerlies, and the polar easterlies. These broad patterns shift with the seasons and may be interfered with in a variety of ways by local winds that result from topographic variations and land-water contrasts, as well as by larger systems like the Asian monsoon. However, the broad patterns are the central interest here.

Following the work of Hadley in the eighteenth century, it became increasingly apparent that wind patterns were more complicated than he had suggested; however, little that was new appeared in the literature until the twentieth century. Following World War II, considerable interest in the general atmospheric circulation developed not only in the work of Rossby (1949) but also in that of a number of other researchers.

One of the first major overviews of new knowledge of the general circulation pattern was provided by V. P. Starr (1948). P. R. Crowe contributed to a better understanding of one portion of the general circulation system, that of the tropical trade winds. During the 1950s research on the general circulation pattern intensified. B. W. Thompson (1951) provided a careful look at the general circulation over the western Pacific. C. S. Ramage (1952) related the general circulation pattern to normal patterns of weather over southern Asia and the South Pacific. V. P. Starr (1953) described the nature of large-scale eddies in the atmosphere, R. Hide (1953) began the development of experimental designs that helped to understand the nature of atmospheric circulation systems, and H. L. Kuo (1953) studied the development of zonal currents in the atmosphere and ways that they were affected by large-scale disturbances.

V. P. Starr (1954) summarized again some of the current contributions that were being made to the study of the general atmospheric circulation. Although brief, his article made it clear just how much new work was being done in the 1950s. E. N. Lorenz (1955) probed the relationship between the amount of potential energy that was available and the maintenance of the general circulation. R. C. Sutcliffe (1956) looked at how water balances could be related to the general circulation system, an important concern. G. B. Tucker (1962) updated earlier summaries of general circulation research, and it was apparent that a fairly complete picture had emerged, the outlines of which today are still discussed in most texts. Another useful summary, especially for geographers, is Crowe (1965). R. G. Barry and R. J. Chorley (1970) provided an overview of the general atmospheric circulation, and similar discussions, though sometimes less detailed, still appear, as in the work of John Oliver and John Hidore (1984).

Hadley identified the underlying cause of the global circulation pattern as differential heating of the earth's surface. Specifically, he looked at the rising air in the equatorial zone as a response to the heating that occurred there. Later, Rossby pointed out that most of the air that rises in the equatorial zone descends not in the polar zone but in the subtropics, forming broad belts of subtropical high pressure in both hemispheres at latitudes of about 25 to 40 degrees, regions that early mariners recognized as the horse latitudes. Because air is descending in these regions, calms are common, and it was these calms that caught the attention of the early mariners. As the story goes, traders engaged in carrying horses from New England to the West Indies would sometimes get caught in these calms and, if they lasted long enough, would run low on water, necessitating that the horses be thrown overboard, hence the "horse latitudes." Specific studies of the trade-wind belt and low-latitude circulation patterns include those of Crowe (1949, 1951) and H. Riehl (1962). A very useful geographic study on tropical circulation in the Pacific is that of Leslie Curry and R. W. Armstrong (1959).

Poleward of the subtropical high-pressure belts we find the westerlies, which cover the latitudes from about 35 to 60 degrees in both hemispheres. These westerly winds result from air moving poleward from the subtropical high-pressure belts into the lower pressures of the polar-front zones. These winds, although they show, statistically, a westerly component, are in a broad belt within which considerable wind variations exist on a daily basis. These wind variations result in large part from the influence of migrating wave cyclones in the westerly belts. Furthermore, in the northern hemisphere the general pattern is broken up by the presence of large land masses. The westerlies have been of considerable interest since the early sailing days, and an excellent account of the westerlies was presented by F. Kenneth Hare (1960). A useful summary of daily disruptions of wind patterns in the westerlies in the northern hemisphere was done by W. H. Klein (1957).

In the high latitudes of the polar and arctic zones we find the polar easterlies, although as Arthur Strahler and Alan Strahler (1984: 84) warned, "The concept is at best greatly oversimplified and is certainly misleading when applied to the northern hemisphere." The polar easterlies that spread out from Antarctica are much more consistent than are those in the northern hemisphere. In both cases, however, the cause of these polar easterlies is a combination of high pressure at the poles and the lower pressure associated with the polar front zone.

References

Barry, R. G., and Chorley, R. J. 1970. *Atmosphere, Weather, and Climate*. New York: Holt, Rinehart and Winston. Remains one of the best books on climate and weather that has been written by geographers.

Crowe, P. R. 1949. "The Trade Wind Circulation of the World." *Transactions of the Institute of British Geographers* 15: 37–56. A thoughtful and detailed look at the trade winds and their geographic patterns.

———. 1951. "Wind and Weather in the Equatorial Zone." *Transactions of the Institute of British Geographers* 17: 23–76. An excellent study of tropical weather patterns.

———. 1965. "The Geographer and the Atmosphere." *Transactions of the Institute of British Geographers* 36: 1–19. Must reading for anyone interested in physical geography, especially climatology.

Curry, Leslie and Armstrong, R. W. 1959. "Atmospheric Circulation of the Tropical Pacific Ocean." *Geografiska Annaler* 41: 245–255. A very useful summary of circulation in that region.

Hare, F. Kenneth. 1960. "The Westerlies." *Geographical Review* 50: 345–367. A classic in the climatology literature.

Hide, R. 1953. "Some Experiments on Thermal Convection in a Rotating Liquid." *Quarterly Journal of the Royal Meteorological Society* 79: 161. One of several early works on the experimental side of circulation studies.

Klein, W. H. 1957. "Principal Tracks and Mean Frequencies of Cyclones and Anticyclones in the Northern Hemisphere." *Research Paper*, no. 40. Washington, D.C.: U.S. Weather Bureau. Remains the major source on storm tracks and frequencies in the United States.

Kuo, H. L. 1953. "On the Production of Mean Zonal Currents in the Atmosphere by Large Disturbances." *Tellus* 5: 475–493.

Lorenz, E. N. 1955. "Available Potential Energy and the Maintenance of the General Circulation." *Tellus* 7: 157–167. An important early study of energy potential as a driving force in atmospheric patterns.

Oliver, John E., and Hidore, John J. 1984. *Climatology: An Introduction*. Columbus, Ohio: Charles E. Merrill Publishing Co. An excellent recent climatology text by two well-known geographers.

Pfeffer, Richard L. 1967. "Atmospheric Circulation—Global." In Rhodes W. Fairbridge, ed., *The Encyclopedia of Atmospheric Sciences and Astrogeology*. Encyclopedia of Earth Sciences Series. Vol. 2. New York: Reinhold Publishing Co., pp. 71–78.

Ramage, C. S. 1952. "Relationships of General Circulation to Normal Weather over Southern Asia and the Western Pacific during the Cool Season." *Journal of Meteorology* 9: 403–408. Useful only as a brief overview.

Riehl, H. 1962. "General Atmospheric Circulation of the Tropics." *Science* 135: 13–22. Good summary of tropical circulation patterns.

Rossby, C-G. 1949. "On the Nature of the General Circulation of the Lower Atmosphere." In G. P. Kuiper, ed., *The Atmosphere of the Earth and Planets*. Chicago: University of Chicago Press, pp. 16–48. Important as an example of Rossby's work.

Starr, V. P. 1948. "An Essay on the General Circulation of the Earth's Atmosphere." *Journal of Meteorology* 5: 39–43.

———. 1953. "Note Concerning the Nature of the Large-Scale Eddies in the Atmosphere." *Tellus* 5: 494–498.

———. 1954. "Commentaries Concerning Research on the General Circulation." *Tellus* 6: 268–272. A brief summary of the work of Starr and others as of the mid-1950s.

Strahler, Arthur N., and Strahler, Alan H. 1987. *Elements of Physical Geography*. 3d ed. New York: John Wiley and Sons. One of the most popular of current physical geography texts.

Sutcliffe, R. C. 1956. "Water Balance and the General Circulation of the Atmosphere." *Quarterly Journal of the Royal Meteorological Society* 82: 385–395. Illustrates the important relationships between circulation patterns and regional water balances.

Thompson, B. W. 1951. "An Essay on the General Circulation over South-East Asia and the West Pacific." *Quarterly Journal of the Royal Meteorological Society* 77: 569–597. A detailed regional survey of circulation patterns.

Tucker, G. B. 1962. "The General Circulation of the Atmosphere." *Weather* 17: 320–340. Concise and useful.

Sources of Additional Information

Basic discussions of the general atmospheric circulation can be found in most introductory texts in physical geography, including the following: Tom L. McKnight, *Physical Geography: A Landscape Appreciation* (Englewood Cliffs, N.J.: Prentice-Hall, 1984), and E. Willard Miller, *Physical Geography: Earth Systems and Human Interactions* (Columbus, Ohio: Charles E. Merrill Publishing Co., 1985). Somewhat better discussions can be found in meteorology and climatology texts, including any of the following: Joe R. Eagleman, *Meteorology: The Atmosphere in Action* (New York: D. Van Nostrand Co., 1980); Paul E. Lydolph, *Weather and Climate* (Totowa, N.J.: Rowman and Allanheld, Publishers, 1985); Glenn T. Trewartha and Lyle H. Horn, *An Introduction to Climate*, 5th ed. (New York: McGraw-Hill Book Co., 1980); and James Hanwell, *Atmospheric Processes* (London: Allen and Unwin, 1980). Even more in-depth treatment of the general circulation system can be found in the following specialized books: J. A. Dutton, *The Ceaseless Wind: An Introduction to the Theory of Atmospheric Motion* (New York: McGraw-Hill Book Co., 1976); E. E. Gossard and W. H. Hooke, *Waves in the Atmosphere* (Amsterdam: Elsevier, 1975); Brian Hoskins and Robert Pearce, eds., *Large-Scale Dynamical Processes in the Atmosphere* (Orlando, FL: Academic Press, 1983); and E. Palmen and C. W. Newton, *Atmospheric Circulation Systems: Their Structure and Physical Interpretation* (New York: Academic Press, 1969). Wind has become important again as an energy source, and this receives thorough discussion in Paul N. Vosburgh, *Commercial Applications of Wind Power* (New York: Van Nostrand-Reinhold Co., 1983). Finally, anyone interested in the landforms associated with wind should look at R. A. Bagnold, *The Physics of Blown Sand and Desert Dunes* (London: Methuen & Co., 1941).

ATMOSPHERIC PRESSURE. The force exerted at any point by the weight of a column of air above that point.

The origin of the term *atmospheric pressure* is unclear, although it was in use, in its current form, at least as early as 1942 and probably well before, as is apparent in the work of L. P. Harrison (1942) and C. W. Barber (1943).

Atmospheric pressure is measured with a barometer, and normal atmospheric pressure at sea level on the earth's surface is 14.7 pounds per square inch, 29.92

inches of mercury, or 1013.25 millibars. According to Joe Eagleman (1980: 11), "The force arises from the activity of the molecules that comprise the atmosphere and this is determined by air temperature and by the pull of gravity on the gaseous molecules, since they have a small amount of mass." The *millibar* is a unit of pressure in the metric system that is one-thousandth of a bar, and it is now the unit that is now generally employed by meteorologists in the production of weather and climatological maps on which isobars are drawn. The millibar is equal to a pressure of 1,000 dynes per square centimeter. An *isobar* is a line on a map that connects all points of equal barometric pressure. Similarly, although not directly related to atmospheric pressure, an *isotherm* is a line on a map that connects all points of equal temperature.

Although average atmospheric pressure over the entire earth's surface remains a constant, because the amount of air is not changing, there are considerable variations in atmospheric pressure from place to place, and in the same place from time to time. Such variations are not random, and their patterns were studied early by Sir Gilbert Walker (1923). Thus there are areas and times when we find pressures that are notably low or high relative to surrounding areas. These centers of low and high pressure are known respectively as cyclones and anticyclones. Regions of high and low pressures are an important part of the ATMOSPHERIC CIRCULATION. One useful overview of cyclones and anticyclones and their movements is found in W. H. Klein (1957).

A cyclone, then, is a center of low pressure, and winds are drawn into the center as they move down the pressure gradient, setting up a rotating circulation pattern around the cyclone. This circulation pattern is counterclockwise in the northern hemisphere and clockwise in the southern hemisphere. Because the air is converging toward the cyclone's center, it must rise in the center and spread out again somewhere in the atmosphere. A good discussion of the development of cyclones is found in the work of R. S. Scorer (1966).

Cyclones are also associated with specific types of storms, in which case they are traveling cyclones that are guided by larger forces in the atmospheric circulation. The wave cyclone forms along the polar front, where warm and cold air masses are in conflict. Thus it is a feature of the middle and high latitudes. The TROPICAL CYCLONE develops in the low latitudes, typically between 8 and 15 degrees latitude. It is more commonly known in the United States as a hurricane.

Conversely, an anticyclone is a center of high pressure, and air flows outward from the center, down the pressure gradient. Again a rotating circulation pattern develops, in this case with clockwise rotation in the northern hemisphere and counterclockwise rotation in the southern hemisphere. Because air is circulating away from the high-pressure center, or diverging, air in the center must be descending. Like cyclones, anticyclones may also migrate, as is apparent in an early study by J. E. Belasco (1948).

References

Barber, C. W. 1943. *An Illustrated Outline of Weather Science.* New York: Pitman Publishing Corp. Useful as an example of where the science of meteorology was during the 1940s.

Belasco, J. E. 1948. "The Incidence of Anticyclonic Days and Spells over the British Isles." *Weather* 3: 233–242. One of many early statistical studies of regional weather patterns.

Eagleman, Joe R. 1980. *Meteorology: The Atmosphere in Action.* New York: D. Van Nostrand Co. One of several good current meteorological texts.

Harrison, L. P. 1942. *Meteorology.* New York: National Aeronautics Council. An early text.

Klein, W. H. 1957. "Principal Tracks and Mean Frequencies of Cyclones and Anticyclones in the Northern Hemisphere." Research Paper, no. 40. Washington, D.C.: U.S. Weather Bureau. A classic study of these phenomena.

Scorer, R. S. 1966. "Origin of Cyclones." *Science Journal* 2: 46–52. An exposition that is still well worth reading.

Walker, Sir Gilbert. 1923. "Correlations in Seasonal Variation of Weather. VIII. A Preliminary Study of World Weather," *Indian Meteorological Memoirs* (Calcutta) 24, Pt. 4.

Sources of Additional Information

Basic discussions of atmospheric pressure, cyclones, and anticyclones can be found in most physical geography textbooks, including the following: Arthur N. Strahler and Alan H. Strahler, *Elements of Physical Geography*, 3d ed. (New York: John Wiley and Sons, 1984); Tom L. McKnight, *Physical Geography: A Landscape Appreciation* (Englewood Cliffs, N.J.: Prentice-Hall, 1984); and E. Willard Miller, *Physical Geography: Earth Systems and Human Interactions* (Columbus, Ohio: Charles E. Merrill Publishing Co., 1985). Somewhat more detailed discussions are typically found in meteorology and climatology texts, including the following examples: Herbert J. Spiegel and Arnold Gruber, *From Weather Vanes to Satellites: An Introduction to Meteorology* (New York: John Wiley and Sons, 1983); John E. Oliver and John J. Hidore, *Climatology: An Introduction* (Columbus, Ohio: Charles E. Merrill Publishing Co., 1984); and Paul E. Lydolph, *The Climate of the Earth* (Totowa, N.J.: Rowman and Allanheld, Publishers, 1985).

B

BAROMETER. An instrument for measuring atmospheric pressure.

The concept of air having weight and exerting pressure (see ATMOSPHERIC PRESSURE) as well as the instrument used to measure this pressure (the barometer) evolved during the seventeenth century. There had been conjecture about air as a heavy body as early as the fourth century B.C. Aristotle tried to measure the weight of air but failed. The lack of precise instruments would not allow him to measure such small masses, nor were the techniques available for such measurement (Crombie, 1953: 238).

Although there is general agreement that a student of Galileo, Evangelista Torricelli, invented the mercury barometer in 1643, there is some speculation that others at least had conceived of the idea before 1643 (Zinszer, 1944: 261). Descartes described what sounds like the mercury barometer to pupils in a letter written in 1631. He was actually the first person to put a scale on the mercury tube and take precise measurements; this was in 1647 (Middleton, 1964: 46). Another possible inventor was Gasparo Berti, an astronomer and mathematician who in 1639 performed barometric experiments. Nonetheless, Torricelli is considered by most to be the actual inventor of the inverted tube, mercury barometer (Frisinger, 1977: 68–71).

The standard name, barometer, did not come about for over a decade after the invention of the instrument. Robert Boyle named the instrument (circa 1669) and used it for many experiments with gases. He discovered Boyle's gas law using the barometer in 1662.

Various modifications to the barometer came within a century of the original invention. Robert Hooke created the wheel barometer in 1665. In 1669 Boyle described a completely portable siphon barometer. Hooke and Descartes both produced two-liquid barometers, which increased the sensitivity of the instrument and made accurate reading easier. In 1685 Hooke also designed a three-liquid barometer, which solved some of the problems encountered with the two-liquid instrument. In 1947 Jean Andre De Luc produced the first good portable siphon instrument (Wolf, 1939: 306).

The aneroid barometer (without liquid) was produced around 1700 by Gottfried Leibniz. In correspondence with Johann Bernoulli, Leibniz described his aneroid barometer. Through subsequent correspondence, defects in design (e.g., the effects of humidity) were corrected (Frisinger, 1977: 76–77).

An accepted system of standard barometric measurements and correction for temperature, humidity, and capillarity were developed in the nineteenth century by several people (Frisinger, 1977: 77).

More recent research has used the basically unchanged barometer to study weather and climate phenomena. Standard units of measure have been developed. Meterologists around the world now use the unit of force, the newton, as their standard. One newton is the force created by accelerating 1 kilogram of mass 1 meter per second squared. At sea level this would be 101,325 newtons per meter squared. In 1940 the U.S. Weather Service adopted the millibar as its standard unit. One millibar equals 100 newtons per meter squared (1013.25 millibars is the standard atmospheric pressure) (Lutgers and Tarbuck, 1986: 157–158).

Two related instruments have been invaluable for research and practical applications. The barograph uses an aneroid barometer to record air pressure continuously over time. This instrument produces a graph of pressure for a given place and a given time interval. The second instrument is the aneroid barometer used as an altimeter in aircraft. Pressure in the atmosphere changes at a calculable rate as altitude increases or decreases. If an aneroid barometer is connected to a scale adjusted for this rate, altitude can be read off the instrument (Lutgens and Tarbuck, 1986: 159).

Except for refinements and increases in accuracy, little contemporary research is actually done with the barometer. Most work using the barometer is in conjunction with weather and climate phenomena. The progress of research using the barometer and air pressure is covered under other concepts such as AIR MASS, FRONTAL SYSTEM, FOEHN, CORIOLIS EFFECT, and ATMOSPHERIC CIRCULATION.

References

Crombie, A. C. 1953. *Augustine to Galileo—The History of Science*. Cambridge: Harvard University Press. This is the classic work on the history of science before and during the Renaissance.

Frisinger, Howard H. 1977. *The History of Meterology*. New York: Science History Publications. The classic work specifically for the history of meterology before 1800.

Lutgens, Frederick K., and Tarbuck, Edward J. 1986. *The Atmosphere: An Introduction to Meteriology*. 3d ed. Englewood Cliffs, N.J.: Prentice-Hall. One of a multitude of good, introductory texts on meteorology.

Middleton, W.E.K. 1964. *The History of the Barometer.* Baltimore: The Johns Hopkins Press. The definitive work on the history of the barometer.

Wolf, A. 1939. *A History of Science, Technology, and Philosophy in the Eighteenth Century.* New York: Macmillan.

Zinszer, Harvey A. 1944. "Meterological Mileposts." *Scientific Monthly* 58: 261–264.

Sources of Additional Information

Examples of atmospheric modeling in which air pressure is important include R. M. Chervin and S. H. Schneider, "On Determining the Statistical Significance of Climate Experiments with General Circulation Models," *Journal of Atmospheric Science* 33 (1976): 405–412, and M. E. Schlesinger and W. L. Gates, *Preliminary Analysis of Four General Circulation Model Experiments on the Role of the Ocean in Climate*, Report no. 25 (Corvallis, Oreg.: Climatic Research Institute, 1981). For the classic work on microclimates and pressure see R. Geiger, *Climate Near the Ground* (Cambridge: Harvard University Press, 1965). For work on global circulation patterns that are inextricably linked with pressure patterns see R. G. Barry and R. J. Chorley, *Atmosphere, Weather, and Climate* (London: Methuen & Co., 1968); J. Bjerknes, "The Possible Response of the Atmospheric Hadley Circulation to Equatorial Anomalies of Ocean Temperature," *Tellus* 4 (1966): 820–829; P. R. Julian and R. M. Chervin, "A Study of the Southern Oscillation and Walker Circulation Phenomenon," *Monthly Weather Review* 106 (1978): 1433–1451; and H. van Loon and J. Williams, "The Connection between Trends of Mean Temperature and Circulation at the Surface," *Monthly Weather Review* 105 (1977): 636–647. For the outstanding general history of scientific discovery including the barometer see D. J. Boorstin, *The Discoverers* (New York: Random House, 1983). For works on climatic history and pressure changes see H. H. Lamb, *Climate: Present, Past, and Future. Vol. I, Fundamentals and Climate Now* (London: Methuen & Co., 1972); H. H. Lamb, *Climate: Present, Past, and Future. Vol. II, Climatic History and the Future* (London: Methuen & Co., 1977); and T.M.L. Wigley, M. J. Ingram, and G. Farmer, eds., *Climate and History* (Cambridge: Cambridge University Press, 1981). For the most authoritative reference on all meterological instruments see F. A. Berry Jr., E. Bollay, and N. R. Beers, *Handbook of Meterology* (New York: McGraw-Hill Book Co., 1973).

BASE LEVEL. The lower limit of stream erosion, normally considered to be sea level, assuming that the stream enters the ocean.

Although the idea of a limit to which erosion can reduce a landmass was recognized by Leonardo da Vinci (Chorley and Beckinsale, 1968), the geologic concept of base level did not receive formal recognition until the work of John Wesley Powell (1875). Not only did Powell introduce the concept into the geologic literature, he also recognized that base level is not necessarily coincident with sea level, that there can also be local base levels, as, for example, when a stream drains into a lake or an interior basin. Such a stream can erode no lower than the lake level or the level of the interior basin.

Although Powell's initial definition of base level seemed straightforward, confusion resulted. William Morris Davis (1902) reviewed Powell's work and sought to clarify the meaning of the term as it was used in somewhat different contexts by Powell. According to Richard Chorley and Robert Beckinsale (1968: 59), interpretations of base level today can be placed in the following four categories: (1) grand base level, (2) temporary base level, (3) base-leveled surface, and (4) local base level.

The notion of a grand base level was introduced by C. A. Malott (1928), who actually used the term *ultimate base level*. It is this definition of base level that typically is found in current textbooks. Tom McKnight (1984: 462), for example, defined base level as "an imaginary surface extending underneath the continents from sea level at the coasts, and indicating the lowest level to which land can be eroded." H. F. Garner (1974) introduced still another terminology problem, because he suggested that base level can be viewed as either regional or local. Thus his regional base level is analogous to what we noted above as the grand base level, which is essentially sea level.

With respect to the idea of a temporary base level, Edgar Spencer (1983: 304) suggested that it exists in situations in which "locally, lakes and enclosed or land-locked depressions may act in a similar way and limit downward erosion of local areas." However, the term *local base level* is also used to describe such a situation, so confusion still exists, at least with respect to the differentiation between temporary and local base levels. As Luna Leopold, M. Gordon Wolman, and John Miller noted, with respect to local and temporary base levels, "Clearly these are not precise terms and on occasion are even used synonymously, or more often to distinguish local levels to which streams flow on their way to the oceans, to which by definition all exterior or unenclosed river systems ultimately flow" (1964: 258). According to Chorley and Beckinsale, temporary base level is a "limit to downward erosion of an ephemeral character imposed headward of a resistant outcrop" (1968: 59).

The idea of a base-leveled surface remains somewhat confusing also, partially because it overlaps with other terms, especially *peneplain* and *graded stream*, both of which have their own usage and literature. Garner wrote, for example, that "the ultimate, low-relief erosion surface resulting from base leveling is the *peneplain* of the Davisian scheme" (1974: 7).

Despite arguments about definitions, the concept of a grand base level has underlain considerable research in geology and physical geography, beginning with the introduction of the cycle of erosion by Davis (1899) and its extension to arid environments by Davis (1905).

Aside from the definitional problems, the base-level concept has received other criticism as well. For example, major rivers are able to erode below sea level, as are turbidity currents. In addition, fluctuations in sea level have confounded attempts to interpret landforms as a result of degradation by rivers to an assumed base level. However, none of these criticisms is sufficient to outweigh the usefulness of the base level concept in modern geomorphology.

References

Chorley, Richard J., and Beckinsale, Robert P. 1968. "Base Level." In Rhodes W. Fairbridge, ed., *The Encyclopedia of Geomorphology*. Encyclopedia of Earth Sciences Series. Vol. 3. New York: Reinhold Book Corp., pp. 59–60.

Davis, William Morris. 1899. "The Geographical Cycle." *Geographical Journal* 14: 481–504. The classic statement on the cycle of erosion.

———. 1902. "Base Level, Grade, and Peneplain." *Journal of Geology* 10: 77–111. Extends and clarifies his earlier work on the cycle of erosion.

———. 1905. "The Geographical Cycle in an Arid Climate." *Journal of Geology* 13: 381–407. Carries the idea of an erosion cycle into arid environments.

Garner, H. F. 1974. *The Origin of Landscapes: A Synthesis of Geomorphology*. New York: Oxford University Press. A monumental text on geomorphology and a basic reference for anyone who seeks information on the subject.

Leopold, Luna B.; Wolman, M. Gordon; and Miller, John P. 1964. *Fluvial Processes in Geomorphology*. San Francisco: W. H. Freeman and Co. Remains the classic work on fluvial processes.

Mallott, C. A. 1928. "Base-level and Its Variations." *Indiana University Studies* 82: 37–59. A major early discussion of base level and problems associated with the concept.

McKnight, Tom L. 1984. *Physical Geography: A Landscape Appreciation*. Englewood Cliffs, N.J.: Prentice-Hall. One of several current texts on physical geography.

Powell, John Wesley. 1875. *Exploration of the Colorado River of the West*. Washington, D.C.: Smithsonian Institution. A classic in the literature of geology in the United States.

Spencer, Edgar W. 1983. *Physical Geology*. Reading, Mass.: Addison-Wesley Publishing Co. An excellent introductory text in physical geology.

Sources of Additional Information

In introductory texts in physical geography the term *base level* is barely introduced. Typical examples include the following: James S. Gardner, *Physical Geography* (New York: Harper's College Press, 1977), and Arthur N. Strahler and Alan H. Strahler, *Elements of Physical Geography* (New York: John Wiley and Sons, 1984). Somewhat better discussions of base level can be found in the following: R. J. Small, *The Study of Landforms: A Textbook of Geomorphology* (Cambridge: Cambridge University Press, 1972); Karl W. Butzer, *Geomorphology from the Earth* (New York: Harper and Row, Publishers, 1976); Arthur L. Bloom, *Geomorphology: A Systematic Analysis of Late Cenozoic Landforms* (Englewood Cliffs, N.J.: Prentice-Hall, 1978); and Darrell Weyman and Valerie Weyman, *Landscape Processes: An Introduction to Geomorphology* (London: Allen and Unwin, 1977).

BEACH. A shoreline feature that consists at least partially of accumulated sediments, primarily sand or coarser materials.

According to James Murray et al. (1933), the term *beach* was first used no later than the first half of the sixteenth century in Sussex and Kent, where it was used to indicate shingles or pebbles that had been worn by the waves along the shoreline of England. Shakespeare used the term *beach* more or less in its current context in *The Merchant of Venice*.

In the geological literature the term appears as early as 1830 in the work of Charles Lyell (1830), when he referred to "raised beaches." Later in the nineteenth century Sir Archibald Geikie (1880: 154) wrote that "the strip of sand, gravel or mud, which is alternately covered and laid bare by the rise and fall of the tidal undulation, is called the beach." The latter half of the nineteenth century witnessed an increase in research on beaches and their materials, forms, and characteristics. Much of this research was done by the Germans, but there were other studies as well, such as the study of beach cusps by M.S.W. Jefferson (1899). Soon afterward N. M. Fenneman (1902) studied profile equilibrium and its development.

The beach is recognized as a major feature of shores, especially ocean shores, and their expression along coastal zones looks different depending on local circumstances. Within the coastal zone, between low- and high-water lines, beach materials are carried along by beach drift, whereas in the nearby breaker zone materials are carried by longshore drift. Because these two processes often operate together to move sand along a beach, the beach has been referred to in popular terms as a "river of sand." Soon after the beginning of the twentieth century enough material had accumulated to produce the first whole book on beaches (Johnson, 1919), a work that helped establish a baseline from which future research needs and objectives could be measured.

Basic ideas about beaches were well established early in the twentieth century, and more detailed research was undertaken in many locations. For example, in the 1920s P. Marshall (1929) worked extensively to classify types of sand beaches in Europe. Furthermore, R. S. Patton (1931) studied the evolution of beaches along the south shore of Long Island, New York. Because of its economic importance, Patton's study was important in determining whether to continue to allow natural shoreline processes to operate in this area near New York City or whether to have engineers and scientists interfere in the process.

During the 1930s a central focus of beach research was on the transportation of beach materials and beach-formation processes. W. V. Lewis (1931), for example, looked at how the configuration of a beach was determined by wave incidence. Kirk Bryan and Robert Nichols (1939) studied the role of wind as an agent of shoreline transportation.

Beach research intensified further in the 1940s and 1950s. Geographers were active participants, as exemplified by the work of B. W. Sparks (1949) and H. A. Wood (1951), among others. Laboratory studies of shoreline processes were improving also, as is apparent in the work of G. M. Watts (1954) and R. L. Rector (1954).

Following this period of active scholarly work on beaches, two important books appeared by Cuchlaine King (1959) and Willard Bascom (1964) and remain basic reference works on shoreline processes during the 1980s, attesting to their quality and to the quality of the research that went before them.

Bascom (1960) summarized beaches at the beginning of the 1960s, and this was followed by some new areas of research, some of which at least partially reflected the changing concerns of society. J. R. Byerly (1963), for example, studied the way in which watershed geology and beach radioactivity were related. Geographers continued to study beaches also, as we see in the work of J. R. Hardy (1964). Another important book appeared at the end of the 1960s (Bird, 1969).

Since 1970 other new books and articles have appeared, although the frequency of appearance of beach-related research seems to have dropped somewhat relative to the two previous decades, perhaps because of the rising attention paid to continental drift, plate tectonics, and sea-floor spreading. Among important books on beaches published subsequent to 1970 would certainly be those of Paul Komar (1976) and Maurice Schwartz (1982).

Beaches are varied in their size, composition, and distribution. Furthermore, they are dynamic. After their initial development, they change seasonally, and they can change under the influence of human activities as well. Dams, for example, rob the ocean of part of its sand supply, requiring adjustments along the coastal zone. Beaches differ in size from small, localized pocket beaches to barrier beaches that go on for hundreds of miles.

References

Bascom, Willard N. 1960. "Beaches." *Scientific American* 203: 81–94. Still well worth reading.

———. 1964. *Waves and Beaches*. New York: Anchor Books, Doubleday and Co. (Revised edition published in 1979.) The classic work on beaches and related phenomena.

Bird, E.C.F. 1969. *Coasts*. Cambridge, Mass.: M.I.T. Press. Detailed study of coasts, including beaches.

Bryan, Kirk, and Nichols, Robert L. 1939. "Wind Deposition Shorelines." *Journal of Geology* 4: 431–435.

Byerly, J. R. 1963. "Naturally Occurring Radioactive Minerals as Littoral Tracers," *Shore and Beach* 31: 20–27.

Fenneman, N. M. 1902. "Development of the Profile of Equilibrium of the Subaqueous Shore Terrace." *Journal of Geology* 10: 1–32. Perhaps the earliest such study of equilibrium profiles.

Geikie, Sir Archibald. 1880. *Physical Geography*. New York: D. Appleton and Co., Science Primers. One of the very earliest texts on physical geography. Well worth looking at for a historical perspective on how the subject has changed.

Hardy, J. R. 1964. "The Movement of Beach Material and Wave Action Near Blakeney Point, Norfolk." *Transactions of the Institute of British Geographers* 34: 53–69. An excellent example of a geographic study of shoreline and beach processes.

Jefferson, M.S.W. 1899. "Beach Cusps." *Journal of Geology* 7: 237–246.

Johnson, D. W. 1919. *Shore Processes and Shoreline Development*. New York: John Wiley and Sons. Probably the earliest book on the topic.

King, Cuchlaine A. M. 1959. *Beaches and Coasts*. London: E. Arnold. (Revised edition published in 1972.) Remains an excellent resource for beach studies.

Komar, Paul D. 1976. *Beach Processes and Sedimentation.* Englewood Cliffs, N.J.: Prentice-Hall. Most useful for specialists.

Lewis, W. V. 1931. "The Effect of Wave Incidence on the Configuration of a Shingle Beach." *Geographical Journal* 78: 129–148.

Lyell, Charles. 1830. *Principles of Geology: Being an Attempt to Explain the Former Changes of the Earth's Surface by Reference to Causes now in Operation.* Vol. 1. London: J. Murray. A classic in the literature of geology.

Marshall, P. 1929. "Beach Gravels and Sands," *New Zealand Institute, Transactions and Proceedings* 60: 324–365.

Murray, James A. H., Bradley, Henry, Craigie, W. A. and Onions, C. T., eds. 1933. *The Oxford English Dictionary.* Oxford: Clarendon Press.

Patton, R. S. 1931. "Moriches Inlet: A Problem in Beach Evolution." *Geographical Review* 21: 627–632. Another well-done geographic study.

Rector, R. L. 1954. "Laboratory Study of Equilibrium Profiles of Beaches." *Beach Erosion Board Technical Memo*, no. 41. Washington, D.C.: Beach Erosion Board.

Schwartz, Maurice L. 1982. *The Encyclopedia of Beaches and Coastal Environments.* Stroudsburg, Pa.: Hutchinson Ross Publishing Co. If you could choose only one book on the subject of beaches, this would be it.

Sparks, B. W. 1949. "Procedures in Studying Shore Erosion." *Geography* 34: 216–220.

Watts, G. M. 1954. "Laboratory Study of Effect of Varying Wave Periods on Beach Profiles." *Beach Erosion Board Technical Memo*, no. 53. Washington, D.C.: Beach Erosion Board.

Wood. H. A. 1951. "Procedures in Studying Shore Erosion." *The Canadian Geographer* 1: 31–37.

Sources of Additional Information

General discussions of beaches abound in most introductory textbooks in physical geography and geology. Good examples include the following: Tom L. McKnight, *Physical Geography: A Landscape Appreciation* (Englewood Cliffs, N.J.: Prentice-Hall, 1984); Arthur N. Strahler and Alan H. Strahler, *Elements of Physical Geography*, 3d ed. (New York: John Wiley and Sons, 1984); and Edgar W. Spencer, *Physical Geology* (Reading, Mass.: Addison-Wesley Publishing Co., 1983). More detailed information on beaches can be found in J. S. Fisher and R. Dolan, eds., *Beach Processes and Coastal Hydrodynamics* (Stroudsburg, Pa.: Dowden, Hutchinson and Ross, 1977); F. P. Shepard, *Submarine Geology*, 3d ed. (New York: Harper and Row, Publishers, 1973); S. P. Leatherman, ed., *Barrier Islands* (New York: Academic Press, 1979); J. L. Davies, *Geographical Variation in Coastal Development* (New York: Hafner Publishing Co., 1973); and V. P. Zenkovich, *Processes of Coastal Development*, translated by D. G. Fry (New York: Wiley-Interscience, 1967).

BIOCYCLES. See BIOSPHERE.

BIOGEOGRAPHY. The geographical study that deals with the distribution of plants and animals and the reason for their distribution.

Plants and animals are irregularly distributed both on the continents and in the OCEANS. According to A. Starker Leopold (1974: 59), "Biogeographic studies are concerned with learning the manner in which living organisms are arranged on the Earth and the causative factors of this arrangement." The study of these living organisms has historically been divided by both biologists and geographers into phytogeography (plant studies) and zoogeography (animal studies), although attention primarily has been focused on plant studies. The main reason for this focus on plants is that most of the total world biomass (volume of living material) is composed of plants. Also, because of their greater mobility and small size, studying animal species is more difficult than studying plants.

The origins of biogeography can be traced to a growing interest and curiosity about the nature of the earth. This field of study originally referred to as "natural history" was concerned with the description and classification of natural phenomena. Early work on these topics was done in the latter half of the eighteenth century by the Swedish scientist Carolinus Linnaeus (1707–1778), who laid the foundations of modern biological study with his discussion of taxonomy and nomenclature (Tivy, 1982: 3).

The nineteenth century saw a significant increase in world exploration and the accompanying discovery of many new kinds of organisms. Among the most significant of these explorers was the German naturalist Alexander von Humboldt (1769–1859). Humboldt's famous research expeditions, particularly his South American expedition with the French naturalist Aime Bonpland, laid the foundations for much of modern biogeography. Humboldt's writings (Humboldt and Bonpland, 1805) pointed out the significant relationship between climate (see CLIMATOLOGY) and the distributions and forms of plant life throughout the world. Of particular significance to Humboldt were the changes in plant life associated with altitudinal changes. Because of this pioneering work by Humboldt he is often referred to as the "father of phytogeography."

During the nineteenth century many scientists further refined the ideas of Humboldt. The Swiss botanist Alphonse de Candolle (1885) studied the distribution of major VEGETATION types throughout the world. Candolle (1882) also was one of the first scientists to look at the origin and distribution of cultivated plants. The latter part of the nineteenth century brought continued attempts to explain the distribution and morphological features of plants. These studies culminated with the classic study of world vegetation patterns, *Plant Geography on a Physiological Basis* by A.W.F. Schimper (1903).

The development of zoogeography can be traced to the interest in animal distribution by Aristotle, but the first scientist to perceive clearly the differences between Old World and New World faunas was Georges Buffon. Another early synthesis of ideas about zoogeography was produced by William Swainson (see Brown and Gibson, 1983: 8) but the first comprehensive global classification of faunal regions (for birds) was developed in the middle of the nineteenth century by P. L. Sclater (1858). The classification and description of animals accelerated in the latter part of the nineteenth century, and by the end of the century at least twenty such classification systems had been developed. Amongst the most important classifications of zoogeographical regions were the works of T. H. Huxley (1868) and A. R. Wallace (1860, 1876).

One cannot discuss biogeography in the nineteenth century without recognizing the contributions of Charles Darwin. Darwin not only studied GEOLOGY and indigenous peoples on his journey aboard H.M.S. *Beagle* but also was very interested in native plants and animals. His observations of tortoises and finches led to his ideas about the role of geographic isolation. When Darwin returned to England after the voyage on the *Beagle*, he developed his theory of evolution through natural selection (Darwin, 1859). His theory of evolution was a very significant scientific advance and had an important impact on biogeography.

Darwin and his followers have influenced the development of biogeography in at least four ways (Stoddart, 1966): through establishing the concept of evolution or change through time, by analyzing the idea of struggle or natural selection, by discussing the concept of randomness or chance variation, and by developing the important concepts of association and organization in nature.

The study of biogeography made giant advances in the twentieth century. In the early part of the twentieth century climate once again was emphasized, and its important role in vegetation complexes was stressed. Wladimir Köppen (1918), a biologist, established climatic regions primarily based on major types of vegetation. About the same time A. J. Herbertson (1905) used vegetation-climate associations to establish his "natural regions of the world."

Another area of increasing interest was the ecological relationship between plants and their environment. The dynamics of vegetation succession was studied by H. C. Cowles (1899), who analyzed plant succession (see PLANT SUCCESSION/CLIMAX) on the sand DUNES near Lake Michigan. However, Cowles's student Frederic Clements (1916) published the major textbook on the topic and also developed the concept of a "climax" or ultimate terminal stage in plant succession. In Britain, Arthur Tansley (1939) challenged some of Clements's ideas by stressing the wide range of environmental factors that control the composition of plants in the final or climax stage. Tansley's most important contribution, however, was the development of the concept and coining of the term ECOSYSTEM.

Since this initial conceptualization by Tansley, the ecosystem concept has played an important role in biography. D. R. Stoddard (1967) said that the ecosystem concept appeals to geographers because of four main basic characteristics of its makeup: (1) it includes the worlds of humans, plants, and animals and the interactions between humans and the living environment; (2) it is a functional approach and is dynamic; (3) through a structural approach it organizes nature in an orderly fashion; and (4) it is a general system. R. J. Chorley argued, however, that the ecosystem concept "may fail as a supposed key to the general understanding of relations between modern society and nature, and therefore as a basis for contemporary geographical studies, because it casts social man in too subordinate and ineffectual a role" (1973: 155).

The past several decades have seen major advances in biogeography. The late 1960s and 1970s saw a renewed interest in island biogeography. The stimulus to this revived interest was the publication of R. H. MacArthur and E. O. Wilson's *Theory of Island Biogeography* (1967). They argued that biota on an island would be in a dynamic equilibrium between the in-migration of new species and the extinction of old species. Further support for this thesis is found in the work of J. Hellier (1976), J. H. Brown (1971), and R. C. Buckley (1985). A recent study of particular interest dealt with biogeography on the caldera of Krakatoa (Flenley and Richards, 1982). These ideas on island biogeography, however, have been challenged in a recent article by F. S. Gilbert (1980).

Continental drift and PLATE TECTONICS theories have revolutionized both geology and historical biogeography. The changes in the relative positions and sizes of landmasses and oceans have had important biogeographic consequences. The significance to biogeography of the new plate-tectonics synthesis is clearly evident in a variety of recent research studies (Coleman, 1980; Stevens, 1980; Stoddart, 1983).

According to Joy Tivy, the past two decades have seen the rapidly increasing use of "statistical analyses and mechanical aids in the collection, storage and processing of data" (1982: 9).

References

Brown, J. H. 1971. "Mammals on Mountain Tops: Non Equilibrium Insular Biogeography." *American Naturalist* 105: 467–478.

Brown, James H., and Gibson, Arthur C., 1983. *Biogeography*. St. Louis: C. V. Mosley Co.

Buckley, R. C. 1985. "Distinguishing the Effects of Area and Habitat Type of Island Plant Species Richness by Separating Floristic Elements and Substrate Types and Controlling Island Isolation." *Journal of Biogeography* 12: 527–535.

Chorley, R. J. 1973. "Geography as Human Ecology." In Chorley, R. J., ed., *Directions in Geography*. London: Methuen & Co., pp. 155–169.

Clements, F. E., 1916. *Plant Succession*. Washington, D.C.: Carnegie Institute. An early study discussing climax vegetation.

Coleman, P. J., 1980. "Plate Tectonics Background to Biogeographic Development in the Southwest Pacific Over the Last 100 Million Years." *Paleogeography, Paleoclimatology, Paleoecology* 31: 105–121.

Cowles, H. C. 1899. "The Ecological Relations of the Vegetation on the Sand Dunes of Lake Michigan." *Botanical Gazette* 27: 95–117, 167–202, 281–308, 361–391. An early study of the relationship between plants and the environment.

Darwin, Charles, 1859. *On the Origin of Species by Means of Natural Selection*. London: J. Murray. The classic book on evolution.

DeCandolle, A. 1882. *Origine des Plantes Cultivees*. Paris: Germer Bailliere. A classic study on the development of cultivated plants.

———. 1885. *Geographie Botanique Raisonne en Exposition des Faits Principaux et des Lois Concernant la Distribution des Plantes d'Epoque Actuelle*. Paris: Massion. A classic study on the worldwide distribution of plants.

Diamond, J. M. 1975. "The Island Dilemma: Lessons of Modern Biogeographic Studies for the Design of Natural Reserves." *Biological Conservation* 7: 129–146.

Flenley, J. R., and Richards, K., eds. 1982. *The Krakatoa Cenenary Expedition: Final Report.* Miscellaneous Series, no. 25. Hull, U.K.: Department of Geography, University of Hull.

Gilbert, F. S. 1980. "The Equilibrium Theory of Island Biogeography: Fact or Fiction?" *Journal of Biogeography* 7: 209–235. A critical evaluation of island biogeography theory.

Hellier, J. 1976. "The Biogeography of Enio Landsnails on the Aegean Islands." *Journal of Biogeography* 3: 281–292.

Herbertson, A. J. 1905. "The Major Natural Regions: An Essay in Systematic Geography." *Geographical Journal* 25: 300–312.

Humboldt, F.H.A. von, and Bonpland, A. 1805. *Essai Sur la Geographie des Plantes.* Paris: Levrault, Schoell and Compagnie. A classic work by one of the founders of modern biogeography.

Huxley, T. H. 1868. "On the Classification and Distribution of the Alectoromorphae and Heteromorphae." In *Proceedings of the Zoological Society of London.* London: Zoological Society. pp. 294–319.

Koppen, W. 1918. "Klassification der Climate nach Temperatur, Niederschlag, und Jahreslauf," *Petermanns Geographische Mittheilungen* 64: 193–203, 243–248. An important climate classification system.

Leopold, A. Starker. 1974. "Biogeography." In *McGraw-Hill Encyclopedia of Environmental Science.* New York: McGraw-Hill Book Co. A short discussion of the basic concepts of biogeography.

MacArthur, R. H., and Wilson, E. O. 1967. *The Theory of Island Biogeography.* Princeton, N.J.: Princeton University Press. A modern interpretation of island biogeography theory.

Schimper, A.W.F. 1903. *Plant Geography on a Physiological Basis* (trans. W. R. Fisher). Oxford: Clarendon Press. An important early study on world vegetation patterns.

Sclater, P. L. 1858. "On the General Geographic Distribution of the Members of the Class Aves." *Journal of the Linnean Society of London* 2: 130–145. One of the first attempts to develop zoogeographical regions.

Stevens, G. R. 1980. "Southwest Pacific Faunal Paleobiogeography in Mesozoic and Cenozoic Times: A Review." *Paleogeography, Paleoclimatology, Paleoecology* 31: 153–196. A recent study on the biogeography associated with plate tectonics.

Stoddart, D. R. 1966. "Darwin's Impact on Geography." *Annals of the Association of American Geographers* 56: 683–698.

———. 1967. "Organism and Ecosystem as Geographical Models." In Chorley, R. J., and Haggett, P., eds., *Models in Geography.* London: Methuen & Co., pp. 511–538.

———. 1983. "Biogeography: Darwin Devalued or Darwin Revalued?" *Progress in Physical Geography* 7, no. 2: 256–264. A good review of recent advances in biogeography.

Tansley, A. G. 1939. *The British Islands and Their Vegetation.* Cambridge: Cambridge University Press.

Tivy, Joy. 1982. *Biogeography: A Study of Plants in the Ecosphere.* 2d ed. London: Longmans, Green and Co. A good biogeography textbook with an excellent discussion of historical developments.

Wallace, A. R. 1860. "On the Zoological Geography of the Malay Archipelago." *Journal of the Linnean Society of London* 14: 172–184.

———. 1876. *The Geographical Distribution of Animals*. 2 vols. London: Macmillan. A classic work on animal distribution.

Sources of Additional Information

Short definitions and discussions of the concept can be found in Philip Gore, ed., *Webster's Third New International Dictionary of the English Language* (Springfield, Mass.: G & C Mirriam Co., 1976); Michael Allaby, ed., *The Oxford Dictionary of Natural History* (Oxford: Oxford University Press, 1985); and Dudley Stemp, ed., *Longmans Dictionary of Geography*. (London: Longmans, Green and Co., 1966). Recent textbooks in biogeography are P. Stott, *Historical Plant Geography* (London: Allen and Unwin, 1984); Christopher Cox, *Biography: An Ecological and Evolutionary Approach* (Oxford: Blackwell Scientific Publications, 1976); R. L. Jones, *Biogeography* (Amersham, Eng.: Hulton Educational Press, 1980); N. Pears, *Basic Biogeography* (London: Longmans, Green and Co., 1977); and J. A. Taylor, ed., *Biogeography: Recent Advances and Future Directions* (Totowa, N.J.: Barnes and Noble Books, 1984). Good discussions of recent advances in biogeography can be found in J. A. Taylor, "Biogeography," *Progress in Physical Geography* 2 (1984): 94–101; idem, "Biogeography," *Progress in Physical Geography* 9 (1985): 104–112; and idem, *Progress in Physical Geography* 10 (1986): 239–248.

BIOMES. See BIOSPHERE.

BIOSPHERE. 1. The part of the earth in which living organisms can be sustained, including portions of the lithosphere, hydrosphere, and atmosphere. 2. The thin film or life zone found at the outer surface zone of the earth, sometimes called the ecosphere.

Although the idea of the biosphere was not introduced into geographical studies until the latter part of the nineteenth century, the study of the earth's life zone can be tracked to antiquity. Much of the early scientific work dealing with the biosphere was called "natural history" and was primarily concerned with the description and classification of natural phenomena. The Swedish scientist Carolus Linnaeus did pioneering work on classification and nomenclature during the latter half of the eighteenth century (Tivy, 82: 3).

The nineteenth century saw many attempts to analyze the biosphere. Foremost among the geographic studies was the work of F.H.A. von Humboldt and Aime Bonpland on the significant relationships between climate (see CLIMATOLOGY) and plant distribution (Humboldt and Bonpland, 1805) and Alphonse de Candolle's work on the distribution of VEGETATION types throughout the world (de Candolle, 1885).

The specific idea of the biosphere, however, was first introduced into science in the latter part of the nineteenth century by the Austrian geologist Eduard Suess. It was Suess who "first used the term in a discussion of the various envelopes of the earth in the last and most general chapter of a short book on the genesis of the Alps published in 1875'' (Hutchinson, 1970: 45). Suess believed that this thin layer of living material supported by the earth's crust plays a very important role in patterns of geological EROSION (Watts, 1971).

After this early discussion of the biosphere concept, little was done for about 50 years until the publication, first in Russian in 1926 and later in French in 1929, of two lectures by the Russian mineralogist Vladimir Vernadosky entitled *La Biosphere*. Vernadsky was primarily interested in the importance of living organisms in geochemical processes and gave much of the credit for the origin of the biosphere concept to the French naturalist Jean Lamarck (Hutchinson, 1970: 45). The importance of these circulatory elemental movements were further emphasized in many later studies (Woodwell, 1970; Young, 1968).

A further subdivision of the biosphere was proposed by Pierre Dansereau in the form of biocycles. Dansereau subdivided the biosphere into three biocycles: salt water, fresh water, and land. According to Dansereau, ''the ecologically determining physical factor which varies here is the density of the environment. Life in the sea, in ponds and streams, and on land requires radically different adjustments'' (1957: 125).

Since the biosphere is such a broad concept, geographers and biologists have made various attempts to make the concept more manageable by subdividing it into smaller categories called biomes. *Biomes* are complex biotic communities that have distinctive plant and animal forms and cover relatively large geographic areas (Durrenberger, 1973). The idea of the biome has gradually developed over the years, but it probably began in 1877 when Karl Möbius called it a ''community of living beings, a collection of species and a massing of individuals which find here everything necessary for their growth and continuance'' (Smith, 1966: 333).

The exact delimitation of biomes is difficult because the parameters or boundaries are less than precise. Biomes include aquatic (freshwater and marine) as well as terrestrial (deserts, tundra, grassland, savanna, chaparral, woodland, coniferous forest, deciduous forest, and tropical forest) categories (Brown and Gibson, 1983). The past several decades have seen much research on the biosphere that has used the biome approach as an organizing principle (Hammond, 1972; Gilluly, 1970).

Recent research dealing with the biosphere was given an impetus with the organization by UNESCO of the ''Man and the Biosphere'' (MAB) program. The establishment of a worldwide network of biosphere reserves was an outgrowth of the MAB program. The objective of the biosphere reserve program is to ''identify and protect representative and unique segments of the world's biotic province as major centers for biotic and genetic preservation, ecological and environmental research, education, and demonstration'' (Franklin, 1977: 267).

Research continues on the biosphere by American as well as international scientists (Taylor, 1986).

References

Brown, James H., and Gibson, Arthur C. 1983. *Biogeography.* St. Louis: C. V. Mosby Co.
Dansereau, Pierre. 1957. *Biogeography: An Ecological Perspective.* New York: The Ronald Press Co.
Durrenberger, Robert W. 1973. *Dictionary of the Environmental Sciences.* Palo Alto, Calif.: National Press Books.
de Candolle, A. 1885. *Geographie Botanique Raisonnee en Exposition des Faits Principaux et des Lois Concernant la Distribution des Plantes d'Epoque Actuelle.* Paris: Massion. A classic study on plant distribution in the biosphere.
Franklin, Jerry F. 1977. "The Biosphere Reserve Program in the United States." *Science* 195: 262–267. A discussion of the "Man and the Biosphere" program of UNESCO.
Gilluly, Richard. 1970. "Ecology: The Biome Approach." *Science News* 98: 204–206.
Hammond, Allen L. 1972. "Ecosystem Analysis: Biome Approach to Environmental Research." *Science* 175: 45–48.
Humboldt, F.H.A. von, and Bonplant, A. 1805. *Essai sur la Geographie des Plantes.* Paris: Levrault, Schoell and Compagnie. A classic work by one of the founders of modern geography.
Hutchinson, G. E. 1970. "The Biosphere." *Scientific American* 223, no. 3: 45–53. An excellent introduction to the biosphere concept.
Smith, Robert Leo. 1966. *Ecology and Field Biology.* New York: Harper and Row, Publishers.
Taylor, J. A. 1986. "Biogeography." *Progress in Physical Geography* 10, no. 2: 239–248. A good discussion of recent research.
Tivy, Joy. 1982. *Biogeography: A Study of Plants in the Ecosphere.* 2d ed. London: Longmans, Green and Co. A good biogeography textbook with an excellent discussion of historical developments.
Vernadsky, W. 1929. *La Biosphere.* Paris: Libraire Felix Alcan. An early discussion of the biosphere emphasizing biogeochemical processes.
Watts, David. 1971. *Principles of Biogeography.* New York: McGraw-Hill Book Co.
Woodwell, G. M. 1970. "The Energy Cycle of the Biosphere." *Scientific American* 223, no. 3: 64–75.
Young, H. E., ed. 1968. *Primary Productivity and Mineral Cycling in Natural Ecosystems.* Portland: University of Maine Press.

C

CHEMICAL WEATHERING. See WEATHERING.

CHINOOK. See FOEHN.

CIRQUE. 1. A usually deep and steep-sided recess that is semicircular in plan and eroded by a firn bank or glacier. 2. A steep-walled recess or hollow usually at the head of a glacial valley. (French, *cirque*; German, *kar*; Norwegian, *botn*; Welsh, *cwm*; Scottish, *corrie*; Swedish, *nisch*.)

The term *cirque* originated with Jean de Charpentier (1823) in his study of the Pyrenees. He did not establish the direct link between cirque morphology and glaciation (see PLEISTOCENE, VALLEY GLACIER, and CONTINENTAL GLACIER) at that time. A. C. Ramsey (1859) made this link relating alpine glacial activity and the cirque form. In 1873 B. Gastaldi (1873: 397) reiterated that glaciers (ice) certainly had the ability to carve out cirques in either soft or hard rock.

Several early writers adamantly stated that cirques were merely water-erosion features without any evidence that glacial ice had any relationship to them (see, for example, Bonney, 1871, 1877; Gregory, 1915). But as Clifford Embleton and C.A.M. King pointed out, these writers "failed to explain, for example, the presence of rock [see ROCK] basins commonly present in the cirque floors or to consider those cirques which have developed at or near the summits of mountains" (1975: 205). W. V. Lewis (1938: 250) cited the lack of eroding power by water so close to mountain summits also, therefore virtually eliminating fluvial erosion as a possible cause for cirques.

A preponderance of the research during the past eight decades has attempted to discover the actual mechanics of ice erosion in cirques. William Morris Davis (1909) admitted that cirques were glacially caused but that no one knew the mechanics of ice erosion that could cause such a LANDFORM. There is still no general consensus about the primary cause of cirque formation, although at least three theories of cirque erosion have been put forward through the years. The "Bergschrund theory" of W. D. Johnson (1904) details diurnal freezing and thawing at the base of the large crevasse (the Bergschrund) at the head of the cirque. Over time this process would weather and erode rock debris, which would then be incorporated into the glacier and transported downvalley. The "meltwater theory" of W. V. Lewis (1938) stresses the importance of the introduction of water into the area of the headwall. There may also be some pressure-release fracturing, which would allow glacial plucking to occur in the cirque. A third theory of cirque development involves the "rotational slip" of the glacier on its bed, thus providing a mechanism for the grinding of rock at the glacier base as well as a mechanism for the formation of the rock lip at the downvalley side of the cirque. Most likely, all of these processes are partially responsible for cirque development along with physical phenomena such as boundary layer interactions between cold ice and isothermal ice acting on the rock face (Fisher, 1963: 518). More recently, Embleton and King (1975: 229–238) cited abrasion, joint-block removal, freeze-thaw action, rotational slipping, and other minor factors as the basis for cirque development.

Considerable research in recent years has involved studying the morphology, elevation, and aspect of cirques. King and M. Gage (1961), E. Derbyshire (1964), J. T. Andrews (1965), P. H. Temple (1965), D. E. Sugden (1969), and E. E. Roberts (1984) all cited aspect as a dominant agent in determining the numbers and sizes of cirques. Others (e.g., Andrews and Dugdale, 1971; Gordon, 1977) looked at morphometry as an important part of cirque research: length/width ratios, length/height ratios, and several other variables have been used for cirque analysis and comparison. A more recent emphasis has been a systems approach to looking at cirque formation factors as seen in the work of G. A. Olyphant (1981).

References

Andrews, J. T. 1965. "The Corries of the Northern Nain-Okak Section of Labrador." *Geographical Bulletin* 7: 129–136.

Andrews, J. T., and Dugdale, R. E. 1971. "Quaternary History of Northern Cumberland Peninsula, Baffin Island, N.W.T.; Part V: Factors Affecting Corrie Glaciation in Okoa Bay." *Quaternary Research* 1: 532–551. A classic in the use of morphometric variables in cirque description.

Bonney, T. G. 1871. "On a Cirque in the Syenite Hills of Skye." *Geological Magazine* 8: 535–540.

———. 1877. "On Mr. Helland's Theory of the Formation of Cirques." *Geological Magazine* 14: 273–277. These two works were cornerstones of the idea of nonglacial cirque formation.

Charpentier, J. de. 1823. *Essai sur la Constitution Geognostique des Pyrenees*. Paris. The original work that used the term *cirque*.
Davis, W. M. 1909. "Glacial Erosion in North Wales." *Quarterly Journal of the Geological Society of London* 65: 281–350.
Derbyshire, E. 1964. "Cirques, Australian Landform Example no. 2." *Australian Geography* 9: 178–179.
Embleton, C., and C.A.M. King. 1975. *Glacial Geomorphology*. New York: John Wiley and Sons. An excellent reference text with an entire chapter devoted to cirques.
Fisher, J. E. 1963. "Two Tunnels in Cold Ice at 4000m on the Breithorn." *Journal of Glaciology* 2: 513–520.
Gastaldi, B. 1873. "On the Effects of Glacier Erosion in Alpine Valleys." *Quarterly Journal of the Geological Society of London* 29: 396–401.
Gordon, J. E. 1977. "Morphometry of Cirques in the Kintail-Affric-Cannich Area of Northwest Scotland." *Geografiska Annaler* 59A: 177–194.
Gregory, J. W. 1915. "The Geology of the Glasgow District." In *Proceedings of the Geological Association* 26: 162.
Johnson, W. D. 1904. "The Profile of Maturity in Alpine Glacial Erosion." *Journal of Geology* 12: 569–578. Proposed the "Bergschrund theory" of cirque development.
King, C.A.M., and Gage, M. 1961. "Note on the Extent of Glaciation in Part of West Kerry." *Irish Geographer* 4: 202–208.
Lewis, W. V. 1938. "A Melt-Water Hypothesis of Cirque Formation." *Geological Magazine* 75: 294–265. Proposed the "melt-water theory" of cirque formation.
———. 1960. "Norwegian Cirque Glaciers." *Royal Geographical Society Research Series* 4: 104 pp.
Olyphant, G. A. 1981. "Interaction among Controls of Cirque Development: Sangre de Cristo Mountains, Colorado, USA." *Journal of Glaciology* 27: 449–458.
Ramsey, A. C. 1859. "The Old Glaciers of Switzerland and North Wales." In Ball, J., ed., *Peaks, Passes, and Glaciers*. London: Longman, Green, Longman and Roberts, pp. 400–466. First proposed the idea that cirques were glacially produced.
Roberts, E. E. 1984. "The Quaternary History of the Falkland Islands." Ph.D. dissertation, University of Aberdeen.
Sugden, D. E. 1969. "The Age and Form of Corries in the Cairngorms." *Scottish Geographical Magazine* 85: 34–46.
Temple, P. H. 1965. "Some Aspects of Cirque Distribution in the West-Central Lake District, Northern England." *Geografiska Annaler* 47A: 185–193.

Sources of Additional Information

For an excellent sequence of works by an individual on the formation processes of cirques see W. V. Lewis, "The Function of Meltwater in Cirque Formation," *Geographical Review* 30 (1940): 64–83; idem, "Valley Steps and Glacial Valley Erosion," *Transactions, Institute of British Geographers* 14 (1947): 19–44; idem, "Glacial Movement by Rotational Slipping," *Geografiska Annaler* 31 (1949): 146–158; and idem, "Pressure Release and Glacial Erosion," *Journal of Glaciology* 2 (1954): 417–422. For a good review of glacial and nival morphology see I. S. Evans, "The Geomorphology and Morphology of Glacial and Nival Areas," in Richard J. Chorley, ed., *Water, Earth, and Man* (London: Methuen & Co., 1969), pp. 369–380. For information on cirques and climatic influences see E. Derbyshire and I. S. Evans, "The Climatic Factor in Cirque Variation," in E. Derbyshire, ed., *Geomorphology and Climate* (New York: John Wiley and Sons, 1976), pp. 447–494. For good general information on cirques see Rhodes W. Fairbridge, ed., *The Encyclopedia of Geomorphology*, Encyclopedia of Earth Sciences Series, vol. 3 (Stroudsburg, Pa.: Dowden, Hutchinson and Ross, 1968), and Richard F. Flint, *Glacial and Quaternary Geology* (New York: John Wiley and Sons, 1971).

CLIMATOLOGY. 1. The study of the collective or aggregate state of the atmosphere in a given area or region over a specified period. 2. The study of the average of weather conditions at a given place usually based on statistics gathered over a long period and including average values as well as deviations from those average conditions.

The factors associated with climate usually include the common weather elements of temperature, humidity, PRECIPITATION, sunshine, and wind speed. These factors are summarized in terms of arithmetic averages for specified periods, usually based on records of 30 years or more. Also included in the science of climatology, or the study of climate, are the factors that determine climate variations. These factors include latitude (see LATITUDE/LONGITUDE), altitude, location, aspect, and wind patterns as well as many other smaller-scale influences.

Observations about weather and climate are as old as recorded history. Early *Homo sapiens* were greatly affected by climate but they had little understanding of climatic mechanisms. To interpret the many mysteries of climate they relied on superstition and pagan religion. Many early civilizations named gods to represent the major climatic elements. For example, Thor was the thunder god in Norse mythology, the north wind was guided by the Greek god Boreas, Ra was the Egyptian sun god, and Jupiter Pluvius was the Roman god of rain (Critchfield, 1974: 5). Even today in some of the arid regions of the southwestern United States Native Americans still hold rain dances.

Climatology originated in ancient Greece, and the origin of the word *climatology* comes from the Greek words *klima*, meaning slope or latitude of the earth, and *logos*, a discourse or study. Important among these early Greek conceptions of climate were the five world climatic zones (torrid, north and south temperate, north and south frigid) of Parninides developed in the fifth century B.C.; the important work on medical climatology *Airs, Waters, and Places* written by Hippocrates about 400 B.C.; and perhaps the most famous of the early Greek writings, *Meteorologica* by Aristotle, written around 350 B.C. (Oliver, 1981: 2).

Few new advances in the scientific study of climate took place after the Greeks until the sixteenth and seventeenth centuries, when instruments were developed to measure weather changes. In 1593 Galileo developed the first thermometer, but perhaps even more important was the discovery of the principle of the mercurial barometer in 1643 by Evangelisto Torricelli, an associate of Galileo at Florence. The development and refinement of these instruments made it possible to compare climatic data from different times and places. By the mid-seventeenth century enough information had been gathered by early explorers so that the British astronomer Edmund Halley could construct the first chart of the winds, which he included in his *Historical Account of the Trade Winds and*

Monsoons (1686). This work was further refined by George Hadley (1685–1768), who wrote a treatise on the trade winds and how they were affected by the rotation of the earth (Bynum, Brown, and Porter, 1981: 265).

The eighteenth century saw an increase in the refinements of climatological instrumentation. Data collection and description were the areas of primary emphasis. The following century, however, saw major advances in the understanding of climatic processes. Alexander von Humboldt developed the concept of the isotherm in 1817 and used isotherms on the first map showing temperature variations. Shortly after this, in 1827, Heinrich Wilhelm Dove developed the idea that storms occur when polar and equatorial air masses converge and that this convergence is an important element in local and regional climates.

Americans became interested in climate phenomena in the early nineteenth century, primarily as it was related to disease control (Brown, 1873). Out of these climatic observations came the first general work on the climate of the United States, *The Climate of the United States and Its Endemic Influences* (1842) by Samuel Forry. Forry relied much on the earlier work by von Humboldt, and the frontispiece of his book contains the first temperature map of the United States.

Climate study in America received an enormous boost in 1849 with the appointment of Joseph Henry as administrator of the Smithsonian Institution. Henry organized a nationwide net of observation stations, and the results and analysis of these observational data were published by the Smithsonian in James H. Coffin's (1854) study of winds in the northern hemisphere. Perhaps the most significant work of the Smithsonian was completed by one of Henry's assistants, Lorin Blodget. Blodget had a great aptitude for statistical analysis, and shortly after being employed by the Smithsonian he began to publish the results of his analysis of climatological data (Blodget, 1853). Blodget's major contribution, however, was his book *Climatology in the United States*. In a discussion of the history of American climatology, John Leighly said that Blodget's Climatology book remains

> the great monument of American climatology in the 19th century. Its author brought together the material he had published elsewhere, revised his maps, and skillfully fashioned the whole into a rounded scientific and literary fabric. He gave his readers, in truth, God's plenty: tables of observations, maps, quotations from traders, forceful verbal characterizations. (1954: 338)

The meteorological activities of the Smithsonian Institution were taken over in 1870 by the National Weather Service, originally under the direction of the Army Signal Corps. Toward the end of the nineteenth century state weather services were also initiated.

Many advancements in climatology have been made in the twentieth century. Of primary importance to geographers have been the development and refinement of climate-classification systems. The most used climate-classification system in its original form or with modifications was the system developed by the German biologist Wladimir Köppen. Köppen's scheme related climate and VEGETATION but also provided an objective, numerical basis for climatic types based on climatic elements. Köppen's original work appeared in 1900, but he later revised it with more emphasis on temperature, rainfall, and their seasonal characteristics. The German climatologist R. Geiger collaborated with Köppen on revised editions of his system (Köppen and Geiger, 1936), and several more recent modifications (Trewartha, 1954; Strahler and Strahler, 1983) have been presented.

Another important pioneer in the development of a climate-classification system was the American climatologist C. W. Thornthwaite, whose first classification of North American climates appeared in 1931 and was extended to the world in 1933. Although Thornthwaite's system was similar to Köppen's, he concentrated on the effectiveness of precipitation and temperature efficiency. In a further refinement of his system, Thornthwaite (1948) proposed a basis for climatic classification based on the concept of potential EVAPOTRANSPIRATION, that is, the amount of moisture that would be evaporated from soil and transpired by plants if it were available.

The last several decades have seen increasing sophistication in measurement techniques (Bennett, 1979) and emphasis on both the synoptic (Court, 1957) and dynamic (Hare, 1957) aspects of climatology. In a recent discussion of climatological research, K. J. Gregory defined these two areas of climatology:

> Synoptic climatology is devoted to obtaining insight into local or regional climates by examining the relationship of weather elements to atmospheric circulation processes. Dynamic climatology deals with the explanatory description of world climates in terms of the circulation or disturbance of the atmosphere. (1985: 53)

In a recent review of research on synoptic and dynamic climatology, P. A. Smithson concluded that "dynamic and synoptic climatology offer a wide range of research activity from theory and principles to applications in time and space" (1986: 106).

Applied climatology is another aspect of climatic study that has received much attention in recent years. Two recent reviews in English (Musk, 1985) and French (Escourrov, 1984) have highlighted recent advances in applied climatology. Among the many recent topics covered in applied climatology are the nuclear-winter debate (Elsom, 1984), climatic hazards like drought and desertification (Hare, 1984), the implications of climatic change (Bach, 1984), and the role that climate plays in a range of issues affecting social and economic development (Biswas, 1984).

References

Bach, W. 1984. *Our Threatened Climate*. Dordrecht: Elsevier. A good discussion of the ways of averting the CO_2 problem through rational energy use.

Bennett, R. J. 1979. "Statistical Problems in Forecasting Long-Term Climate Change." In N. Wrigley, ed., *Statistical Applications in the Spatial Sciences*. London: Pion, pp. 242–256. A good article summarizing recent statistical advances in the study of climate.

Biswas, A. K. 1984. *Climate and Development*. Dublin: Tycody International Publishers. A good overview of the role of climate in development issues.

Blodget, Lorin. 1853. "On the Distribution of Precipitation in Rain and Snow on the North American Continent." In *Proceedings of the American Association for the Advancement of Science* 7: 101–108.

Brown, H. E. 1873. *The Medical Department of the United States Army from 1775 to 1873*. Washington, D.C.: U.S. Army.

Bynum, W. F., Brown, E. G., and Porter, R. 1981. *Dictionary of the History of Science*. Princeton, N.J.: Princeton University Press. A short discussion of the evolution of climate study.

Coffin, James H. 1854. *Winds of the Northern Hemisphere*. Smithsonian Contributions to Knowledge, no. 6. Washington, D.C.: Smithsonian Institution.

Court, A. 1957. "Climatology: Complex, Dynamic, and Synoptic." *Annals of the Association of American Geographers* 47: 125–136.

Critchfield, Howard J. 1974. *General Climatology*. 3d ed. Englewood Cliffs, N.J.: Prentice-Hall. A good textbook on climatology.

Elsom, D. M. 1984. "Climatic Change Induced by a Large Nuclear Exchange." *Weather* 39: 268–271. A short discussion of the idea of a nuclear winter.

Escourrov, G. 1984. "La Climatologie Appliquee en France." *Annales de Geographie* 516: 249–253. A review of recent research in applied climatology.

Forry, Samuel. 1842. *The Climate of the United States and Its Endemic Influences*. New York.

Gregory, K. J. 1985. *The Nature of Physical Geography*. London: E. Arnold. A recent book that traces the history of developments in physical geography.

Hare, F. K. 1957. "The Dynamic Aspects of Climatology." *Geografiska Annaler* 39: 87–104.

———. 1984. "Climate, Drought, and Desertification." *Nature and Resources* 20: 2–8. A short discussion of drought problems in the Sahel region of West Africa.

Köppen, W. 1931. *Grundriss der Klimakunde*. Berlin: Walter DeGruyter Company. A classic study and development of a climate-classification system.

Köppen, W., and Geiger, R. 1936. *Handbuch der Klimatologie*. Berlin: Gebruder Borntraeger.

Leighly, John. 1954. "The Growth of Climatology in the United States." In Preston E. James and Clarence F. Jones, eds., *American Geography: Inventory and Prospect*. Syracuse, N.Y.: Syracuse University Press. A good discussion of the historical evolution of climate study in the United States.

Musk, Leslie F. 1985. "Applied Climatology." *Progress in Physical Geography* 9: 442–453. A review of recent research in applied climatology. Contains a good bibliography.

Oliver, John E. 1981. *Climatology: Selected Applications*. London: V. H. Winston and Sons. A recent textbook on applied climatology.

Smithson, P. A. 1986. "Synoptic and Dynamic Climatology." *Progress in Physical Geography* 10: 100–109. A review of recent research in synoptic and dynamic climatology with a good bibliography.

Strahler, Arthur N., and Strahler, Alan H. 1983. *Modern Physical Geography*. 2d ed. New York: John Wiley and Sons. A good analysis of climate classification is presented with descriptions of each climate category.

Thornthwaite, C. W. 1931. "The Climates of North America According to a New Classification." *Geographical Review* 21: 633–655.

———. 1933. "The Climates of the Earth." *Geographical Review* 23: 433–440.

———. 1948. "An Approach Toward a Rational Classification of Climate." *Geographical Review* 38: 55–94. A classic paper on the concept of potential evapotranspiration and its application to climate classification.

Trewartha, Glenn T. 1954. *An Introduction to Climate*. 3d ed. New York: McGraw-Hill Book Co. A modified version of the Köppen classification system is presented.

Sources of Additional Information

For good general descriptions of climate principles see Helmut E. Landsberg, "Climatology," in Rhodes Fairbridge, ed., *The Encyclopedia of Atmospheric Sciences and Astrogeology*, Encyclopedia of Earth Sciences Series, vol. 2 (New York: Reinhold Publishing Co., 1967); W. D. Sellers, "Climatology," in *McGraw-Hill Encyclopedia of Science and Technology*, 5th ed. (New York: McGraw-Hill Book Co., 1982); P. R. Crowe, *Concepts in Climatology* (London: Longmans Green and Co., 1971); G. T. Trewartha, *An Introduction to Weather and Climate*, 2d ed. (New York: Holt, Reinhart and Winston, 1977). Much excellent material has been written relating climate to other environmental factors. For more information see J. E. Aronin, *Climate and Architecture* (New York: Reinhold Publishing Co., 1956); R. Claiborne, *Climate, Man, and History* (New York: W. W. Norton and Co., 1970); M. J. Tooley and G. M. Sheail, eds., *The Climatic Scene* (London: Allen and Unwin, 1985); J. G. Lockwood, "The Influence of Vegetation on the Earth's Climate," *Progress in Physical Geography* 7 (1983): 81–89; and Jane F. Gillooly and J.R.E. Lutjeharms, "The Ocean and Climate," *South African Journal of Science* 80 (1984): 36–40.

CLOUD. A visible mass of condensed water vapor or ice particles suspended in the atmosphere at elevations up to 45,000 or more feet above sea level.

According to James Murray et al. (1933), the term *cloud* was used in its modern sense as early as the beginning of the fourteenth century, although scientific studies of clouds and cloud physics have occurred primarily since the end of World War II. Cloud studies are now a fundamental component of meteorology.

"Clouds form," according to B. J. Mason, "primarily as the result of vertical motions in the atmosphere, the study of which is one of the most difficult and fundamental in meteorology" (1962: 9). He suggested four classes of clouds: (1) layer clouds that are formed by widespread, regular ascent; (2) layer clouds that result from irregular stirring motions; (3) convective clouds; and (4) clouds that form as a result of orographic disturbances.

Water vapor (see PRECIPITATION), a gaseous form of water that has entered the atmosphere primarily as a result of evaporation (see LAPSE RATE), in clouds condenses on minute particles known as condensation nuclei, primarily dust, ash, and salt particles that have suspended in the atmosphere. Such condensation begins to occur when a rising air mass reaches its dew point (see LAPSE RATE) temperature. (See also ADIABATIC HEATING AND COOLING.) The study of condensation nuclei is difficult, although a thorough theoretical discussion can be found in the 1957 work of Mason and a more generalized discussion in his 1962 work. Without the presence of condensation nuclei, water-droplet formation would occur only in highly supersaturated air. These water droplets range in size from 0.02 to 0.06 millimeters in diameter.

The most widely used cloud-classification systems are descriptive, and they are based on the early work of Luke Howard in 1803 (Fairbridge, 1967). The current system is that used by the World Meteorological Organization (1956). It is also used in most meteorology and climatology texts, as well as in introductory physical geography texts such as that of Arthur Strahler and Alan Strahler (1984).

Cloud types are classified according to a combination of two criteria, form or appearance and altitude above sea level. On the one hand, there are two major cloud-form classes, cumuliform and stratiform, which may be described, respectively, as globular and layered (Strahler and Strahler, 1984: 97).

On the other hand, these two forms may be cross-classified by height into three categories: (1) high clouds, which are those at or above 20,000 feet, (2) middle clouds, which are those found between 10,000 and 20,000 feet, and (3) low clouds, which are those found below 10,000 feet. High clouds include cirrus, cirrocumulus, and cirrostratus. Middle clouds are comprised primarily of altocumulus and altostratus. Low clouds include stratus, nimbostratus, stratocumulus, fair weather cumulus, and cumulus. In addition, there is the towering cumulonimbus, a vertically developed cloud that may extend from very low elevations upward to 20,000 feet or more. This cloud is associated with THUNDERSTORMS. Within these cloud groups or families, there are considerable variations, and a list of specific terms that describe major variations can be found in the work of Rhodes Fairbridge (1967: 235).

Clouds are important as visual indicators of weather conditions. They also play a major role in atmospheric heating and cooling. During daylight hours clouds reflect incoming solar radiation back into the upper atmosphere, whereas at night they act to trap outgoing terrestrial radiation. Thus, all else being equal, clouds reduce daytime temperatures below what they would be otherwise and raise nighttime temperatures above what they would be otherwise.

Fog should be considered here as well, because it is really only a cloud layer that has developed in contact with either a land or sea surface. Two types of fogs are typically noted by meteorologists. *Radiation fog* typically forms at night under temperature inversions when the air at and near the surface cools below the dew point. *Advection fog*, on the other hand, results from movement of warm moist air over a cool surface. Strahler and Strahler (1984: 99) provided a brief discussion of these two fog types.

References

Fairbridge, Rhodes W. 1967. "Clouds." In Rhodes W. Fairbridge, ed., *The Encyclopedia of Atmospheric Sciences and Astrogeology*. Encyclopedia of Earth Sciences Series. vol. 2. New York: Reinhold Publishing Corp., pp. 234–236.

Fletcher, N. H. 1962. *The Physics of Rain Clouds.* Cambridge: Cambridge University Press. A major early work on cloud physics.

Mason, B. J. 1957. *The Physics of Clouds.* Oxford: Clarendon Press. The premier source for studies of cloud physics.

———. 1962. *Clouds, Rain, and Rainmaking.* Cambridge: Cambridge University Press. A classic work on the subject, still cited as an authority by virtually all meteorologists.

Murray, James A. H., et al., eds. 1933. *The Oxford English Dictionary.* Oxford: Clarendon Press.

Strahler, Arthur N., and Strahler, Alan H. 1984. *Elements of Physical Geography.* 3d ed. New York: John Wiley and Sons. Probably the most widely used of all current physical geography texts and an excellent beginning point for anyone interested in that topic.

World Meteorological Organization. 1956. *The International Cloud Atlas.* Geneva: World Meteorological Organization. The major single source of information for identifying cloud types and distributions.

Sources of Additional Information

Discussions of clouds and cloud types can be found in most introductory physical geography texts, including the following: Tom McKnight, *Physical Geography: A Landscape Appreciation* (Englewood Cliffs, N.J.: Prentice-Hall, 1984), and E. Willard Miller, *Physical Geography: Earth Systems and Human Interactions* (Columbus, Ohio: Charles E. Merrill Publishing Co., 1984). More detailed discussions can be found in the following: Joe R. Eagleman, *Meteorology: The Atmosphere in Action* (New York: D. Van Nostrand Co., 1980), and Paul E. Lydolph, *The Climate of the Earth* (Totowa, N.J.: Rowman and Allanheld, Publishers, 1985). A concise look at cloud physics is provided in W. G. Durbin, "An Introduction to Cloud Physics," *Weather* 16 (1961): 71–82, 113–125.

CONDENSATION. See LAPSE RATE.

CONDENSATION NUCLEI. See CLOUD.

CONTINENTAL DRIFT. See PLATE TECTONICS.

CONTINENTAL GLACIER. 1. An ice sheet of very great size, large enough to cover a subcontinent or continent. 2. An ice sheet emanating from some lowland source region.

Much of the discussion on the concept of VALLEY GLACIER is also relevant to continental glaciers and ice sheets. Alpine and continental glacier landforms, processes, and ice characteristics have much in common. However, as stated by H. F. Garner:

> there are differences in magnitude, genesis and consequences that are important. At the peak of Pleistocene [see PLEISTOCENE] glacial advances, it is estimated that some 30% of the earth's land surfaces were under ice. Probably no more than 15% to 20% of that total was in the form of alpine glaciers. Also, though one can argue

that any glacier requires a starting point, continental ice sheets apparently reflect more of a geographically localized hydrographic condition than a topographic shelter. The major centers of continental glacier spreading probably reflect regional areas of low temperature and ready access to precipitation [see PRECIPITATION]. (1974: 497)

The study of the cyclical nature of major glaciations dependent upon climatic influences dealt initially and most profoundly with the continental ice sheets of the Pleistocene. The seminal work detailing this theory was *The Great Ice Age* by James Geikie (1874). It was a systematic account of the glacial epoch and paid particular attention to climate variation. The idea of multiple glacial advances was strongly implied by this work. Several decades later the work of Albrecht Penck and Eduard Bruckner (1909) confirmed Geikie's theory with their exhaustive study of the four Alpine glaciations in the Alps.

Between 1900 and the 1960s, the major research thrust was the mapping of glacial ice sheets and the classification of specific LANDFORMS. Most mapping of the marginal limits of ice advances, "especially those resulting from deposition, is slow and the delineation of the marginal positions of the ice sheets depends on an accurate assessment of their meaning, in terms of the marginal processes of glacial retreat" (Tinkler, 1985: 191). Because of the obvious technical problems involved, most of this work was done on postglacial landscapes and little, if any, done during actual glaciation. The classification of landforms was done using residual landforms not currently evolving forms. R. F. Flint's (1929) work distinguished active and dead ice and the landforms of glacial retreat. Others, such as S. A. Andersen (1931) in Denmark, soon followed. Most research was regional and included the work on drift boundaries (Flint, 1943), the delimitation of isostatically deformed shorelines (Hough, 1958), and the definitive work on till fabric by C. D. Holmes (1941). The problems of this research are explained well by K. J. Tinkler:

> A particular problem here was that it was much easier to study glacial behavior at existing ice margins than the glacier bed hundreds of kilometers behind the ice front. Consequently, the emphasis of work until well into the 1960s was upon the progress of deglaciation, partly because of this marginal bias, and partly because the evidence of glacial retreat is, for obvious reasons, the best preserved. Yet, at the height of glaciation, ice covered some 30% of the earth's surface and much erosion and deposition must have occurred which cannot be attributed to the processes of deglaciation. (1985: 206–207)

An obvious need in continental glacier studies was large-scale analysis of active ice sheets. Before World War II there were no significant scientific investigations in either Greenland or the Antarctic (the two areas of active continental glaciation) (Tinkler, 1985: 193). Significant results and interpretation on glacial landform formation and ice movement began to appear in the late 1950s and 1960s (see Schytt, 1956; Bishop, 1957; Hollin, 1962; Zotikov, 1963; Hansen and Langway, 1966).

The 1970s and early 1980s saw a renewed effort to reconstruct the chronology of the major ice advances of the Pleistocene. For example, D. E. Sugden (1977, 1978) has worked on the morphology, dynamics, and other characteristics of the Laurentide ice sheet. G. S. Boulton et al. (1977) have done similar work for Britain. More recently the CLIMAP project has resulted in the comprehensive volume of the chronology of the Pleistocene, *The Last Great Ice Sheets* (Denton and Hughes, 1982). Although this last modeling effort dealt only with static ice conditions, it is the best attempt thus far at explaining Pleistocene glacial history.

A current topic of considerable interest is ice-sheet modeling with particular emphasis on isostacy. The mapping of deformed late-glacial and postglacial shorelines has been the basis for most of this work. J. T. Andrews (1970) gave an excellent summary of much of this research. Other studies in the areas of isostacy, sea-level changes, and ice-sheet/bedrock relationships are those of G. de Q. Robin et al. (1977) and the 1984 Binghamton Symposium on tectonic GEOMORPHOLOGY.

References

Andersen, S. A. 1931. "The Waning of the Last Continental Glacier in Denmark as Illustrated by Varved Clays and Eskers." *Journal of Geology* 39: 609–624.

Andrews, J. T. 1970. *Post-Glacial Uplift in Arctic Canada*. Special Publication, no. 2. London: Institute of British Geographers. An excellent summary of work on isostacy and shoreline changes.

Bishop, B. C. 1957. *Shear Moraines in the Thule Area, Northwest Greenland*. Snow, Ice, and Permafrost Research Establishment Research Report, no. 17. Washington, D.C.: U.S. Army, Corps of Engineers.

Boulton, G. S.; Jones, A. G.; Clayton, K. M.; and Kenning, M. J. 1977. "A British Ice Sheet Model and Patterns of Glacial Erosion and Deposition in Britain." In F. W. Shotten, ed., *British Quaternary Studies*. Oxford: Clarendon Press, pp. 231–246.

Denton, G. H., and Hughes, T. J., eds., 1982. *The Last Great Ice Sheets*. New York: John Wiley and Sons. An indispensable text for work in reconstruction of Pleistocene glacial events.

Flint, R. F. 1929. "The Stagnation and Dissipation of the Last Ice Sheet." *Geographical Review* 19: 256–289. Seminal work on the landforms of glacial retreat.

———. 1943. "Growth of the North American Ice Sheet during the Wisconsin Age." *Geological Society of America Bulletin* 54: 325–362.

Garner, H. F. 1974. *The Origin of Landscapes*. New York: Oxford University Press.

Geikie, J. 1874. *The Great Ice Age*. London: Dolby and Ibister. Master work initially detailing the advances of Pleistocene ice due to climate changes.

Hansen, B. L., and Langway, C. C. 1966. "Deep Core Drilling in Ice and Core Analysis at Camp Century, Greenland, 1961–1966." *Antarctic Journal of the U.S.* 1: 207–208.

Hollin, J. T. 1962. "On the Glacial History of Antarctica." *Journal of Glaciology* 4: 173–195.

Holmes, C. D. 1941. "Till Fabric." *Geological Society of America Bulletin* 52: 1299–1354.

Hough, J. L. 1958. *Geology of the Great Lakes*. Urbana: University of Illinois Press.
Penck, A., and Bruckner, E. 1909. *Die Alpen in Eiszeitalter*. Leipzig: Tauchnitz. This work confirms the theory of multiple glacial events through work in the European Alps.
Robin, G. de Q.; Drewry, D. J.; and Meldrum, D. T. 1977. "International Studies of the Ice Sheet and Bedrock." *Philosophical Transactions of the Royal Society of London*, Series B, 279: 185–196.
Schytt, V. 1956. "Lateral Drainage Channels along the Northern Side of the Moltka Glacier, Northern Greenland." *Geografiska Annaler* 38: 64–77.
Sugden, D. E. 1977. "Reconstruction of the Morphology, Dynamics, and Thermal Characteristics of the Laurentide Ice Sheets at Its Maximum." *Arctic and Alpine Research* 9: 21–47.
———. 1978. "Glacial Erosion by the Laurentide Ice Sheet." *Journal of Geology* 20: 367–391.
Tinkler, K. J. 1985. *A Short History of Geomorphology*. Totowa, N.J.: Barnes and Noble Books. An excellent reference on the history of geomorphic research including glacial geomorphology.
Zotikov, I. A. 1963. "Bottom Melting of the Central Zone of the Ice Shield of the Antarctic Continent." *International Association of Scientific Hydrology Bulletin* 8: 36–43.

Sources of Additional Information

Additional work on continental glaciers and their relationship with bedrock and sedimentation can be found in J. B. Anderson, D. D. Kurtz, E. W. Domack, and K. M. Balshaw, "Glacial and Glacial Marine Sediments of the Antarctic Continental Shelf," *Journal of Geology* 88 (1980): 399–414; D. J. Drewry, S. R. Jordan, and E. Jankowski, "Measured Properties of the Antarctic Ice Sheet: Surface Configuration, Ice Thickness, Volume, and Bedrock Characteristics," *Annals of Glaciology* 3 (1982): 83–91; J. Gordon, "Ice-Scoured Topography and Its Relationships to Bedrock Structure and Ice Movement in Parts of Northern Scotland and West Greenland," *Geografiska Annaler* 63A (1981): 55–65; and T. J. Kemmis, "Importance of the Regulation Process to Certain Properties of Basal Tills Deposited by the Laurentide Ice Sheet in Iowa and Illinois, USA," *Annals of Glaciology* 2 (1981): 147–152. A general collection of the classic works on continental glaciers and Pleistocene geology include Clifford Embleton and Cuchlaine A. M. King, *Glacial Geomorphology* (New York: John Wiley and Sons, 1975); Richard Foster Flint, *Glacial Geology and the Pleistocene Epoch* (New York: John Wiley and Sons, 1947); Richard Foster Flint, *Glacial and Pleistocene Geology* (New York: John Wiley and Sons, 1957); Richard Foster Fling, *Glacial and Quaternary Geology* (New York: John Wiley and Sons, 1971); D. E. Sugden and B. S. John, *Glaciers and Landscape: A Geomorphological Approach* (London: E. Arnold, 1976); W. B. Wright, *The Quaternary Ice Age* (London: Macmillan, 1914); and H. E. Wright and D. G. Frey, eds., *The Quaternary of the United States* (Princeton, N.J.: Princeton University Press, 1965).

CONVECTION. Within a liquid or gas, convection is a mass movement of fluid elements that results from differential heating of the liquid or gas and leads to changes in temperature and density within the fluid. Within the atmosphere, it refers specifically to updrafts of warm air.

Heat energy can be transferred in three ways: radiation, conduction, and convection. According to R. G. Barry and R. J. Chorley, convection "occurs in fluids (including gases) which are able to circulate internally and distribute heated parts of the mass" (1970: 14). James Murray et al. (1933) argued that although the term *convection* was used as early as 1623, its current use as a term in physics relating to a mechanism for heat transfer dates from the work of W. Prout in 1834 and A. Parkes in 1869. Prout discussed the term only as a mechanism for heat transfer within water bodies, whereas Parkes used the term more broadly, in a way that would include atmospheric convection.

Within the atmosphere, convection serves two important functions. First, it is one way that heat is transferred from the earth's surface into the lower atmosphere, and, as such, it serves as part of the complex system that maintains the atmosphere's heat balance. Second, convection is one way that air is forced to rise, thus creating one of the necessary conditions to initiate precipitation. In this role it generates strong updrafts of air that develop into a convection cell.

Updrafts of air are common over heated surfaces, creating "thermals." Hawks, eagles, and even glider pilots often take advantage of these updrafts to gain elevation effortlessly. As the rising column of air cools, however, it may reach the dew-point temperature, at which time a CLOUD will begin to form, normally a cumulus cloud. If the air is unstable, large cumulus-cloud development may take place, creating a cumulonimbus cloud. Such clouds may reach altitudes of as much as 40,000 feet, and within them a mixture of updrafts and downdrafts develops. At this point thunderstorm development begins, and hail may follow as well. (see THUNDERSTORM.)

Convection cells and thunderstorms are not the only result of convection. As discussed under ATMOSPHERIC CIRCULATION, convection plays a major role in the development of earth's major wind and pressure systems.

References

Barry, R. G., and Chorley, R. J. 1970. *Atmosphere, Weather, and Climate*. New York: Holt, Rinehart and Winston. Despite its age, this book remains an excellent reference for discussion of atmospheric processes.

Murray, James A. H., et al., eds. 1933. *The Oxford English Dictionary*. Oxford: Clarendon Press.

Sources of Additional Information

Convection is introduced and discussed to some extent in the following introductory physical geography texts: Arthur N. Strahler and Alan H. Strahler, *Elements of Physical Geography*, 3d ed. (New York: John Wiley and Sons, 1984); E. Willard Miller, *Physical Geography: Earth Systems and Human Interactions* (Columbus, Ohio: Charles E. Merrill Publishing Co., 1985); and John E. Oliver, *Physical Geography: Principles and Applications* (North Scituate, Mass.: Duxbury Press, 1979). More detailed discussions can be found in James D. Hanwell, *Atmospheric Processes* (London: Allen and Unwin, 1980); Paul E. Lydolph, *The Climate of the Earth* (Totowa, N.J.: Rowman and Allanheld, Publishers, 1985); John E. Oliver and John J. Hidore, *Climatology: An Introduction*

(Columbus, Ohio: Charles E. Merrill Publishing Co., 1984); and John E. Oliver, *Climate and Man's Environment: An Introduction to Applied Climatology* (New York: John Wiley and Sons, 1973). A detailed discussion of convective adjustment can be found in A. S. Monin, *An Introduction to the Theory of Climate* (Boston: D. Reidel Publishing Co., 1986).

CORIOLIS EFFECT. The deflecting motion in the atmosphere and other fluids caused by the rotation of the reference system of the earth's surface.

"It is a curious fact that all things which move over the surface of the earth tend to sidle from their appointed paths—to the right in the Northern Hemisphere, to the left in the Southern Hemisphere" (McDonald, 1952: 72). This tendency to sidle is called the Coriolis effect. It is not a force in strict Newtonian physics terms since it is due, in simplistic terms, only to the rotation of the earth out from under a moving object.

As early as 1735 Hadley noticed the effects of the curving of the atmosphere in motion and the curving movement of the ocean currents (Hadley, 1735). The curving movements of these fluids were not explained, however, for another 100 years. G. G. Coriolis, a French mathematician, made the first complete analysis of this effect in 1835 (Coriolis, 1835). Almost a decade later he published a full mathematical explanation of the effect (Coriolis, 1844).

Coriolis was a mathematician and did not relate his work directly to the global fluid movements. This was soon done by W. Ferrel (1889: 78), who produced Ferrel's law, which is paraphrased in the opening sentence, above. He also estimated the magnitudes of these global movements and publicized the Coriolis parameter by which the magnitude of the effect is calculated. The Coriolis parameter is:

$$f = 2 \omega \sin \phi$$

where f is the parameter, ω is the earth's angular velocity, and $\sin \phi$ is the sine of the latitude where the measurement is being taken. R. G. Barry and A. H. Perry (1973: 31) and A. J. Simmons and L. Bengtsson (1984: 41) gave a thorough mathematical explanation for this calculation.

The Coriolis effect is a complex mathematical manipulation when put into the context of global circulation patterns. But there are simple, graphic explanations that help the nonmathematician to understand it. One must first understand the rotation of the earth in its west to east movement. Herbert Riehl expressed this vividly: "A person standing near the pole turns round on his heels once in 24 hours, while a person on the equator makes one somersault in space" (1965: 121). If a person were standing at the North Pole and shot a rocket directly south along the 0° meridian, and the rocket took one hour to reach the equator, the projectile would hit the equator at 15°W. The earth has rotated out from under the rocket and makes the rocket appear to curve to the right even though it is actually going straight with a reference in space.

The significance of this apparent movement is very real. In the earth's atmosphere there are movements of air (wind) from high to low pressure. This wind does not go directly from the high to the low, however, because of the Coriolis effect. Air in the northern hemisphere moves to the right, in the southern hemisphere to the left. Without other interference such as surface friction, the winds would actually circle the high and low pressure centers and be parallel to the isobars. This kind of wind is called a geostrophic wind and occurs only at high altitudes away from surface interference. Wind at lower altitudes is affected not only by the Coriolis effect but also by other interference and does not parallel the isobars. Recent research deals not directly with the Coriolis effect but indirectly by looking at circulation-generation modeling, modeling of the boundary-layer areas between upper and surface winds, and global circulation-pattern models (e.g., Marchuk et al., 1979; Simmons and Bengtsson, 1984). The Coriolis effect remains a basic and integral part of this new thrust toward atmospheric modeling.

References

Barry, R. G., and Perry, A. H. 1973. *Synoptic Climatology—Methods and Applications.* London: Methuen & Co.

Coriolis, G. G. de. 1835. "Memoire sur les Equations du Mouvement Relatif des Systemes de Corps." *Journal de L'Ecole Polytechnique* 15: 142-154. The original complete explanation of the effect.

———. 1844. *Traite de la Mecanique de Corps Solides.* Paris. An expanded, mathematical version of the 1835 paper.

Ferrel, W. 1889. *A Popular Treatise on the Winds.* London: Macmillan. Puts the Coriolis effect into the context of atmospheric circulation for the first time.

Hadley, G. 1735. "Concerning the Cause of the General Trade Winds." *Philosophical Transactions of the Royal Society* 39: 58-62. Hadley first recognized the phenomenon but did not explain it.

Marchuk, G. I.; Dymnikov, V. P.; Lykosov, V. N.; Galin, V. Ya.; Bobyleva, I. M.; and Perov, V. L. 1979. "A Global Model of the General Atmospheric Circulation." *Atmospheric and Oceanic Physics* 15: 321-331.

McDonald, J. E. 1952. "The Coriolis Effect." *Scientific American* 186: 72-78. A very good review article explaining the Coriolis effect.

Riehl, H. 1965. *Introduction to the Atmosphere.* New York: McGraw-Hill Book Co.

Simmons, A. J., and Bengtsson, L. 1984. "Atmospheric General Circulation Models: Their Design and Use for Climate Studies." In Houghton, J. T., ed., *The Global Climate.* Cambridge: Cambridge University Press, pp. 37-62.

Sources of Additional Information

Overviews and discussions of atmospheric circulation and the Coriolis effect can be found in P. R. Crowe, *Concepts in Climatology* (New York: St. Martin's Press, 1971); and F. K. Lutgens and E. J. Tarbuck, *The Atmosphere: An Introduction to Meteorology,* 3d ed. (Englewood Cliffs, N.J.: Prentice-Hall, 1986). For an excellent discussion of Coriolis effect and the ocean-atmospheric system see A. H. Perry and J. M. Walker, *The Ocean-Atmosphere System* (New York: Longman, 1977). For more on definitions and short explanations see R. H. March, "Coriolis Acceleration," in *Encyclopedia of Physics*

(Reading, Mass.: Addison-Wesley Publishing Co., 1981), p. 152; S. P. Parker, ed., *McGraw-Hill Dictionary of Scientific and Technical Terms*, 3d ed. (New York: McGraw-Hill Book Co., 1984); and J. M. Pike, "Coriolis Effect," in *The Encyclopedia of Physics*, 3d ed. (New York: Van Nostrand-Reinhold Co., 1985).

CYCLONE. See ATMOSPHERIC PRESSURE.

D

DELTA. 1. An alluvial deposit at the mouth of a river or stream, often triangular and resembling the Greek letter delta. 2. A deposit of sediment found at the mouth of a stream or tidal inlet.

When a river flows into the sea or any body of standing water such as a pond or lake, it deposits its load of debris. The form that this debris takes depends upon the quantity and character of the load and the waves and currents of the body of water into which the debris is deposited. As material is deposited, it becomes easier for the river to divide and flow to each side; thus new branches form through the division of the main channel, and this alluvial deposit grows outward in the shape of a fan or triangle (Lapedes, 1978: 142).

The term *delta* originated with Herodotus (circa 430 B.C.), who described the shape of the Nile River where it empties into the Mediterranean (Schwartz, 1982: 358). This river deposit resembled the shape of the Greek letter delta (Δ). He referred to Egypt (meaning the Nile delta) as "an acquired country, the gift of the river" (Moore and Asquith, 1971: 2563). Another description of the Nile delta by Diodorus resembles a description for a geometry text: "Now where the Nile in its course through Egypt divides into several streams it forms the region which is called from its shape the Delta. The two sides of the Delta are described by the outermost branches, while its base is formed by the sea which receives the discharge from the several outlets of the river. It empties into the sea by seven mouths" (Oldfather, 1960: 5).

The modern study of deltas started in the latter part of the nineteenth century. G. R. Credner (1878) looked at the morphology, distribution, and growth of deltas, but the pioneering work at that time was done by G. K. Gilbert (1884) in his study of Lake Bonneville. Gilbert recognized that the principal charac-

Table 1
Ten Largest Deltas

River	Receiving Basin	Area $km^2 \times 10^3$
Amazon	Atlantic Ocean	467.1
Ganges-Brahmaputra	Bay of Bengal	105.6
Mekong	South China Sea	93.8
Yangtze-Kiang	East China Sea	66.7
Lena	Laptev Sea	43.6
Hwang Ho	Yellow Sea	36.3
Indus	Arabian Sea	29.5
Mississippi	Gulf of Mexico	28.6
Volga	Caspian Sea	27.2
Orinoco	Atlantic Ocean	20.6

teristic of a deltiac STREAM deposit is the prograding nature of the sediment pile into the standing body of water, leading to a succession of bottomset, foreset, and topset deposits (Morgan, 1970: 5). He succinctly described the universal process characteristic of all river mouths when he said (Gilbert, 1884: 104): "the process of delta formation depends almost wholly on the following law: the capacity and competence of a stream for the transportation of detritus are increased and diminished by the increase and diminution of the velocity." When a stream deposits its load into a receiving basin, its momentum is dispersed by its interaction with the ambient water. The result is a loss of transporting ability and deposition occurs.

In 1899 F. P. Gulliver classified deltas based on the ratio of activity between the river and the sea and the general configuration of deltas as seen from TOPOGRAPHIC MAPS. This work on classification was further refined by D. W. Johnson (1919). A more recent and comprehensive classification of deltas was developed in a book by I. Samojlov (1956). Table 1 lists the world's ten largest deltas.

The decades of the 1930s, 1940s, and 1950s saw an increasing interest in the study of deltas. Of particular geographic significance were the studies by Richard Russell (1942: 54) and H. N. Fisk (1944: 52). Fisk's studies primarily concerned the lower Mississippi River and its delta. He followed up and expanded upon the pioneering work by S. H. Lockett, who mapped the "birds' foot" shape of the lower delta of the Mississippi (Russell, 1967: 14). Among Fisk's detailed reports is "Geological Investigations of the Alluvial Valley of the Lower Mississippi River," which is one of the classics of geological literature. Through physiographic analysis of maps and airphotos (Fisk, 1955), study of numerous borings, and carbon-14 dating (Morgan, 1970: 5), it has been possible to reconstruct the chronological development of the deltaic plain and the progradational sequence of the Mississippi delta.

Recent years have seen an increasing interest in delta research primarily because of the role deltas may play in accommodating the world's energy needs. According to A. J. Scott and W. L. Fisher "In regard to mineral fuels, probably no depositional units are of more significance than those developed as parts of ancient delta systems" (1969: 10–30).

Many models of delta development have been discussed in the recent literature (Broussard, 1975; Roy, 1983; Wang, 1984); yet global variations in the relative intensities of delta-molding forces produce a wide array of different morphologies and depositional sequences. A better understanding of these complex processes could lead to important economic discoveries.

References

Bates, C. C. 1953. "Rational Theory of Delta Formation." *American Association of Petroleum Geologists Bulletin* 37: 2119–2162. An important paper using water density as a classification method.

Broussard, M. L., ed. 1975. *Deltas: Models for Exploration.* Houston: Houston Geological Society. An excellent analysis of deltas and deltaic environments.

Credner, G. R. 1878. *Die Deltas, Ihre Morphologie, Geographische Verbreitung, und Entstehungs Bedingungen.* Petermann's Geographischen Mittheilungen, Erganzungsband 12, Erganzungsheft 56.

Fisk, H. N. 1944. "Geological Investigation of the Alluvial Valley of the Lower Mississippi River." Vicksburg, Miss.: U.S. Army Corps of Engineers, Mississippi River Commission. A classic paper on the development of the Mississippi River delta.

———. 1952. "Geological Investigation of the Atchafalaya Basin and Problems of Mississippi River Diversion." Vicksburg, Miss.: U.S. Army Corps of Engineers, Mississippi River Commission.

———. 1955. "Sand Facies of Recent Mississippi Delta Deposits." In *Proceedings, 4th World Petroleum Congress.* Rome: Sec. 1, pp. 377–398.

Gilbert, G. K. 1884. *The Topographical Features of Lake Shores.* U.S. Geological Survey Annual Report, no. 5. Washington, D.C.: U.S. Geological Survey, pp. 104–108. One of the first studies to analyze delta development.

Gulliver, F. P. 1899. "Shoreline Topography." *Proceedings of the American Academy of Arts and Sciences* 34: 151–258. An early paper on shoreline classification.

Johnson, D. W. 1919. *Shore Processes and Shoreline Development.* New York: Hafner Publishing Co.

Jopling, A. V. 1965. "Hydraulic Factors and the Shape of Laminae." *Journal of Sedimentary Petrology* 35: 777–791.

Lapedes, Daniel N., ed. 1978. *McGraw-Hill Encyclopedia of Geological Sciences.* New York: McGraw-Hill Book Co. An excellent overview of delta development.

Moore, G. T., and Asquith, D. Q. 1971. "Delta: Term and Concept." *Geological Society of America Bulletin* 82: 2563–2568.

Morgan, James P., ed. 1970. *Deltaic Sedimentation, Modern and Ancient.* Tulsa, Okla.: Society of Economic Paleontologists and Mineralogists.
Oldfather, C. H. 1960. *Diodorus of Sicily.* Book I, 33: 5–7. London: W. Heinemann.
Roy, P. S. 1983. "Holocene Sedimentation Histories of Estuaries in Southwestern Australia." In E. Hodgkin, ed., *Man's Impact on the Estuarine Environment.* Department of Conservation and Environment, Western Australia.
Russell, Richard J. 1942. "Geomorphology of the Rhone Delta." *Annals of the Association of American Geographers* 32: 149–254. An early analysis of deltas by an important geographer.
———. 1954. "Alluvial Morphology of Anatolian Rivers." *Annals of the Association of American Geographers* 32: 149–154.
———. 1967. *River Plains and Sea Coasts.* Berkeley: University of California Press. A book by an important geographer who contributed much to the development of delta research.
Samojlov, I. 1956. *Die Flussmundungen.* Gotha: Veb Hermann Haack.
Schwartz, Maurice L., ed. 1982. *The Encyclopedia of Beaches and Coastal Environments.* Stroudsburg, Pa.: Hutchinson Ross Publishing Co. An excellent summary of deltas and delta research.
Scott, A. J., and Fisher, W. L. 1969. "Delta Systems and Deltaic Deposition." In W. Fisher, W.L.F. Brown, A. J. Scott, and J. H. McGowen, eds., *Delta Systems in the Exploration for Oil and Gas.* Austin: University of Texas, Bureau of Economic Geology, pp. 10–30. A good discussion of the economic development of deltas.
Wang, F. C. 1984. "The Dynamics of a River-Bay-Delta System." *Journal of Geophysical Research* 89: 8054–8060.
Wright, L. D.; Coleman, J. M.; and Erickson, M. W. 1974. *Analysis of Major River Systems and Their Deltas: Morphologic and Process Comparisons.* Technical Report, no. 156. Baton Rouge: Louisiana State University Coastal Studies Institute.

Sources of Additional Information

Brief discussions of deltas can be found in Tjeerd H. Van Andel, *McGraw-Hill Encyclopedia of Science and Technology,* 5th ed. (New York: McGraw-Hill Book Co., 1982), pp. 77–80; Karl W. Butzer, *Geomorphology from the Earth* (New York: Harper and Row, Publishers, 1976), pp. 160–162; and Dale F. Ritter, *Process Geomorphology,* 2d ed. (Dubuque, Ia.: W. C. Brown, 1986), pp. 294–301. Two thorough and excellent discussions of deltas are Richard A. Davis, Jr., ed., *Coastal Sedimentary Environments* (New York: Springer-Verlag, 1985), Chapter 1, and James M. Coleman, *Deltas: Processes of Deposition and Models for Exploration,* 2d ed. (Minneapolis: Burgess Publishing Co., 1981). A good discussion of coastal inlets can be found in J. L. Davies, *Geographical Variation in Coastal Development* (New York: Hafner Publishing Co., 1973), Chapter 11. For recent studies of deltas in glaciated regions see Jim Bogen, "Morphology and Sedimentology of Deltas in Fjord and Fjord Valley Lakes," *Sedimentary Geology* 36 (1983): 245–267; G.S.P. Thomas, "Sedimentation of a Sub-Aqueous Esker-Delta at Strabathie, Aberdeenshire," *Scottish Journal of Geology* 20 (1984): 9–20; and James P. Syvitski and George E. Farrow, "Structures and Processes in Bayhead Deltas: Knight and Bute Inlet, British Columbia," *Sedimentary Geology* 36 (1983): 217–244.

DEW POINT. See LAPSE RATE.

DRAINAGE BASIN. 1. An area of land drained by a particular river or stream. 2. An area limited by the drainage divide whereby the upper reaches supply the lower reaches with water and sediment. 3. An open process-response system dealing with the movement of water and sediment.

A simple drainage-basin classification system was proposed by E. de Martonne (1950) and modified by R. W. Fairbridge (1968: 284). This is a four-class system: (1) *exorheic*, where the basin empties directly or eventually into an ocean; (2) *endorheic*, where the basin has no outlet to an ocean; (3) *arheic*, where there are no obvious basins; and (4) *cryptorheic*, where there is no obvious basin as in KARST topography. Considerable work has also been done classifying drainage basins according to the pattern of the STREAM network on the land surface. These patterns are at least partially determined by structure, lithology, tectonic history, vegetation, climate, and basin morphology. Good synopses of these classification patterns are in the work of W. D. Thornbury (1969: 121–122) and D. S. Way (1973).

The drainage basin, especially the exorheic basin, has long been recognized as a basic unit for geomorphological thought. As early as 1802 John Playfair (1802) wrote of the adjustment of tributaries to the main stream in drainage basins. G. K. Gilbert (1877) wrote about interdependent relationships within the basin system creating a dynamic equilibrium that affects all streams and slopes within a basin. William Morris Davis (1899) wrote that the river was the dominant influence not only on the channel but also on the entire basin including the slopes. This implies the important relationship between the basin and the processes of landscape development. Throughout the nineteenth and early twentieth centuries, the drainage basin became the most useful unit for landscape analysis, but analytical methods had not yet been established to allow comprehensive study of the basin quantitatively.

Robert E. Horton began rectifying this deficiency. Horton's (1945) powerful work on the hydrophysical aspects of the drainage basin allowed the morphometric measurement of basins, showed how these measurements were related, and attempted to rationalize the basin characteristics with hydrological processes (Chorley et al., 1984: 316). Horton's previous extensive publications (e.g., 1933) were found in the civil engineering and the SOIL science literature and, therefore, were less well known to physical geographers (see PHYSICAL GEOGRAPHY) and geomorphologists until the 1945 article was published.

E. Derbyshire et al. (1979: 46–47) presented a useful historical synopsis of basin studies enumerating six general approaches to drainage-basin inquiry that have been developed during the four decades since Horton's initial impact in the mid-1940s. The first is Horton's approach, which has already been discussed. The second approach also came out of Horton's work and has a theme of drainage-basin processes:

The basis for this second theme was that *drainage basin processes* are the consequence of the way in which drainage basin characteristics of relief, soil, land use and rock (see ROCK] type modify inputs of water and radiation. The objective of such studies was the development of equations which could express the relations between inputs, basin characteristics and output, so that outputs could be predicted for areas with no records of river flow or sediment production. (Derbyshire et al., 1979: 46)

This line of research was promoted extensively by Arthur Strahler (1950, 1952) and M. A. Melton (1958, 1960). These studies, however, did little to show how the basin process worked. This problem was attacked with vigor in several of the remaining approaches. The third approach concentrated on the hydraulic geometry of streams themselves. This was a narrower topic but showed the desire to quantify conditions for prediction. An example of this work is in the book by L. B. Leopold and T. Maddock (1953). The fourth approach, an offshoot of the hydraulic geometry studies, dealt with river-channel patterns and their effect on human activity such as irrigation. The fifth approach developed rapidly during the 1960s and 1970s as statistical methods of analysis improved and physical principles were incorporated in more theoretical approaches to basin analysis (e.g., Gregory, 1976). This allowed for theory to be developed based upon strict physical principles. The sixth was the paleo approach as defined by S. A. Schumm (1977) in which the primary concern is with water and sediment production over time. Therefore, the past, present, and future processes and morphology are examined. This last approach necessarily leads into more applied work as seen in publications such as those of R. G. Craig and J. L. Craft (1982) and R. U. Cooke et al. (1983).

References

Chorley, R. J.; Schumm, S. A.; and Sugden, D. E. 1984. *Geomorphology*. New York: Methuen & Co. This is an excellent reference for all topics in geomorphology including drainage basins.

Cooke, R. U.; Doornkamp, J. C.; Brunsden, D.; and Jones, D.K.C. 1983. *Urban Geomorphology in Drylands*. New York: Oxford University Press.

Craig, R. G., and Craft, J. L., eds. 1982. *Applied Geomorphology*. London: Allen and Unwin.

Davis, W. M. 1899. "The Geographical Cycle." *Geographical Journal* 14: 481–504.

Derbyshire, E.; Gregory, K. J.; and Hails, J. R. 1979. *Geomorphological Processes*. Boulder, Colo.: Westview Press. Another very good reference in geomorphology, especially the historical development of research.

Fairbridge, R. W. 1968. "Drainage Basin." In R. W. Fairbridge, ed., *The Encyclopedia of Geomorphology*. Encyclopedia of Earth Sciences Series. Vol. 3. Stroudsburg, Pa.: Dowden, Hutchinson and Ross. This has a good section on drainage basins and drainage patterns.

Gilbert, G. K. 1877. *Report on the Geology of the Henry Mountains*. Washington, D.C.: U.S. Department of the Interior.

Gregory, K. J. 1976. "Changing Drainage Basins." *Geographical Journal* 142: 237–247. Initiated the work on paleo analysis of drainage basins.

Horton, R. E. 1933. "The Role of Infiltration in the Hydrological Cycle." *Transactions, American Geophysical Union* 14: 446–460.

———. 1945. "Erosional Development of Streams and Their Drainage Basins: Hydrophysical Approach to Quantitative Morphology." *Geological Society of America Bulletin* 56: 275–370. The seminal work on quantitative drainage basin analysis.

Leopold, L. B., and Maddock, T. 1953. "The Hydraulic Geometry of Stream Channels and Some Physiographic Implications." *U.S. Geological Survey Professional Paper*, no. 252. Washington, D.C.: U.S. Geological Survey.

Martonne, E. de. 1950–1955. *Traite de Geographie Physique*. 3 vols. Paris: A. Colin.

Melton, M. A. 1958. "Correlation Structure of Morphometric Properties of Drainage Systems and Their Controlling Agents." *Journal of Geology* 67: 442–460.

———. 1960. "Intravalley Variation in Slope Angles Related to Microclimate and Erosional Environment." *Geological Society of America Bulletin* 71: 133–144.

Playfair, J. 1802. *Illustrations of the Huttonian Theory of the Earth*. Edinburgh: William Creech.

Schumm, S. A. 1977. *The Fluvial System*. New York: John Wiley and Sons. An excellent text for all topics in fluvial geomorphology.

Strahler, A. N. 1950. "Equilibrium Theory of Erosional Slopes Approached by Frequency Distribution Analysis." *American Journal of Science* 248: 673–696, 800–814.

———. 1952. "The Dynamic Basis of Geomorphology." *Geological Society of America Bulletin* 63: 923–938.

Thornbury, W. D. 1969. *Principles of Geomorphology*. 2d ed. New York: John Wiley and Sons.

Way, D. S. 1973. *Terrain Analysis*. Stroudsburg, Pa.: Dowden, Hutchinson and Ross.

Sources of Additional Information

For excellent works on the form of basins and geomorphic activity see M. A. Carson and M. J. Kirkby, *Hillslope Form and Process* (Cambridge: Cambridge University Press, 1971); K. J. Gregory and D. E. Walling, *Drainage Basin Form and Process* (London: E. Arnold, 1973); and M. E. Morisawa, *Fluvial Geomorphology* (Binghamton, N.Y.: SUNY Binghamton, 1973). For work on drainage basins, fluvial systems, and humans see R. J. Chorley, ed., *Water, Earth, and Man* (London: Methuen & Co., 1969); W. F. Graf, "The Impact of Suburbanization on Fluvial Geomorphology," *Water Resources Research* 2 (1977): 690–692; and K. J. Gregory, "Drainage Basin Adjustments and Man," *Geographica Polonica* 34 (1976): 155–173. For drainage basin responses within systems theory see R. J. Chorley, *Geomorphology and General Systems Theory*, U.S. Geological Survey Professional Paper, no. 500-B (Washington, D.C.: U.S. Geological Survey, 1962); R. J. Chorley and B. A. Kennedy, *Physical Geography: A Systems Approach* (London: Prentice-Hall, 1971); and G. H. Dury, *Introduction to Environmental Systems* (Portsmouth, N.H.: Heinemann Educational Books, 1981). For general work on fluvial systems including drainage basins see J. Lewin, ed., *British Rivers* (London: Allen and Unwin, 1981), and M. Morisawa, ed., *Fluvial Geomorphology* (Winchester, Mass.: Allen and Unwin, 1981).

DUNE. 1. An accumulation of windblown, sand-sized particles. 2. A low ridge or hill of drifted sand, primarily deposited and moved by the wind and found either in deserts or along the coastline of large water bodies.

All desert areas of the world have dunes, but the most extensive area of dune formation occurs in the great Saharan and Arabian deserts, where approximately 30 percent of the surface is covered with sand. Dune sand covers only about 1 percent of the surface of the North American desert (Davis, 1982).

Sand dunes are also found in coastal areas along many lake shores and seashores. In Europe large dune areas are found along the coasts of France and the Low Countries. In North America dunes are found along the coasts of both the Atlantic and Pacific oceans as well as along some of the shoreline regions of the Great Lakes.

For dunes to develop, several factors are important: (1) a fairly continuous supply of sand; (2) a constant wind strength and direction; and (3) some sort of obstacle to trap the sand. According to W. G. Moore (1978:72), the process is relatively simple: "The sand particles are carried along by the wind and piled into a heap, which gradually increases in size till it becomes a mound or ridge; the dune is often commenced where an obstacle of some kind exists, and the sand is heaped against this until it is covered and the sand falls over on the leeward side."

Early literature about dunes comes from many different sources. The dunes of the Sahara and the Egyptian Desert were first described by romantic adventurers in search of lost civilizations or treasures. In the United States, early records of dune formations were made by explorers and naturalists such as John Wesley Powell (1878), who was commissioned by the U.S. government to explore and report on the resources (see NATURAL RESOURCE) of the arid regions of the West. U.S. government surveyors and geologists such as G. K. Gilbert (1875), in his work west of the 100th meridian, and J. G. Parke (1857), exploring possible railroad routes west, also mentioned dunes in their reports.

Scientific study of dunes began in the late nineteenth century. Europe was actively involved in colonialism, and thus Europeans were exposed to vast dune regions. Naturally, Europeans were leaders in dune study during that era, particularly V. Cornish (1914), W. T. Blandford (1877), and Sir H. Frere (1870) in England; K.J.V. Steensrup (1894) in Denmark; and G.B.M. Flammand (1899) in France. Americans were also involved in dune research. F. P. Gulliver (1899) was doing work on coastal dunes, and W. C. Cowles (1899) concentrated on the dunes of Lake Michigan. As might be expected, most of the early studies were mostly descriptive, with observations primarily concerned with movement of sand bodies and external form of dunes. Attempts to classify dune types advanced slowly, but basic dune forms were recognized in relation to factors such as wind direction and physical barriers (McKee, 1979).

The French and Americans have been deeply involved in dune research in the twentieth century with a lesser volume of work coming from the British and Russians. Permanent research organizations are now established in desert areas. They include research centers at the University of Arizona, the Negev Institute for Arid Zone Research, L'Institute Fondemental de l'Afrique Noire, the Turkmanistan Academy of Sciences, and the Academy of Sciences of the People's Republic of China.

Without a doubt, the greatest contribution to the study of dunes has come from Ralph Bagnold of Great Britain. He has written many books on dunes and dune processes, and his *Physics of Blown Sand and Desert Dunes* (1941) is considered a classic in the field. Almost every author who has written on the subject of dunes since the 1930s makes reference to Bagnold's pioneering work.

Dune research covers many different areas of concern. The topics of study can be divided into four areas: dune formation, dune patterns and classification, dune composition, and dune ecology. There has been a changing emphasis on these areas of research over time.

Dune formation was a popular topic of early works and continues to be popular among researchers. Most scholars agree that EROSION/deposition and changing climate play important roles in dune development, but the extent of each is a primary source of controversy. Contributions in the field of dune formation have come from W. T. Blandford (1877), Leon Aufrere (1928), R. Capot-Rey (1957), and S. F. Hanna (1969). A more recent study by H. Tsoar (1978) attempts to resolve some of the conflicting ideas about dune formation.

A second area of research has been that of dune patterns and classification. The past 100 years have seen the development of many classification schemes. Many names in these classification systems originate from local languages, and some names of features, such as barchan, have achieved near-universal acceptance. According to Ronald Greeley and James Iversen, "It seems that each time someone completes a study of dunes in a given area, the existing nomenclature and classification schemes do not quite apply, and the investigator feels obliged to devise a new, or at least modified, scheme" (1985: 160).

One of the earliest and most accepted classification schemes was the result of research in northern Arizona completed by J. T. Hack (1941). Hack defined three primary types of dunes: (1) longitudinal dunes, oriented parallel to the prevailing wind direction; 2) transverse dunes, oriented perpendicular to the prevailing wind direction; and 3) parabolic dunes, which are U-shaped dunes pointing downwind. Hack related the development of these dune types to sand supply, wind, and VEGETATION. In 1979 Carol Breed and Teresa Gow developed a correlation table of dune types and useful references for comparing the same dune types in different areas. A very useful classification system that looks at internal structures of dunes as well as providing a uniform comparison of dunes using Landsat images is presented by E. D. McKee (1979).

Much research has also been done on the composition of dunes. The texture of eolian sand grains has been the subject of research for a long time (Dake, 1921; Wentworth, 1932). Ralph Bagnold (1941) has probably done the most research in the area of dune composition, concentrating on grain size and type. For the past 30 years and more, McKee (1966) and his colleagues at the U.S. Geological Survey have conducted much research on the internal features and sedimentary structures of sand dunes. Dune composition has also become a popular topic because of its relationship to petroleum reserves, and work has been done by D. A. Holm (1960) in the Arabian Penninsula. Also, the most advanced technology such as infrared photography and satellite imagery (see REMOTE SENSING) is now being used in hopes of discovering petroleum reserves in dune areas (Stanley and Rhoad, 1967; McKee and Breed, 1977).

The "ecology movement" of the 1960s brought about a greater awareness of human influence on dune environments. Thus dune ecology has been of increasing importance in the past two decades and continues to be studied. Coastal dunes are particularly susceptible to human activities, and a lot of research is being done on the ecology of coastal regions. Recently, the impact of off-road vehicles on the dune ecosystem (Leatherman and Steiner, 1979) and the role of man-made structures on dune dynamics (Nordstrom and McCluskey, 1982) have been studied.

The study of dunes is a fascinating topic, and although there is a vast amount of literature now available, there are still new areas to be explored and new challenges to meet.

References

Aufrere, Leon. 1928. "L'Orientation des Dunes et la Direction des Vents." *(Paris) Academie des Sciences Comptes Renous* 5, no. 187: 833–835. An early dune study by a French scientist.

Bagnold, Ralph, 1941. *The Physics of Blown Sand and Desert Dunes.* London: Methuen & Co. A classic book on dune processes.

Blandford, W. T. 1877. "Geographical Notes on the Great Desert between Sino and Rajputana." *Geological Survey of India Records* 10: 10–21. One of the early dune studies by a British scientist.

Breed, Carol, and Gow, Teresa, 1979. "Morphology and Distribution of Dunes in Sand Seas Observed by Remote Sensing." In McKee, E. D., ed., "A Study of Global Sand Seas." U.S. Geological Survey Professional Paper, no. 1052. Washington, D.C.: U.S. Geological Survey, pp. 253–308. A recent study using remote sensing.

Capot-Rey, R. 1957. "Sur une Forme D'erosion Eolienne dans le Sahara Francais." *Tyd. Kon. Nederland Aardnjsk Gen.* 74: 242–247.

Cornish, V. 1914. *Waves of Sand and Snow.* London: Fisher-Unwin.

Cowles, W. C. 1899. "The Ecological Relations of the Vegetation on the Sand Dunes of Lake Michigan." *Botanical Gazette* 27: 95–117, 167–202, 281–308, 361–391. An early ecological study of dunes.

Dake, C. L. 1921. "The Problem of St. Peter Sandstone." *Missouri University School of Mines and Metallurgy Bulletin* 6, no. 1.

Davis, John D. 1982. *McGraw-Hill Encyclopedia of Science and Technology.* 5th ed. Vol. 4. New York: McGraw-Hill Book Co., pp. 412–416. An excellent general discussion of dune types, formation, and distribution.

Flammand, G.B.M. 1899. "La Traversee de l'erg Occidental (Grand Dunes du Sahara Oranasis)." *Annals de Geographie* 9: 231–241. An early study in the Sahara by a French scientist.

Frere, Sir H. 1870. "On the Rann of Cutch and Neighboring Regions." *Journal of the Royal Geographical Society* 40: 181–207.

Gilbert, G. K. 1875. *Geographical and Geological Explorations and Surveys West of the 100th Meridian*. Washington, D.C.: U.S. Army, Engineers Department. One of the first observations of dunes by an American geologist.
Greeley, Ronald, and Iversen, James. 1985. *Wind as a Geological Process on Earth, Mars, Venus, and Titan*. Cambridge: Cambridge University Press. A recent textbook that looks at terrestrial and extraterrestrial dunes.
Gulliver, F. P. 1889. "Shoreline Topography." *Proceedings of the American Academy of Arts and Sciences* 34: 151–258. An early study of coastal dunes.
Hack, J. T. 1941. "Dunes of the Western Navajo Country." *Geographical Review* 31: 240–263. A classic study on dune classification.
Hanna, S. F. 1969. "The Formation of Longitudinal Sand Dunes by Large Helical Eddies in the Atmosphere." *Journal of Applied Meteorology* 8: 874–883.
Holm, D. A. 1960. "Desert Geomorphology in the Arabian Penninsula." *Science* 132: 1369–1379.
Leatherman, S. P., and Steiner, A. J. 1979. "Recreational Impacts on Foredunes: Assateague Island National Seashore." In *Proceedings, Second Scientific Conference on the National Parks*. San Francisco, Calif. A recent study dealing with human impact on the ecology of dune areas.
McKee, E. D. 1966. "Structures of Dunes at White Sands National Monument, New Mexico." *Sedimentology* 7: 1–69.
———, ed. 1979. "A Study of Global Sand Seas." *Geologic Professional Paper*, no. 1052, 1, 19.16: 1052. A very thorough study of dunes by one of America's foremost experts. It has an excellent bibliography.
McKee, E. D., and Breed, C. S. 1977. "Desert Sand Seas." In Carr, G. P., ed., "Skylab Explores the Earth," National Aeronautics and Space Administration Special Paper, no. 380, Washington, D.C.: National Aeronautics and Space Administration, Chapter 2, pp. 5–48.
Moore, W. G. 1978. *A Dictionary of Geography: Definitions and Explanations of Terms Used in Physical Geography*. New York: Barnes and Noble Books. A short definition and description of dune formation.
Nordstrom, K. F., and McCluskey, J. M. 1982. *The Effects of Structures on Dune Dynamics at Fire Island National Seashore*. Final Report, Center for Coastal and Environmental Studies. New Brunswick, N.J.: Rutgers University.
Parke, J. G. 1857. *Report of Explorations for Railroad Routes*. U.S. Congress, House Executive Document 91, 33rd Congress, 2d Session. One of the first dune studies by an American.
Powell, John W. 1878. *Report on the Lands of the Arid Region of the United States* (ed. W. Stegner). Cambridge: Harvard University Press. One of the first studies of arid regions in the United States.
Stanley, D. J., and Rhoad, D. C. 1967. "Dune Sands Examined by Infrared Photography." *American Association of Petroleum Geologists Bulletin* 51: 424–430.
Steenstrup, K.J.V. 1894. "On Klitterns Vandrung." *Meddelelser, Dansk, Geologisk Forening, Copenhagen* 1: 1–14.
Tsoar, H. 1978. "The Dynamics of Longitudinal Dunes." Ph.D. thesis, Ben-Gurion University of the Negev.
Udden, J. A. 1898. *The Mechanical Compositions of Wind Deposits*. Augustana Library Publication, no. 1.
Wentworth, C. K. 1932. "The Mechanical Composition of Sediments in Graphic Form." *Iowa University Studies in Natural History* 14, no. 3: 7–14.

Sources of Additional Information

Short definitions and discussions of dunes can be found in F. J. Monkhouse, *A Dictionary of Geography*, 2d ed. (Chicago: Aldine Publishing Co., 1970), p. 115; Stella Stiegeler, ed., *A Dictionary of Earth Sciences* (New York: Pico Press, 1976), p. 301; Daniel L. Lapedes, ed., *McGraw-Hill Encyclopedia of the Geological Sciences* (New York: McGraw-Hill Book Co., 1978); and John Whitton, *Dictionary of Physical Geography* (New York: Penguin Books, 1984), p. 157. Excellent discussions of dune development and classification can be found in J. A. Mabbutt, *Desert Landforms* (Cambridge, Mass.: The M.I.T. Press, 1977), and Ronald V. Cooke and Andrew Warren, *Geomorphology in Deserts* (Berkeley: University of California Press, 1973). Detailed examinations of coastal dune processes can be found in E.C.F. Bird, *Coasts* (Canberra: Australian National University Press, 1968); Victor Goldsmith, "Coastal Dunes," in Richard A. Davis, Jr., ed., *Coastal Sedimentary Environments*, 2d ed. (New York: Springer-Verlag, 1985), pp. 27–34; J. L. Davies, *Geographical Variation in Coastal Development* (New York: Hafner Publishing Co., 1973), Chapter 10; and B. Greenwood and R. A. Davis, Jr., eds., *Hydrodynamics and Sedimentation in Wave-Dominated Coastal Environments* (New York: Elsevier, 1984).

E

EARTHQUAKE. 1. A vibration of the earth's surface caused by a series of shocks or tremors associated with subterranean crustal movements. 2. A sudden motion or movement of the earth's crust caused by the abrupt release of energy along active faults or in areas of volcanic activity.

Earthquakes have always been a topic of interest from ancient cultures to the present. In many countries legends or myths attributed earthquake activity to supernatural forces. The first systematic and nonmystical analysis of earthquake activity, however, can be traced to the Greeks. A number of Greek philosophers starting in the fifth century B.C. took an interest in earthquake research. According to Thales, earthquakes were attributed to the composition of the earth's interior. He believed the earth floated on water and cracked and buckled in response to its fluid core (Fried, 1973: 25). Anaxagoras, a friend of Pericles, believed a large subterranean fire within certain mountains caused tremors and aftershocks (Svenson, 1984: 1). Strabo noticed that earthquakes were more frequently associated with coasts than inland areas. Even Aristotle, the most renowned of all Greek philosophers, took an interest in earthquake research and believed that the wind on the earth's surface entered underground caves and, after combining with underground gases, exploded, causing the tremors (Bolt, 1978: 3). The modern science of seismology can be traced to its Greek origins. The word *seismology* is formed from the Greek *seismos*, meaning earthquake, and logos, meaning science.

Although seismology first became an independent science around the turn of the twentieth century, much of its theoretical foundations can be traced to earlier times. The first mathematician who studied problems of stress and strength of materials was Galileo in 1638. Following Galileo's work, Robert Hooke developed the general idea in 1660 that the deformation of a body is directly proportional to the applied stress (Bath, 1973).

By the mid-eighteenth century the study of earthquake-wave propagation was being intensively studied by John Mitchell. He proposed, in 1760, that earthquakes bring about wave motions in the earth and the center of an earthquake could be determined mathematically by recording the time at which the motions are felt.

About 100 years after Mitchell's pioneering work, the first maps of the distribution of earthquakes were made by the Englishman Robert Mallet. To measure earthquake effects more accurately in a quantitative way, intensity scales were introduced in the 1870s. In 1874 Bruno Rossi developed a scale of earthquake intensity that was modified by Giuseppe Mercalli in 1902 and later used by C. F. Richter in his work (Bath, 1973).

In 1878 R. Hoernes developed a classification of earthquakes that is still used. He divided earthquakes into three categories: collapse earthquakes, volcanic earthquakes, and tectonic earthquakes (Bath, 1973: 19).

During the twentieth century many advances in the study of earthquakes have been made. The development of the PLATE TECTONICS theory has been responsible for much new thinking about the distribution and causes of earthquakes (Cox, 1972). There has also been much work in recent years on the prediction of earthquakes. Forecasting future events is both complex and fascinating. Large earthquakes can inflict enormous damage, and better prediction methods can have great benefits to society (Mogi, 1985).

In the past few decades geographers have become increasingly concerned with the consequences of earth movements in urban areas (Berlin, 1980) and the NATURAL HAZARDS associated with earthquake activity (Hart, 1977; U.S. Geological Survey, 1974).

References

Berlin, G. Lennis. 1980. *Earthquakes and the Urban Environment*. Vol. 1. Boca Raton, Fla.: CRC Press. A good geographical analysis of earthquake problems in urban areas.
Bolt, Bruce A. 1978. *Earthquakes, A Primer*. San Francisco: W. H. Freeman and Co. An easy-to-read introduction to earthquakes.
Cox, A., ed. 1972. *Plate Tectonics and Geomagnetic Reversals*. San Francisco: W. H. Freeman and Co. A very technical discussion of plate tectonics.
Fried, John J. 1973. *Life along the San Andreas Fault*. New York: Saturday Review Press.
Hart, Earl. 1977. *Fault Hazard Zones in California*. Special Publication, no. 42. Sacramento: California Division of Mines and Geology.
Mogi, Kiyoo. 1985. *Earthquake Prediction*. New York: Academic Press. A recent analysis of advancements in earthquake prediction. Many examples from Japan.
Svenson, Arthur G. 1984. *Earthquakes, Earth Scientists, and Seismic-Safety Planning in California*. New York: University Press of America. A discussion of the politics of earthquake planning in California.
U.S. Geological Survey. 1974. "Seismic Hazards and Land-Use Planning." *Geologic Survey Circular*, no. 690. Washington, D.C.

Sources of Additional Information

Many good books on the nature of earthquakes for the nonspecialist are available, including G. A. Eiby, *Earthquakes* (London: F. Muller, 1957); B. A. Bolt, *Nuclear Explosions and Earthquakes: The Parted Veil* (San Francisco: W. H. Freeman and Co., 1976); G. B. Oakeshott, *Volcanoes and Earthquakes* (New York: McGraw-Hill Book Co., 1976); and Robert Iacopi, *Earthquake Country* (Menlo Park, Calif.: Lane Book Co., 1976). For more information on earthquake prediction see T. Utsu, "Probabilities Associated with Earthquake Prediction and Their Relationships," *Earthquake Prediction Research* 2 (1983): 105–114; D. W. Simpson and P. G. Richards, eds., *Earthquake Prediction* (Washington, D.C.: American Geophysical Union, 1981).

ECOSYSTEM. 1. A complex of living and nonliving materials that functions as a natural unit and through which the flow of energy and matter can be traced. **2.** A biological unit comprised of all organisms in a given area, interacting with the physical environment so that energy flows can be traced through trophic structures and material cycles.

The term *ecosystem* was first proposed by the British ecologist A. G. Tansley (1935), but the concept is not of recent origin. Allusions to the interaction of organisms and the physical environment can be found far back in written history. Geographers have long been concerned with the nature and development of ecosystems. George Perkins Marsh's *Man and Nature* (1864) is a classic treatise on the concept. Marsh analyzed the causes for the decline of ancient civilizations and forecast a similar doom for modern societies unless humans took what we would today call an "ecosystems" approach to man and nature. Other early writers interested in the ecosystems concept were the German Karl Möbius (1877), who studied oyster reefs, and the American S. A. Forbes (1887), who wrote an important essay, "The Lake as a Microcosm."

Early contributions to the ecosystem concept came primarily from geographers interested in the distribution of plants and animals. The development of the idea of a life zone was first introduced in 1898 by C. H. Merriam. He divided North America into climatic zones, each of which was categorized by certain biota. A. D. Hopkins (1938) and L. R. Holdridge (1947) further developed the idea of a life zone.

F. E. Clements (1916) and H. C. Cowles (1901) refined and clarified the ecosystem concept. Clements reintroduced the idea of plant succession and climax vegetation (see PLANT SUCCESSION/CLIMAX), and Cowles developed physiographic ecology, observing the relationship between vegetation changes and LANDFORMS. Although Clements and Cowles had an important influence on the geography of ecosystems, Forrest Shreve (Shreve and Livingston, 1921) had a major impact on geographers. His "clear geographic reasoning and his insight into the nature and meaning of regions placed his works among the most important ever to appear in America" (A. W. Kuchler, 1954: 433). Shreve was interested primarily in the effect of climate on the VEGETATION complex.

The distribution of animal species, a part of the ecosystem concept, is of interest to geographers. V. E. Shelford (1911) and W. C. Allee and K. P. Schmidt (1937) studied the relationship between organisms and their environment.

The geography of human ecology is also part of the ecosystem concept. According to H. H. Barrows, "Geographers in increasing numbers define their subjects as dealing solely with the mutual relations between man and his natural environment" (1923: 2). The Berkeley School of Cultural Geography, which developed in the 1920s, has a partial emphasis on the role of the ecosystem. Centered around Carl Sauer (1925), the Berkeley School focused on the earth-as-the-home-of-man, stressing human effect on the environment.

The publication of the symposium proceedings *Man's Role in Changing the Face of the Earth* (Thomas, 1956) renewed interest in the ecosystem concept. In another symposium proceedings, *Man's Place in the Island Ecosystem* (Fosberg, 1963), geographers and anthropologists presented a framework for the ecosystem concept. Geographer David Stoddart's (1965, 1967) analyses of the ecosystem concept were attempts to integrate the concept with Ludwig Von Bertanffy's general systems theory.

The past decade has seen a proliferation of books and articles by geographers on the ecosystem concept. Like Stoddart, some geographers have tried to integrate the ecosystem idea into other, related perspectives. James P. Clarkson (1970) examined the relationship between ecology and spatial analysis, and Larry Grossman (1977), in a similar study, analyzed the contributions of anthropology and geography to human-environmental relations. Fosberg (1976) discussed the development of the ecosystem concept in BIOGEOGRAPHY.

A thorough discussion of the ecosystem concept in geography is presented in the book *Human Environments and Natural Systems* by Ned H. Greenwood and J.M.B. Edwards (1973). In addition, the Association of American Geographers prepared a study (Manners and Mikesell, 1974) dealing with environmental systems.

Another recent trend among geographers is to apply the ecosystem concept to urban areas. The components of the *urban ecosystem* interact like the elements of a biological system. Instead of the interrelationships between the atmosphere, hydrosphere, biosphere, and lithosphere associated with most biological systems, the urban ecosystem studies the manner in which economic, social, political, and decision-making processes work together to produce a city (Stearns and Montag, 1974).

References

Allee, W. C., and Schmidt, K. P. 1937. *Ecological Animal Geography.* New York: Schuman.
Barrows, H. H. 1923. "Geography as Human Ecology." *Annals of the Association of American Geographers* 13: 1–14. A discussion of ecology as the central feature of geographic study.
Clarkson, James P. 1970. "Ecology and Spatial Analysis." *Annals of the Association*

of American Geographers 60: 700–716. A discussion of the role of ecological analysis in geography.
Clements, F. E. 1916. *Plant Succession*. Carnegie Institution Publication, no. 242. Washington, D.C.: Carnegie Institution. An early formulation of the ecosystem concept with regard to plant succession.
Cowles, H. C. 1901. "The Physiographic Ecology of Chicago and Vicinity." *Botanical Gazette* 31: 73–108, 145–182. An early discussion of the relationship between landforms and plant associations.
Forbes, S. A. 1887. "The Lake as a Microcosm." Reprinted in *Illinois Natural History Survey Bulletin* 15: 537–550 (1925). An early study of the ecosystem concept relative to a lake.
Fosberg, F. R. 1976. "Geography, Ecology, and Biogeography." *Annals of the Association of American Geographers* 66: 117–128. A discussion of the ecosystem concept in biogeography.
———, ed. 1963. *Man's Place in the Island Ecosystem*. Honolulu: Bishop Museum Press. An excellent collection of readings using the ecosystem concept as its conceptual framework.
Greenwood, Ned H., and Edwards, J.M.B. 1973. *Human Environments and Natural Systems*. North Scituate, Mass.: Duxbury Press. A thorough discussion of the ecosystem concept in geography.
Grossman, Larry. 1977. "Man-Environment Relationship in Anthropology and Geography." *Annals of the Association of American Geographers* 67: 126.
Holdridge, L. R. 1947. "Determination of World Plant Formations from Simple Climatic Data." *Science* 105: 367–368.
Hopkins, A. D. 1938. *Bioclimatics*. Miscellaneous Publications, no. 280. Washington, D.C.: U.S. Department of Agriculture. A discussion of the life-zone concept.
Kuchler, A. W. 1954. "Plant Geography." In Preston James and Clarence Jones, eds., *American Geography: Inventory and Prospect*. Syracuse, N.Y.: Syracuse University Press, pp. 428–441. A good discussion of the evolution of plant geography.
Manners, Ian, and Mikesell, Marvin, eds. 1974. *Perspectives on Environment*. Washington, D.C.: Association of American Geographers.
Marsh, George P. 1864. *Man and Nature*. Reprint (1965). Cambridge, Mass.: Harvard University Press. A classic study on the interaction of people and the environment.
Merriam, C. H. 1898. "Life Zones and Crop Zones of the United States." *Biological Survey Bulletin*, no. 10. Washington, D.C.: U.S. Department of Agriculture. The initial development of the life-zone concept.
Mobius, Karl. 1877. "Die Auster und Die Austernwirtschaft." Translated in *Report, U.S. Fish Commission* (1880), pp. 683–751.
Sauer, C. O. 1925. *The Morphology of Landscape*. University of California Publications in Geography. Vol. 2. Berkeley: Department of Geography.
Shelford, V. E. 1911. "Physiological Animal Geography." *Journal of Morphology* 22: 551–618. The ecosystem concept as applied to animal geography.
Shreve, F., and Livingston, B. E. 1921. *The Distribution of Vegetation in the United States, as Related to Climatic Conditions*. Carnegie Institution Publication, no. 284. Washington, D.C.: Carnegie Institution. An early attempt to tie together regions and ecosystems.
Stearns, F., and Montag, T. 1974. *The Urban Ecosystem: A Holistic Approach*. Strouds-

burg, Pa.: Dowden, Hutchinson and Ross. An application of the ecosystem concept to urban areas.

Stoddart, D. R. 1965. "Geography and the Ecological Approach: The Ecosystem as a Geographic Principle and Method." *Geography* 50: 242–251.

———. 1967. "Organism and Ecosystem as Geographical Models." In R. J. Chorley and P. Haggett, eds. *Models in Geography*. London: Methuen & Co., 1967, pp. 164–193. A good discussion of the ecosystem model in geography.

Tansley, A. G. 1935. "The Use and Abuse of Vegetational Concepts and Terms." *Ecology* 16: 284–307. The initial discussion of the term *ecosystem*.

Thomas, William L., ed. 1956. *Man's Role in Changing the Face of the Earth*. Chicago: University of Chicago Press. An excellent and extensive collection of readings by geographic scholars and others on the ecosystem concept.

Sources of Additional Information

For a short definition of the concept, see Paul Sarnoff, *Encyclopedic Dictionary of the Environment* (New York: Quadrangel, 1971), p. 96, and Daniel N. Lapedes, ed., *Dictionary of the Life Sciences* (New York: McGraw-Hill Book Co., 1976), p. 264. For extended discussions of the biological aspects of an ecosystem see W. B. Clapham, Jr., *Natural Ecosystems* (New York: Macmillan, 1973); P. Walton Purdom and Stanley H. Anderson, *Environmental Science* (Columbus, Ohio: Charles E. Merrill Publishing Co., 1980), pp. 166–175; and Charles H. Southwick, *Ecology and the Quality of Our Environment* (New York: Van Nostrand-Reinhold Co., 1972), pp. 103–157. An analysis of ecosystems from a worldwide geographic perspective can be found in Charles F. Bennett, Jr., *Man and Earth's Ecosystems: An Introduction to the Geography of Human Modification of the Earth* (New York: John Wiley and Sons, 1975). The concept of the urban ecosystem is briefly outlined in Robert P. Larkin, Gary L. Peters, and Christopher H. Exline, *People, Environment, and Place: An Introduction to Human Geography* (Columbus, Ohio: Charles E. Merrill Publishing Co., 1981), pp. 315–321.

ELECTROMAGNETIC RADIATION. 1. Energy in the form of an advancing interaction between electric and magnetic fields propagated through space or some other material. 2. Energy in the various wavelengths including cosmic rays, gamma rays, and X-rays and ultraviolet, visible, infrared, micro, and radio waves. 3. The emission or propagation of such energy.

The concept of electromagnetic radiation (EMR) as we know and understand it today was first proposed by James C. Maxwell in 1864 and experimentally verified by H. Hertz in 1884. Maxwell's theory essentially stated that EMR was propagated in a wave form by the oscillation of an electric charge whereby a system of electric and magnetic fields moved outward from the region where the electric charges were oscillated or accelerated (Nunnally, 1973: 20–25).

EMR travels at the speed of light (3×10^8 m/sec) in a vacuum and somewhat less through other material. The different categories of EMR are considered to be on a spectrum of differing wavelengths, frequencies, and energy levels. EMR is defined by the general equation

$$C = f\lambda$$

where $C = 3 \times 10^8$ m/sec, f = frequency, and λ = wavelength.

Since C is essentially a constant, as frequency increases, wavelength must decrease. The highest frequency (and shortest wavelength) part of the spectrum are the cosmic rays followed in order by gamma rays, X-rays, ULTRAVIOLET light, visible light, INFRARED energy, microwaves, and radio waves.

EMR also exhibits quantum or particle behavior in its propagation and transport. In 1900 Max Planck recognized the discrete (as opposed to the continuous-wave) nature of EMR. He showed that EMR was given off in separate amounts (quanta) and only in these amounts. The total amount of energy propagated was in direct proportion to the frequency, so the higher frequencies had the higher energy. Both the wave and quantum theories are valid and useful under specific circumstances. Under averaging conditions and very large numbers of waves, the two theories are completely compatible (Nunnally, 1973; Suits, 1983).

Knowledge of the various mechanisms for generating EMR is useful to us in designing instruments to record the EMR of different wavelengths. For the shorter wavelength (X-rays, ultraviolet, and visible), the energy is propagated by the movement of electrons within molecules. As an electron goes from a higher energy electron shell (outer) to a lower energy shell (lower), energy is emitted from that molecule. For the middle range of the spectrum (visible, infrared, lower microwave), energy is propagated from the movement (vibration and rotation) of the molecules themselves. The longer wavelengths (microwave and radio waves) require the oscillation of an electric current to propagate EMR (Dearholt and McSpadden, 1973). Using this information various REMOTE SENSING systems can acquire information about the EMR emitting or reflecting objects.

Currently, the portions of the spectrum of EMR being used in remote sensing go from X-rays through radio waves. The state-of-the-art in using EMR in remote sensing is probably the laser (light amplification by stimulated emitted radiation). A laser can produce a coherent, almost monofrequency energy radiation in a precise direction. This permits the radar techniques and applications to be used in the infrared, visible, and ultraviolet spectral ranges (Suits, 1983: 59–60). Thus most sensing systems can now be active rather than passive with precise control over amount and wavelength of EMR on any scene or object being sensed (see REMOTE SENSING for other information on the EMR).

References

Dearholt, D., and McSpadden, W. 1973. *Electromagnetic Wave Propagation*. New York: McGraw-Hill Book Co.

Nunnally, Nelson R. 1973. "Introduction to Remote Sensing: The Physics of Electromagnetic Radiation." In Holz, Robert K., ed., *The Surveillant Science*. Boston: Houghton Mifflin Co., pp. 18–27. This is a most lucid and succinct article on those factors of the physics of EMR most needed by remote sensing users.

Suits, Gwynn H., 1983. "The Nature of Electromagnetic Radiation." In Colwell, Robert N., eds. *Manual of Remote Sensing*. 2d ed. Falls Church, Va.: American Society of Photogrammetry, pp. 37–60. An excellent, up-to-date synopsis of the nature of EMR and some of the characteristics of the energy for use in remote sensing.

Sources of Additional Information

For a more thorough discussion of electromagnetic fields and the wave theory see V. Rojansky, *Electromagnetic Fields and Waves* (New York: Dover Publications, 1979), and A. Shadowitz, *The Electromagnetic Field* (New York: McGraw-Hill Book Co., 1974). For a good, popular account of the physics of EMR see Isaac Azimov, *Understanding Physics: The Electron, Proton, and Neutron* (New York: New American Library, 1966). The leading text on optics and EMR is F. A. Jenkins and H. E. White, *Fundamentals of Optics*, 4th ed. (New York: McGraw-Hill Book Co., 1976). The two best sources of information on EMR and remote sensing are Robert N. Colwell, *Manual of Remote Sensing*, 2d ed. (Falls Church, Va.: American Society of Photogrammetry, 1983), and R. N. Colwell, W. Brewer, G. Landis, P. Langley, J. Morgan, J. Rinker, J. M. Robinson, and A. L. Sorem, "Basic Matter and Energy Relationships Invol ed in Remote Reconnaissance," *Photogrammetric Engineering* 29 (1963): 761–799.

ELECTROMAGNETIC SPECTRUM. See ELECTROMAGNETIC RADIATION.

ENVIRONMENTAL PERCEPTION. See NATURAL HAZARD.

EQUINOX. The precise point in time when the sun appears directly overhead at the equator and the circle of illumination passes through both the north and south poles.

According to James Murray et al (1933) the term *equinox* was used by Chaucer as early as 1391 to describe the two points in time that are characterized by equal periods of day and night at all places on the globe. Thus from its earliest use the definition of *equinox* has been the same, although it may occasionally be put into somewhat different words. For example, F. J. Monkhouse and John Small wrote that the equinox is "one of the 2 points of intersection of the sun's path during the year with the plane of the terrestrial Equator" (1978: 107). It is also true, as Arthur Strahler and Alan Strahler noted, that "at equinox, the sun rises at a point due east on the horizon and sets at a point due west on the horizon. This fact holds true at all latitudes except at the two poles" (1987: 29).

An equinox occurs twice each year. In March the spring, or vernal, equinox occurs on the twentieth or twenty-first day of the month, and in September the autumnal equinox occurs, usually on the twenty-second day of the month. These terms and dates are appropriate for those who live in the northern hemisphere; in the southern hemisphere the dates are the same, but the terms are opposite. Whereas the September 22 equinox represents the beginning of fall in the northern hemisphere, it represents the beginning of spring in the southern hemisphere.

In climatology the equinoxes are important relative to the distribution of incoming solar radiation. Because major wind and pressure systems migrate with the apparent migration of the sun overhead, the equinoxes serve as major reference points. Between the Tropic of Cancer, which lies 23.5 degrees north latitude, and the Tropic of Capricorn, which lies at 23.5 degrees south latitude, the sun makes its apparent annual migration. Within these tropical latitudes, between the Tropics of Cancer and Capricorn, the sun passes overhead twice during the year. Near the equator there are double maximums and minimums within the annual temperature cycle, and they reflect the sun's two overhead passages. Toward the edges of the tropics the two overhead periods grow closer and closer together in time, so that the double maximum and minimum temperatures are blurred together. The Tropics of Cancer and Capricorn represent the northernmost and southernmost points, respectively, to which the subsolar point migrates.

Aside from the obvious climatological interest in equinoxes and seasonality, there is another interest as well, and that is in equinoxial storms. At high latitudes the most vigorous atmospheric circulations occur during the equinoxial periods. "The equinoxial storms of the subpolar ocean south of Cape Horn at the southern tip of South America are legendary," wrote Paul Lydolph, "as they are along the Arctic coast of Western Siberia and in many other subpolar areas of the world" (1985: 357).

References

Lydolph, Paul E. 1985. *The Climate of the Earth*. Totowa, N.J.: Rowman and Allanheld, Publishers. An excellent and up-to-date climatology text.
Monkhouse, F. J., and Small, John. 1978. *A Dictionary of the Natural Environment*. New York: John Wiley and Sons. A very worthwhile reference book for anyone with an interest in studying physical geography or related fields in the natural sciences.
Murray, James A. H., et al., eds. 1933. *The Oxford English Dictionary*. Oxford: Clarendon Press.
Strahler, Arthur N., and Strahler, Alan H. 1987. *Modern Physical Geography*. 3d ed. New York: John Wiley and Sons. Probably the best of current physical geography texts that are on the market.

Sources of Additional Information

Equinoxes are discussed in most introductory physical geography texts, including the following: Tom L. McKnight, *Physical Geography: A Landscape Appreciation* (Englewood Cliffs, N.J.: Prentice-Hall, 1984), and E. Willard Miller, *Physical Geography: Earth Systems and Human Interactions* (Columbus, Ohio: Charles E. Merrill Publishing Co., 1985). More details on equinoxes and climatology can be found in John E. Oliver and John J. Hidore, *Climatology: An Introduction* (Columbus, Ohio: Charles E. Merrill Publishing Co., 1984).

EROSION. 1. The geologic agencies of movement that secure and remove mineral debris and organic matter from the earth's surface. 2. The wearing away or denudation of land masses. 3. A segment of the overall theory of landscape evolution and development in conjunction with material transportation and deposition.

The concept of erosion is a broad and multifaceted one that is inherently involved in many other concepts in this book such as GEOMORPHOLOGY, BASE LEVEL, DRAINAGE BASIN, GEOLOGY, KARST, PHYSICAL GEOGRAPHY, STREAM OR RIVER, and WEATHERING. Erosion has been an important topic from the beginning of the study of geology and geomorphology and is entrenched in geomorphic literature.

James Hutton (1788 and 1795) incorporated erosion into his views on the processes of surficial decay and the analogy of a machine grinding down the surface of the earth. He saw evidence of erosion and subsequent deposition all around him, yet could not find "any information on erosion rates as a basis for estimating 'the age of the present earth,' or . . . the length o' a cycle" (Tinkler, 1985: 49). The term *cycle* referred to the cycle of uplift, erosion, deposition, and uplift of earth materials and became an important term in erosion studies through the mid-twentieth century.

Erosion as a component of modern geomorphology was developed through the work of G. K. Gilbert. His work on the geology of the Henry Mountains in Utah (Gilbert, 1877) provided the basis for studying landscape development through erosion, transportation, and deposition of weathered detritus. He developed the concept of grade whereby progressive downward erosion ceases when transport capacity of the system is reached. He also provided insights into the role erosion takes in producing land surface isostacy (Gilbert, 1890). In spite of Gilbert's considerable impact on the science of geomorphology, he was never able to organize his ideas into a theoretical synthesis upon which other geomorphologists could build. This integrated view was to be put forward by a contemporary of Gilbert's: William Morris Davis (Tinkler, 1985: 144).

Geology and geomorphology as sciences had matured enough by the end of the nineteenth century to accept a broad, integrated, and very theoretical scheme for explaining the development of landscapes. Such a scheme was devised by W. M. Davis who proposed his ideas in a cursory way in 1889 (Davis, 1889). Succinctly put, he stated that subaerial erosion wears down the earth's surface through a definable and predictable series of stages to the point where little or no relief stands above base level. Only geologic accidents such as tectonic uplifts, major glacial advances, or significant climate change would interfere with this "cycle of erosion" (also called the geographical cycle). His fully documented and formulated theory was published in 1899 (Davis, 1899). He defined in great detail his landform stages: youth, maturity, and old age, which inevitably lead to peneplanation of the land surface. During the next three decades, he wrote dozens of articles on specific aspects of the erosion cycle and applied it to various regions around the world.

Davis's work spread rapidly abroad, where it was wholeheartedly accepted by the academic communities in countries such as Great Britain, Australia, New Zealand, and France. In fact, it was still prevalent in textbooks in these countries well into the 1950s and 1960s. The acceptance of the theory was no less rousing in Eastern Europe and in particular in Germany (Tinkler, 1985: 150–151). Although the German Albrecht Penck was sympathetic to the new terminology presented by Davis, Penck's son Walther developed his own views on landscape evolution, published posthumously (Penck, 1924). Much of the rancor between Penck and Davis arose because the only English interpretation of Penck's work was from Davis himself (Davis, 1932). Even though the two theories have much in common, language became a barrier to understanding. Only in 1953 did a translation of Penck's ponderous German work emerge.

During the 1940s and 1950s, much of the research concerning erosion stressed climate-induced or modified erosion processes. Penck pointed out that many LANDFORMS occur under varying climate and at differing rates (Tinker, 1985: 167). Other researchers took this new idea of climatic geomorphology and expanded its influence. L. C. Peltier (1950), for example, devised a system of morphogenetic regions based primarily on climate. L. C. King (1953) produced an alternative cycle of erosion that had climatic elements and contained four basic components of hillslope formation. He also coined the term *pediplain* to refer to an erosion surface in arid regions.

For the past few decades, erosion studies have concentrated on erosion process. F. D. Hale (1968: 318–320) enumerated the six principal agents of erosion: running water, groundwater, waves and currents, wind, glaciers, and gravity. This process (agent) orientation in erosion studies is epitomized in M. A. Carson and M. J. Kirkby's (1972) *Hillslope Form and Process* and S. A. Schumm's (1977) *The Fluvial System*.

A very practical and applied approach to erosion studies has evolved from the work by the U.S. Department of Agriculture, Soil Conservation Service, particularly that done by W. H. Wischmeier and his colleagues (e.g., 1959; Wischmeier and associates, 1971; and Wischmeier and Smith, 1978). Although this work recognizes the role of erosion through natural processes, it stresses the accelerated erosion that takes place due to human involvement in the natural system. An additional aspect of accelerated erosion is the increase in undesirable sedimentation that takes place after erosion occurs (Jansen and Painter, 1974). The growing concern about accelerated erosion and sedimentation can be seen in several volumes on the topic that have been recently published including those of S. A. El-Swaify et al. (1985) and M. J. Kirkby and R. P. C. Morgan (1980).

References

Carson, M. A., and Kirkby, M. J. 1972. *Hillslope Form and Process*. Cambridge: Cambridge University Press.

Davis, W. M. 1889. "The Rivers and Valleys of Pennsylvania." *National Geographic Magazine* 1: 183–253. This was the first presentation of Davis's erosion cycle theory.

———. 1899. "The Geographical Cycle." *Geographical Journal* 14: 481–504. This was a more formal and thorough explanation of the cycle.

———. 1932. "Piedmont Benchlands and *Primarrumpfe*." *Geological Society of America Bulletin* 43: 399–440. For many years this was the only place where Penck's theories were available in English.

El-Swaify, S. A.; Moldenhauer, W. C.; and Lo, A. 1985. *Soil Erosion and Conservation*. Ankeny, Ia.: Soil Conservation Society of America.

Gilbert, G. K. 1877. *Report on the Geology of the Henry Mountains*. Washington, D.C.: U.S. Geographic and Geologic Survey. This report really provided the groundwork for the science of geomorphology in the United States.

———. 1890. *Lake Bonneville*. Monograph. Vol. 1. Washington, D.C.: U.S. Geological Survey. Gilbert proposed the relationship between the ideas of isostacy and erosion.

Hale, F. D. 1968. "Erosion." In R. W. Fairbridge, ed. *The Encyclopedia of Geomorphology*. Encyclopedia of Earth Science Series. Vol. 3. Stroudsburg, Pa.: Dowden, Hutchinson and Ross.

Hutton, J. 1788. "Theory of the Earth." *Royal Society of Edinburgh Transactions* 1: 209–304. First to incorporate erosion and sedimentation into a theory of the earth's geology.

———. 1795. *The Theory of the Earth*. 2 vols. Edinburgh: William Creech.

Jansen, J. M. L., and Painter, R. B. 1974. "Predicting Sediment Yield from Climate and Topography." *Journal of Hydrology* 21: 371–380.

King, L. C. 1953. "Canons of Landscape Evolution." *Geological Society of America Bulletin* 64: 721–752. King provides an alternative erosion cycle model using climatic inputs and slope segment models.

Kirkby, M. J., and Morgan, R. P. C., eds. 1980. *Soil Erosion*. London: John Wiley and Sons.

Peltier, L. C. 1950. "The Geographical Cycle in Periglacial Regions as It Is Related to Climatic Geomorphology." *Annals of the Association of American Geographers* 40: 214–236. This work is an excellent example of the idea of climatic geomorphology.

Penck, W. 1924. *Die Morphologische Analyse*. Stuttgart: Geogr. Abhandlungungen (trans. H. Czech and K. C. Boswell, [1953]. *Morphological Analysis of Landforms*. London: Macmillan). This is the most widely accepted alternative to Davis's cycle theory.

Schumm, S. A. 1977. *The Fluvial System*. New York: John Wiley and Sons.

Tinkler, K. J. 1985. *A Short History of Geomorphology*. Totowa, N.J.: Barnes and Noble Books. This work gives an extensive history of erosion theory from Hutton to the present.

Wischmeier, W. H. 1959. "A Rainfall Erosion Index for a Universal Soil Loss Equation." In *Soil Science Society of America, Proceedings* 23: 246–249.

Wischmeier, W. H.; Johnson, C. B.; and Cross, B. V. 1971. "A Soil Erodibility Nomograph for Farmland and Construction Sites." *Journal on Soil and Water Conservation* 26: 189–193.

Wischmeier, W. H., and Smith, D. C. 1978. *Predicting Rainfall Erosion Losses—A Guide to Conservation Planning*. Handbook, no. 537. Washington, D.C.: U.S. Department of Agriculture.

Sources of Additional Information

The historical aspects of W. M. Davis and the erosion-cycle theory can be seen in R. J. Chorley, R. P. Beckinsale, and A. J. Dunn, *The History of the Study of Landforms. Vol. 2, The Life and Work of W. M. Davis* (London: Methuen & Co., 1973), and R. P. Beckinsale, "The International Influence of William Morris Davis," *Geographical*

Review 66 (1976): 448–466. For work on the principles and processes of soil erosion see L. B. Leopold, M. G. Wolman, and J. P. Miller, *Fluvial Processes in Geomorphology* (San Francisco: W. H. Freeman and Co., 1964); R. W. Fairbridge, ed., *The Encyclopedia of Geomorphology*, Encyclopedia of Earth Sciences Series, vol. 3 (Stroudsburg, Pa.: Dowden, Hutchinson and Ross, 1968); American Society of Civil Engineers, *Sedimentation Engineering*, Manual, no. 54 (Stroudsburg, 1975); and R. W. Fairbridge and C. W. Finkle, Jr., eds., *The Encyclopedia of Soil Science, Part 1*, Encyclopedia of Earth Science Series, vol. 12 (Stroudsburg, Pa.: Dowden, Hutchinson and Ross, 1979). For additional information on soil erosion and conservation see R. P. C. Morgan, *Soil Conservation: Problems and Prospects* (New York: John Wiley and Sons, 1981), and Food and Agriculture Organization, *World Soil Charter* (Rome: United Nations, 1982).

EROSION CYCLE (GEOGRAPHIC CYCLE). See EROSION.

EVAPORATION. See LAPSE RATE.

EVAPOTRANSPIRATION. 1. The process by which water is removed from the soil surface by evaporation and by transpiration of plants. 2. The quantity of water so removed from the soil surface. 3. The total transfer of water from the earth back to the atmosphere.

The compound word *evapotranspiration* is derived from the two words *evaporation* and *transpiration*. *Evaporization* is the vaporization of liquid (see PRECIPITATION) and/or solid water. The evaporization of solid water without the water going through the liquid phase is called *sublimation*. *Transpiration* is the process by which water is taken into plants, used for plant functions, and released to the air as a vapor.

The ancient Greeks were concerned with the HYDROLOGIC CYCLE, which includes evaporation as a mechanism for resupply of water to the atmosphere. Evaporation was studied more closely when Edmond Halley (1694) estimated evaporation rates from the Mediterranean Sea. Transpiration was first explained by Stephen Hales in his study of tree-sap movement. The discovery of osmosis as a partial mechanism for transpiration was made by Datrochet (Bynum and Browne, 1981: 424).

An invaluable extension of the study of both evaporation and transpiration is the concept of potential evapotranspiration (PE). *PE*, a useful term in CLIMATOLOGY for decades, is the theoretical maximum loss of water in an area to evapotranspiration when the surface is completely covered by vegetation and water is plentiful. When calculated for a given region, it indicates the potential water use within the local environment; and when compared with actual evapotranspiration (AE), it will indicate any water shortage or need for irrigation (Hordon, 1967: 373).

Some research in the past five decades has emphasized agriculture and plant reactions to various PE/AE environments. This emphasis is evident in works by R. K. Schofield and H. L. Penman (1948), E. W. Russell (1973), and J. E.

Begg and N. C. Turner (1976). The main thrust of work dealing with evapotranspiration, however, is in climatology and climate-classification systems.

In 1915 the U.S. Weather Bureau established a set of field stations around the country to measure evaporation. This was the first time a large, systematic plan for evaporation measurement was attempted (Sellers, 1965: 156–157). This system was easy to use and gave very good estimates for evaporation at a site. There is no easy way, however, to measure transpiration that is of equal importance in the total evapotranspiration calculation. In 1931 (later revised in 1948) C. W. Thornthwaite developed a mathematical model that would estimate total evapotranspiration at a site and classify regional climates according to water surpluses and deficits (Thornthwaite, 1931 and 1948).

The Thornthwaite climate-classification system is similar to the one developed by Wladimir Köppen in that they both are quantitative and inherently take natural vegetation classes into account. Thornthwaite's system differs because it expresses temperature efficiency and PRECIPITATION effectiveness in the context of PE/AE and the regional propensity to support vegetation. The key to this system is the ability to calculate PE. Thornthwaite's method is empirical and involves the summation of monthly calculations. PE is calculated by:

$$PE = 16(10\ T/I)^a mm\ mo^{-1}$$

where a = a cubic function of I, T = mean temperature, and I = the sum of twelve monthly heat indices.

Tables and monographs have been published for ease in calculation of PE for most parts of the world (e.g., Thornthwaite and Mather, 1957; Palmer and Havens, 1958). Although this approach to climate classification seems relatively easy, the basic relationships between potential water loss and temperature are extremely complex (Crowe, 1971: 100). Some researchers (e.g., Chang, 1959) make very strong cases against using the Thornthwaite system stating that it is too complex and that some aspects of climate are ignored. Others (e.g., Carter and Mather, 1966) strongly defend the system and attempt to put the Thornthwaite system into the natural evolution of climate classification (Oliver 1973: 181). Despite the possible pitfalls of Thornthwaite's system, it is a useful tool that provides understanding not possible with other climate-classification systems.

Putting the Thornthwaite controversy aside, the concept of evapotranspiration is a very useful tool and even more relevant to global climates than previously thought (Shukla and Mintz, 1982). The environmental limitations and capacities of virtually all regions today depend upon the hydrologic cycle including evapotranspiration. This is obvious from regional modeling applications as seen in G. D. Tasker (1982) and I. Simmers (1984). Hydrologic and evapotranspiration modeling will become increasingly important in the future as climate changes become a more critical concern.

References

Begg, J. E., and Turner, N. C. 1976. "Crop Water Deficits." *Advances in Agronomy* 28: 161–217.
Bynum, W. F., and Browne, E. J. 1981. *Dictionary of the History of Science*. Princeton, N.J.: Princeton University Press.
Carter, D. B., and Mather, J. R. 1966. *Climatic Classifications for Environmental Biology*. Publication, no. 19(4). Elmer, N.J.: C. W. Thornthwaite Associates, Laboratory of Climatology. A strong defensive statement about the usefulness of the Thornthwaite system of climate classification.
Chang, J-Hu. 1959. "An Evaluation of the 1948 Thornthwaite Classification." *Annals of the Association of American Geographers* 49: 24–30. An equally strong statement against Thornthwaite's system.
Crowe, P. R. 1971. *Concepts in Climatology*. New York: St. Martin's Press. Contains an excellent section on evapotranspiration and the historical context of climate classification using evapotranspiration.
Halley, E. 1694. "An Account of the Evaporation of Water." *Philosophical Transactions, Royal Society, London* 18: 183–190. An early experimental work on water evaporation.
Hordon, R. M. 1967. "Evapotranspiration." In R. W. Fairbridge, ed., *The Encyclopedia of Atmospheric Sciences and Astrogeology*. Encyclopedia of Earth Science Series. Vol. 2. New York: Reinhold Publishing Co., pp. 372–373.
Oliver, J. E. 1973. *Climate and Man's Environment*. New York: John Wiley and Sons. Good sections on evapotranspiration and use of the Thornthwaite system.
Palmer, W. C., and Havens, A. V. 1958. "A Graphical Technique for Determining Evapotranspiration by the Thornthwaite Technique." *Monthly Weather Review* 86: 123–128.
Penman, H. L. 1963. "Vegetation and Hydrology." *Technical Communication*, no. 53. Bucks, Eng.: Commonwealth Bureau of Soils.
Russell, E. W. 1973. *Soil Conditions and Plant Growth*. London: Longmans, Green and Co.
Schofield, R. K., and Penman, H. L. 1948. "The Principles Governing Transpiration by Vegetation." In *Proceedings, Conference on Biology and Civil Engineering*. pp. 75–98.
Sellers, William D. 1965. *Physical Climatology*. University of Chicago Press, Chicago.
Shukla, J., and Mintz, Y. 1982. "Influence of Land-Surface Evaporation on the Earth's Climate." *Science* 215: 1498–1506.
Simmers, I. 1984. "A Systematic Problem-Oriented Approach to Hydrological Data Regionalization." *Journal of Hydrology* 73: 71–87.
Tasker, G. D. 1982. "Comparing Methods of Hydrologic Regionalization." *Water Resources Bulletin* 18: 965–970.
Thornthwaite, C. W. 1931. "The Climates of North America." *Geographical Review* 21: 633–655. This is the first version of Thornthwaite's climate-classification system.
———. 1948. "An Approach Towards a Rational Classification of Climate." *Geographical Review* 38: 55–94. The second version of Thornthwaite's system, which included the complex methodology for determining potential evapotranspiration.

Thornthwaite, C. W., and Mather, J. R. 1957. *Instructions and Tables for Computing Potential Evapotranspiration and the Water Budget.* Publications in Climatology. Vol. 10, no. 3. Centerton, Ill.: Drexel Institute of Technology.

Sources of Additional Information

Evapotranspiration in the larger context of hydrology is well covered in many sources, including, for example, F. J. Veihmeyer, "Evapotranspiration," in J. T. Chow, ed., *Handbook of Applied Hydrology* (New York: McGraw-Hill Book Co., 1964), chapter 11; W. D. Sellers, *Physical Climatology* (Chicago: University of Chicago Press, 1965); and R. C. Ward, *Principles of Hydrology* (New York: McGraw-Hill Book Co., 1975). The interrelationships of evapotranspiration and plant growth are well covered by the following: R. M. Hager and J. I. Stewart, "Water Deficits, Irrigation Design and Programming," *Journal of the Irrigation and Drainage Division, Proceedings of the American Society of Civil Engineers* 98 (1972): 215–235; W. K. Smith and G. N. Geller, "Leaf and Environmental Parameters Influencing Transpiration: Theory and Field Measurements," *Oecologia* 46 (1980): 308–313; and C. L. Wiegand, P. R. Nixon, H. W. Gausman, L. M. Namken, R. W. Leamer, and A. J. Richardson, *Plant Cover, Soil Temperature, Freeze, Water Stress, and Evaporation Conditions* (Green Belt, Md.: Goddard Space Flight Center, 1981). General references to evapotranspiration and hydrological processes include R. W. Fairbridge, ed., *The Encyclopedia of Atmospheric Sciences and Astrogeology*, Encyclopedia of Earth Science Series, vol. 2 (New York: Reinhold Publishing Co., 1967); and K. J. Gregory and D. E. Walling, eds., *Man and Environmental Processes* (Boulder, Colo.: Westview Press, 1979).

F

FAULTING. See TECTONICS.

FLOOD PLAIN. 1. That portion of a river valley inundated by water during flooding. 2. An alluvial plain adjacent to a river or a stream that has been built from unconsolidated depositional sediment.

Flood plains are very dynamic topographic surfaces that are created and shaped by factors that are dependent on the entire DRAINAGE BASIN system. They derive their characteristics from the STREAM- or river-deposited material, and they in turn affect the stream or river and the surrounding LANDFORMS. As stated by T. H. Schmudde, "the flood plain can be properly thought of as both a product and a functional part of the whole stream environment, and it plays a necessary role in maintaining the overall adjustment that a stream system makes to the variable quantities of water, solubles, and solid particles imposed on it" (1968: 359).

Hydrologists and geomorphologists have never agreed on the defined extent of the flood plain. Some believe that the area covered by the 100-year flood is the flood plain, whereas others, such as M. G. Wolman and L. B. Leopold (1957: 89–91), believe that the active flood plain is the area subject to flooding by the annual flood. There is still no accepted definition of *flood plain* extent.

Flood-plain deposits have been recognized for centuries. Even Leonardo de Vinci wrote about river planform and channel characteristics (Alexander, 1982). G. K. Gilbert defined and described the deposits that comprise the flood plain. He stated that "the width of the channel remains the same while its position is shifted, and every part of the valley which it has crossed in its shiftings comes to be covered by a deposit which does not rise above the highest level of the water. The surface of this deposit is hence appropriately called the flood-plain of the stream" (1880: 121).

In 1936 F. A. Melton classified flood plains according to the deposit type and stream-channel pattern that makes up the deposit. According to K. J. Gregory and D. E. Walling, "the dependence of the flood plain upon the stream channel pattern is apparent from this [Melton's] classification and the features of flood plains are of two main types, firstly those due to lateral accretion and secondly those formed by overbank deposits" (1973: 261). Leopold and Wolman (1957: 44–48) found that 80 to 90 percent of flood-plain deposits result from lateral accretion and channel deposition, and the remaining 10 to 20 percent are from overbank deposits. On the other hand, some researchers such as S. C. Happ et al. (1940: 26) believe that vertical accretion is the dominant process in flood-plain evolution. Flood-plain aggradation and degradation remain important research topics today. For summaries of this work see E. Derbyshire et al. (1979), N. G. Bhowmik (1984), and S. L. Dingman (1984).

Another very important aspect of research concerning flood plains is the hazard of flooding and inundation of the flood plain. Federal, state, and local governments have made flood plains and flood-hazard management important policy issues. Within this context of study, there are numerous research and policy foci. They include, but are not limited to, issuance of flood-plain land-use regulations, the National Flood Insurance Program, the structural control of floods by organizations like the U.S. Army Corps of Engineers, and various state and local flood-plain laws and ordinances. R. B. Platt (1980) produced an excellent work on the varied intergovernmental relationships involved in the management of flood plains. A case study of flood-plain management can be seen in P. H. Rahn's (1984) work concerning flood plains in Rapid City, South Dakota, following the disastrous flood of 1972 in that city. In the future this policy-oriented work will be at least as important in research agendas as the geologic and geomorphic work previously mentioned.

References

Alexander, D. 1982. "Leonardo de Vinci and Fluvial Geomorphology." *American Journal of Science* 282: 735–755.

Bhowmik, N. G. 1984. "Hydraulic Geometry of Flood Plains." *Journal of Hydrology* 68: 369–401.

Derbyshire, E.; Gregory, K. J.; and Hails, J. R. 1979. *Geomorphological Processes*. Boulder, Colo.: Westview Press. This is a good general reference for flood plains, fluvial processes, and other geomorphological processes.

Dingman, S. L. 1984. *Fluvial Hydology*. New York: W. H. Freeman and Co.

Gilbert, G. K. 1880. *Geology of the Henry Mountains*. Washington, D.C.: U.S. Geographical and Geological Survey of the Rocky Mountain Region. A seminal work in fluvial geomorphology.

Gregory, K. J., and Walling, D. E. 1973. *Drainage Basin Form and Process*. New York: John Wiley and Sons. This is an excellent reference for any topic in fluvial geomorphology.

Happ, S. C.; Rittenhouse, G; and Dobson, G. C. 1940. "Some Principles of Accelerated

Stream and Valley Sedimentation." Technical Bulletin, no. 695. Washington, D.C.: U.S. Department of Agriculture.

Leopold, L. B., and Wolman, M. G. 1957. "River Channel Patterns—Braided, Meandering, and Straight." U.S. Geological Survey Professional Paper, no. 282-B. Washington, D.C.: U.S. Geological Survey. One of the best research papers on channel patterns ever written.

Melton, F. A. 1936. "An Empirical Classification of Flood Plain Stream." *Geographical Review* 26: 593–609.

Platt, R. B. 1980. *Intergovernmental Management of Flood Plains*. Monograph, no. 30. Boulder, Colo.: Institute of Behavioral Science, University of Colorado.

Rahn, P. H. 1984. "Floodplain Management Program in Rapid City, South Dakota." *Geological Society of America Bulletin* 95: 838–843.

Schmudde, T. H. 1968. "Flood Plain." In R. W. Fairbridge, ed., *The Encyclopedia of Geomorphology*. Encyclopedia of Earth Science Series. Vol. 3. Stroudsburg, Pa.: Dowden, Hutchinson and Ross, pp. 359–362.

Wolman, M. G., and Leopold, L. B. 1957. *River Flood Plains: Some Observations on Their Formation*. U.S. Geological Survey Professional Paper, no. 282-C. Washington, D.C.: U.S. Geological Survey. This is an authoritative reference for floodplain-deposit studies.

Sources of Additional Information

For additional work on flood plains and general fluvial processes see R. J. Chorley, *Introduction to Fluvial Processes* (Bungay, Eng.: Methuen & Co., 1969); K. J. Gregory, "River Channel Forms, Processes, and Metamorphosis," in D. J. Briggs and R. S. Waters, eds., *Studies in Quaternary Geomorphology* (Norwich, Conn.: Geobooks, 1983), pp. 19–30; D. Knighton, *Fluvial Forms and Processes* (London: E. Arnold, 1984); and M. E. Morisawa, *Rivers, Form, and Process* (London: Longmans, Green and Co., 1985). Examples of case-study research on flood plains and related landforms are M. G. Wolman, *The Natural Channel of Brandywine Creek, Pa.*, U.S. Geological Survey Professional Paper, no. 271 (Washington, D.C.: U.S. Geological Survey, 1955); W. R. Osterkamp and E. R. Hedman, *Perennial Streamflow Characteristics Related to Channel Geometry and Sediment in Missouri River Basin*, U.S. Geological Survey Professional Paper, no. 1242 (Washington, D.C.: U.S. Geological Survey, 1982); and N. D. Smith and D. G. Smith, "William River: An Outstanding Example of Channel Widening and Braiding Caused by Bed-Load Addition," *Geology* 12 (1984): 78–82. Work dealing with flood plains only includes T. H. Schmudde, "Some Aspects of Land Forms of the Lower Missouri River Flood Plain," *Annals of the Association of American Geographers* 53 (1963): 60–73; S. A. Schumm, *Channel Widening and Flood Plain Construction along Cimarron River in Southwestern Kansas*, U.S. Geological Survey Professional Paper, no. 352-D (Washington, D.C.: U.S. Geological Survey, 1963); J. R. L. Allen, "A Review of the Origin and Characteristics of Recent Alluvial Sediments," *Sedimentology* 5 (1965): 89–191; and N. G. Bhowmik and M. Demissie, "Carrying Capacity of Flood Plains," *American Society of Civil Engineers, Journal of Hydraulics Division* 108 (1982): 443–452. For examples of work done on modeling flood plains see P. E. Ashmore, "Laboratory Modelling of Gravel Braided Stream Morphology," *Earth Surface Processes and Landforms* 7 (1982): 201–225, and H. H. Chang, "Modelling of River Channel Changes," *Journal of Hydraulic Engineering* 110 (1984): 157–179.

FOEHN. The name in the European Alps given to a warm, dry wind that descends, usually at high speed, the leeward side of a mountain or mountain range during a specific set of meteorological conditions. (North American Cordillera, *Chinook*; Andes, *Zonda*; New Zealand, *North-wester*; California, *Santa Ana*; South Africa, *Berg*; Iran, *Samun*.)

A foehn is characterized by gusty winds, high temperatures, low humidity, and very clear air. This strong wind can be very dramatic and at times frightening as evidenced by the following quote from J. von Hann:

> A few brief gusts announce the arrival of the foehn. These gusts are cold and raw at first, especially in winter, when the wind has crossed vast fields of snow. Then there is a sudden calm, and all at once the hot blast of the foehn bursts into the valley with tremendous violence, often attaining the velocity of a gale, which lasts two or three days with more or less intensity, bringing confusion everywhere; snapping off trees, loosening masses of rock; filling up the mountain torrents; unroofing houses and barns—a terror in the land. (1903: 346)

For centuries the foehn has been known in mountain regions, especially the European Alps, where it was first named. The local wisdom described the foehn as hot and dry air from northern Africa flowing north over the Alps and descending down the leeward slopes. It was discovered, however, that there were usually rainstorms on the windward side of the Alps during the foehn, which would not allow the suspected warm air from Africa to flow north over the mountains. Therefore, another explanation was needed.

The thermodynamic theory of the foehn, first proposed by Hann in 1886 and expanded by him in 1903 (Hann, 1886, 1903), states that a low-pressure system sets up on the leeward side of the mountain and a high-pressure system sets up on the windward side. Air forced from the high-pressure side rises over the mountain and cools at the moist adiabatic LAPSE RATE, and precipitation occurs. This drier air then continues toward the low-pressure system, descends the mountain, and is warmed at the dry adiabatic lapse rate, which is higher than the moist adiabatic lapse rate (see LATENT HEAT). This produces warmer and drier air on the leeward side (Crowe, 1971; Price, 1981). Other early studies confirmed this theory at least for most of the Alpine foehn winds (Billwiller and Quervain, 1912). The very warm and dry air of the foehn is dramatized by the case quoted in Hann (1903: 346) in which 65 centimeters of snow were sublimated in twelve hours at Grindelwald by a foehn. In his review article on the foehn, W. A. R. Brinkman (1971) stressed that a true foehn must have an absolute heat increase over and above the invaded, ambient AIR MASS.

During the past four decades, at least three other subtypes of foehn have been described that have meteorological conditions somewhat different from those described by Hann. These subtypes include one type in which no PRECIPITATION falls on the windward side of the mountain due to certain stability conditions on the windward side. In a second type nocturnal cooling is prevented from occurring by surface turbulence, and therefore, temperatures rise. A third type involves cold-air replacement by a warm air mass on the leeward side of the mountain (Cook and Topil, 1951; Glenn, 1961; and Beran, 1967).

People, animals, and vegetation are all affected by the foehn. Sometimes there are advantages to the foehn such as when deep snow is suddenly melted, exposing forage for animals, or when the foehn breaks a cold spell and relieves the stress of winter (Ashwell and Marsh, 1967). However, most of the effects of the foehn are negative. When the snow is melted or sublimated, it does not normally replenish the soil moisture. There have also been a number of studies linking foehns and other meteorological events to health (Berg, 1950; Winstanley, 1972; and World Health Organization, 1972). These health hazards include depression to the point of suicide, heart palpitations, headaches, and lung disorders. However, these physical and psychological maladies are very complex, and weather patterns may be only a small contributing factor to an illness. An interesting sidelight is that few, if any, of these reactions to the foehn (the chinook) have been described in the North American Cordillera or other mountain areas; they have almost all been described in the Alps.

Recent research has focused primarily on the mechanisms for the descent of the wind down the leeward side of the mountain. Mountain waves and lee waves, along with the attendant lenticular and roll clouds, have been seen as the principal products of the foehn (e.g. Nicholls, 1973; Smith, 1976; Whiteman and McKee, 1982). Work is continuing along this line of research.

References

Ashwell, I. Y., and Marsh, J. S. 1967. "Moisture Loss under Chinook Conditions." In *Proceedings, First Canadian Conference on Micrometeorology*: 307–310.
Beran, D. W. 1967. "Large Amplitude Lee Waves and Chinook Winds." *Journal of Applied Meteorology* 6: 865–877. Shows relationship between lee waves and foehns or chinooks.
Berg, H. 1950. "Der Einfluss des Föhns auf den Menschen." *Geofisica Pura e Applicata* 17: 104–111. Connects foehns with suicide attempts.
Billwiller, R., and Quervain, A. de. 1912. "Registrierballon Aufsteige in einem Föhntale." *Meteorologische Zeitschrift* 29: 249–251. Confirmed Hann's theory of foehn development.
Brinkman, W. A. R. 1971. "What Is a Foehn?" *Weather* 26: 230–239. Discussed the importance of temperature in foehn definition.
Cook, A. W., and Topil, A. 1951. "Some Examples of Chinooks East of the Mountains in Colorado." *Bulletin of the American Meteorological Society* 33: 42–47.
Crowe, P. R. 1971. *Concepts in Climatology*. New York: St. Martin's Press.
Glenn, C. L. 1961. "The Chinook." *Weatherwise* 14: 175–182. A good discussion of the chinook phenomenon.
Hann, J. 1886. "Zur Frage über den Ursprung des Föhn." *Zeitschrift Osterr. Gesch. für Meteorologie* 1: 257–263. First explained the thermodynamic theory of foehn development.
———. 1903. *Handbook of Climatology. Part I* (trans. R. De C. Wood). London: Macmillan. Expanded his explanation of foehn development.
Nicholls, J. M. 1973. "The Airflow Over Mountains: Research, 1958–1972." Technical

note, no. 127. Geneva: World Meteorological Organization. A good synopsis of the work done for a fifteen-year period.

Price, L. W. 1981. *Mountains and Man.* Berkeley: University of California Press. Has a good section on foehn research over the years.

Smith, R. B. 1976. "The Generation of Lee Waves by the Blue Ridge." *Journal of Atmospheric Science* 33: 507–519.

Whiteman, C. D., and McKee, T. B. 1982. "Breakup of Temperature Inversions in Deep Mountain Valleys. Part II: Thermodynamic Model." *Journal of Applied Meteorology* 21: 290–302.

Winstanley, D. 1972. "Sharav." *Weather* 27: 146–160. Looks at weather and disease.

World Health Organization. 1972. *Health Hazards of the Human Environment.* Geneva. Looks at health problems and weather phenomena.

Sources of Additional Information

Good general explanations of foehns can be found in H. Riehl, *Introduction to the Atmosphere*, 3d ed. (New York: McGraw-Hill Book Co., 1978); J. F. Griffiths and D. M. Driscoll, *Survey of Climatology* (Columbus, Ohio: Charles E. Merrill Publishing Co., 1982); and A. H. Strahler and A. N. Strahler, *Modern Physical Geography*, 3d ed. (New York: John Wiley and Sons, 1987).

FOG. See CLOUD.

FOLDING. See TECTONICS.

FRONTAL SYSTEM. 1. The system including the sharp transition zone between air masses of differing temperature and humidity and the air masses near this zone. 2. The warm-air boundary of the maximum-temperature gradient identified as the actual frontal surface. 3. The contact zone between two differing air masses that moves with the airflow.

The concept of a frontal system evolves logically from the concept of AIR MASSES, each of which has relatively homogeneous characteristics (e.g., temperature and humidity). These air masses, moved by general and local circulation patterns and complex fluid dynamics interactions, inevitably come into contact with air masses containing different characteristics, thus causing a confrontation. The zone of contact where this confrontation occurs is the frontal zone.

The classical notion of fronts comes principally from the work of Vilhelm Bjerknes (1920) and his son Jacob Bjerknes (1930; Bjerknes and Solberg, 1921, 1922). The term *front* actually was coined during World War I, when opposing forces on the Western Front in France reminded Vilhelm Bjerknes of the opposing masses of air in conflict over any given piece of ground. P. R. Crowe stated the idea very well:

> Even if the battle never involves more than a fraction of the opposing forces, upon its outcome the stability of the whole line may ultimately depend. If we replace the opposing armies by two air masses differing in depth and temperature (and

hence in density) and thrown into mutual relationships by a converging circulation, then the release of their potential energy, the battle, is represented by the development of a weather system along the front and the analogy is seen to be remarkably apt. (1971: 352)

During the 1930s and the 1940s, the analysis of air masses and frontal systems was inseparable and dominated synoptic weather forecasting research. Detailed delimitation of given air masses, however, was difficult and led to interpretation problems (Barry and Perry, 1973: 181). W. L. Godson (1950) devised a system by which air-mass identification, and therefore frontal definition, was described in classical Norwegian terms. His system included five air-mass types with four frontal types at the contact zones between them. The frontal types included the inter-Arctic, Arctic, interpolar, and polar. Several researchers revised this model for winter in North America (the most active frontal weather area on earth) (Anderson, Boville, and McClellan, 1955; Galloway, 1958; McIntyre, 1958). This modified system had four air-mass types and three frontal types including the continental Arctic front, maritime Arctic front, and polar front. Two criteria were adopted for defining the front in this system: a three-dimensional zone of pronounced baroclinicity and a space and time continuity that is displaced with the airflow (Barry and Perry, 1973: 59).

A main research focus during the past few decades has been the quest for analysis tools. Questions include: what are fronts really like, how do they differ, what parameters can be used to predict their actions, and basically, how do fronts develop? The study of "frontogenesis" or "confrontation" is a complex field that has made strides in addressing the above questions but has not satisfactorily answered them as yet. An early attempt to quantify this area of research was done by Sir Napier Shaw, who dealt with this problem in his *Manual of Meteorology* (1931) while discussing the calculus of meteorology in relation to pressure and wind. More recent attempts to identify fronts with quantitative analysis have used relative temperature (Kirk, 1966a), potential temperature (Renard and Clarke, 1965), and wet-bulb potential temperature (Creswick, 1967). Kirk (1966b) also considered frontal zone location identification to be at least a partial function of the change in geostrophic vorticity.

More subjective, yet still valid, methods of frontal zone identification have been devised by several researchers including R. J. Reed (1960), H. Van Loon (1965), and M. Yoshimura (1969).

In addition to other recent work, a most interesting idea has developed, the concept of "ocean fronts." These are contact zones between masses of water (much like masses of air) in the earth's oceans that act and react in ways somewhat similar to the fluid atmosphere (Perry and Walker, 1977: 108). This line of research has expanded and created a determined effort to connect the various factors or regions involved in global climate interactions as seen in the work of J. T. Houghton (1984) and Y. H. Pan and A. H. Oort (1983) for example.

References

Anderson, R.; Boville, B. W.; and McClellan, D. E. 1955. "An Operational Frontal Contour-Analysis Model." *Quarterly Journal of the Royal Meterological Society* 81: 588–599. This article deals with mid-latitude frontal systems.

Barry, Roger G., and Perry, Allen H. 1973. *Synoptic Climatology: Methods and Applications*. London: Methuen & Co. This is an excellent reference work on frontal systems and all other climate phenomena.

Bjerknes, Jacob, 1930. "Practical Examples of Polar Front Analysis over the British Isles." Geophysical Memo, no. 50 (M.O. 307j), HMSO.

Bjerknes, Jacob, and Solberg, H. 1921. "Meteorological Conditions for the Formation of Rain." *Geofysical Publikations* 2, no. 3.

———. 1922. "The Life Cycle of Cyclones and the Polar Front Theory of Atmospheric Circulation." *Geofysical Publikations* 3, no. 1: 1–18.

Bjerknes, Vilhelm. 1920. "The Structure of the Atmosphere When Rain Is Falling." *Quarterly Journal of the Royal Meteorological Society* 46: 119–140. The above four references established the concept of the frontal system and the classical Norwegian school of frontal analysis.

Creswick, W. S. 1967. "Experiments in Objective Frontal Analysis." *Journal of Applied Meteorology* 6: 774–781.

Crowe, P. R. 1971. *Concepts in Climatology*. New York: St. Martin's Press. This is another excellent reference for all concepts in climatology.

Galloway, J. L. 1958. "The Three-Front Model: Its Philosophy, Nature, Construction, and Use." *Weather* 13: 395–403. This reference defined in detail the idea behind the three-front model for the mid-latitudes.

Godson, W. L. 1950. "The Structure of North American Weather Systems." In *Centennial Proceedings of the Royal Meteorological Society*: 89–106.

Houghton, J. T., ed. 1984. *The Global Climate*. Cambridge: Cambridge University Press.

Kirk, T. H. 1966a. "A Parameter for the Objective Location of Frontal Zones." *Meteorological Magazine* 94: 351–353.

———. 1966b. "Some Aspects of the Theory of Fronts and Frontal Analysis." *Quarterly Journal of the Royal Meteorological Society* 92: 374–381.

McIntyre, D. P. 1958. "The Canadian Three-Front, Three-Jet Stream Model." *Geophysica* 6: 309–324.

Pan, Y. H., and Oort, A. H. 1983. "Global Climate Variations Connected with Sea Surface Temperature Anomalies in the Eastern Equatorial Pacific Ocean for the 1958–73 Period." *Monthly Weather Review* 111: 1244–1258.

Perry, A. H., and Walker, J. M. 1977. *The Ocean-Atmosphere System*. New York: Longman. This is a good reference for looking at ocean or atmosphere relationships.

Reed, R. J. 1960. "Principal Frontal Zones of the Northern Hemisphere in Winter and Summer." *Bulletin of the American Meteorological Society* 41: 591–598.

Renard, R. J., and Clarke, L. C. 1965. "Experiments in Numerical Objective Frontal Analysis." *Monthly Weather Review* 93: 547–556.

Shaw, W. N. 1931. *Manual of Meteorology. Vol. IV, Meteorological Calculus; Pressure and Wind*. Cambridge: Cambridge University Press. This was one of the first attempts at quantitative analysis of weather systems in general and frontal systems in particular.

Van Loon, H. 1965. "A Climatological Study of the Atmospheric Circulation in the Southern Hemisphere during the IGY, Part 1: 1 July 1957–31 March 1958." *Journal of Applied Meteorology* 4: 479–491.

Yoshimura, M. 1969. "Annual Change in Frontal Zones in the Northern Hemisphere." *Japanese Progress in Climatology*: 69–71.

Sources of Additional Information

Most introductory physical geography and climatology texts discuss frontal systems in some detail; for example, see J. F. Griffiths and D. M. Driscoll, *Survey of Climatology* (Columbus: Charles E. Merrill Publishing Co., 1982); F. K. Lutgens and E. J. Tarbuck, *The Atmosphere*, 3d ed. (Englewood Cliffs, N.J.: Prentice-Hall, 1986); and A. N. Strahler and A. H. Strahler, *Modern Physical Geography* (New York: John Wiley and Sons, 1983). An excellent reference work on definitions in climatology is R. W. Fairbridge, *The Encyclopedia of Atmospheric Sciences and Astrogeology* (New York: Reinhold Publishing Co., 1967). A good review of the mathematical models of frontal analysis (and climate in general) is J. T. Houghton, ed., *The Global Climate* (Cambridge: Cambridge University Press, 1984). Works dealing with frontogenesis include C. L. Godske, T. Bergeron, J. Bjerknes, and R. C. Bundgaard, *Dynamic Meteorology and Weather Forecasting* (Boston: American Meteorological Society, 1957); C. W. Newton, "Severe Convective Storms," *Advances in Geophysics* 12 (1966): 257–308; and S. Pettersen, D. L. Bradbury, and K. Pederson, "The Norwegian Cyclone Models in Relation to Heat and Cold Sources," *Geophysical Publik* 24 (1962): 248–280. Old and new general and comprehensive works on climate and frontal systems include F. A. Berry, Jr., E. Bollay, and N. R. Beers, *Handbook of Meteorology* (New York: McGraw-Hill Book Co., 1973); H. E. Landsberg, ed., *World Survey of Climatology*, 1 vol./year (Amsterdam: Elsevier, 1969-present); and Sir N. Shaw, *Manual of Meteorology*, 4 vols. (Cambridge: Cambridge University Press, 1919–1936).

G

GEOGRAPHICAL CYCLE. See EROSION; GEOMORPHOLOGY.

GEOLOGIC TIME SCALE. A sequence or arrangement of geologic events determined by absolute and/or relative dating techniques.

The geologic time scale combines traditional relative methods of dating rock strata based on the evolution of plants and animals and depositional sequences with numerical TIME based on radiometric age measurements. Early attempts to establish a geologic time scale based on the time rates associated with geologic processes primarily focused on placing limits in years on the subdivisions of geologic time and establishing the age of the earth.

In the seventeenth century, Nicolaus Steno (1669) made a very important geological discovery. He demonstrated that ROCK layers are deposited in a definite sequence with the oldest layers on the bottom and the youngest on top. This "law of superposition" was critical in determining the order in which geologic events take place.

Efforts to create classification systems based on this law of superposition began around the middle of the eighteenth century by Giovanni Ardiuno (Rodolico, 1969). He proposed in 1760 that the earth's rocks be divided into three groups: Primary (first), Secondary (second), and Tertiary (third). Another group, the Quaternary (fourth) was added and included the very youngest rock materials like soils and ALLUVIUM that were not solidified into rock. In 1766 Johan Lehman developed a simple relative time scale based on three main classes of mountains: those formed since the flood, those formed at the time of the flood, and those formed at the creation of the earth (Schneer, 1969: 10).

The nineteenth century saw many advances in the concept of the geologic time scale. In 1791–1793 the Englishman William Smith was involved in surveying for the building of English canals. While working on canals he observed the constant relationship between certain strata and changes in fossils. In his *Geologic Map of England and Wales with Part of Scotland* (1815), Smith's idea of fossil succession was demonstrated as a valid scientific principle, and since that time it has become a widely accepted procedure for studying and dating sedimentary rock formations (Eyles, 1969). This work of Smith was very important and according to Hallam (1983: 25), it "laid the basis for modern geology [see GEOLOGY] by allowing the establishment of a relative time-scale which could be used across the world." Several years after Smith's publication, Baron Georges Cuvier, a French naturalist, and the French geologist Alexandre Brongniart published a book that described the rocks and fossils in the Paris region. Like Smith, they realized that certain fossils were found in specific strata, and they could trace these strata throughout the Paris basin (Cuvier, 1825). By the 1850s the principles of stratigraphic and faunal succession had played a dominant role in establishing three major time divisions, the Cenozoic (recent life), Mesozoic (middle life), and Paleozoic (ancient life) eras. These eras are designated to indicate the characteristic stage of development of fossils they contain.

Much of the credit for the idea of faunal succession must go to Charles Darwin (1859). Darwin's idea of organic evolution stemmed from geological and biological observations during his five-year voyage to the South Pacific. Darwin's work had a major impact on geological ideas and according to Brainerd Mears: "Today the scientific evidence, much of it from the fossil record, overwhelmingly supports the past occurrence of organic evolution. Geologically we now recognize that organic evolution accounts for the faunal succession that made fossils an indispensable tool in unravelling and assembling a proper geologic time sequence" (1970: 86).

As presently constituted, the geologic time scale is divided into these large divisions called eras; they in turn are subdivided into periods that are further subdivided into epochs. Although there is general agreement as to the nature and time span of the eras and periods, there is still disagreement about the epochs (see Table 2).

While many scientists were concerned with dating geologic strata and establishing a relative time scale, others were interested in determining an absolute age for rocks and MINERALS. An early effort to determine an absolute data for the origin of the earth was undertaken by Charles Darwin. Darwin was a supporter of the British geologist Charles Lyell, and in Darwin's classic work *On the Origin of Species*, he stated: "He who can read Sir Charles Lyell's grand work on the Principles of Geology and yet does not admit how incomprehensibly vast has been the pat periods of time, may at once close this volume" (1859: 282). Darwin went on to confirm Lyell's ideas about the earth's age by calculating how long it would have taken to erode the Weald region in southeast England. He concluded that the age of the earth was approximately 300 million years (Hallam, 1983: 83).

Another person who estimated the age of the earth was the Scottish physicist

Table 2
Table of Geologic Time

ERA	Period	Epoch
Cenozoic	Quaternary	Holocene (Recent)
		Pleistocene
		Pliocene
		Miocene
	Tertiary	Oligocene
		Eocene
		Paleocene
Mesozoic	Cretaceous	
	Jurassic	
	Triassic	
Paleozoic	Permian	
	Carboniferous	
	Devonian	
	Silurian	
	Ordovician	
	Cambrian	
Pre-Cambrian Time		

William Thomason, better known as Lord Kelvin. Kelvin (1862) in his early work estimated the earth to be between 100 million and 500 million years old but later stated that it was between 20 million and 40 million years old. Kelvin believed his estimates were based on "known physical laws" and were therefore more accurate than previous estimates.

With the discovery of the phenomenon of radioactivity in 1896 and the pioneering work on radioactive elements in rocks by Rutherford (Eve, 1939), a new quantitative geologic time scale was developed. This quantitative scale was the result of work by many scientists, but foremost among the earth scientists was Arthur Holmes. He outlined how the age of rocks could be determined based on principles of radioactive decay in conjunction with geological information on rock thickness (Holmes, 1911, 1959).

The geologic time scale is undergoing continual revision and refinement as intensive research goes on but "it is unlikely that the broad structure . . . will be greatly altered as a result of future measurements" (Kulp, 1982: 188).

References

Cuvier, G. 1825. *Essays on the Theory of the Earth*. Edinburgh: Blackwood.

Darwin, Charles. 1859. *On the Origin of Species*. London: J. Murray. The classic study of organic evolution.

Eve, A. S. 1939. *Rutherford*. New York: Macmillan.

Eyles, Joan M. 1969. "William Smith: Some Aspects of His Life and Work." In Schneer, Cecil J., ed. *Toward a History of Geology*. Cambridge, Mass.: The M.I.T. Press, pp. 142–158. A short study dealing with the scientific contributions of William Smith.

Hallam, A. 1983. *Great Geological Controversies*. Oxford: Oxford University Press. An excellent book on the major geological controversies with a chapter on the age of the earth.

Holmes, Arthur. 1911. "The Association of Lead with Uranium in Rock Minerals and Its Application to the Measurement of Geological Time." In *Royal Society of London Proceedings*, Series A, 85: 248–256.

———. 1959. "A Revised Geological Time Scale." *Edinburgh Geological Society Transactions* 17: 183–216.

Kelvin, Lord. 1862. "On the Age of the Sun's Heat." *Macmillans Magazine* 5: 288–375.

Kulp, J. Laurence. 1982. "Geological Time Scale." In *McGraw-Hill Encyclopedia of Science and Technology*. 5th ed. Vol. 6. New York: McGraw-Hill Book Co.

Mears, Brainerd Jr. 1970. "The History and Development of Geology." In Mears, Brainerd, Jr., ed., *The Nature of Geology: Contemporary Readings*. New York: Van Nostrand-Reinhold Co., pp. 5–18. An excellent review of the development of geological thinking.

Rodolico, F. 1969. "Arduino, G." In *Dictionary of Scientific Biography*. New York: Scribners.

Schneer, Cecil J., ed. 1969. *Toward a History of Geology*. Cambridge, Mass.: The M.I.T. Press.

Steno, Nicolaus. 1669. *The Prodromus of Nicolaus Steno's Dissertation* (trans. J. Winter). New York: Macmillan, 1916.

Sources of Additional Information

For a nontechnical discussion of radioactive dating techniques see Ruth Moore, "Kulp: The Age of the Earth," in Brainerd Mears, Jr., *The Nature of Geology* (New York: Van Nostrand Reinhold-Co., 1970), pp. 73–81. For detailed information on all aspects of the geologic time scale see George V. Cohee, Martin F. Glaessner, and Hollis D. Hedberg, *The Geologic Time Scale* (Tulsa, Okla.: The American Association of Petroleum Geologists, 1978), and W. B. Harland et al., *A Geologic Time Scale* (Cambridge: Cambridge University Press, 1982). For a good discussion on geologic time as it relates to the evolution-creation controversy see National Association of Geology Teachers, *Journal of Geological Education* 31 (1983).

GEOLOGY. 1. The science that deals with the earth and its composition, structure, processes, and historical evolution. 2. One of the major earth sciences including the history of the earth, geophysics, physical geology, geomorphology, stratigraphy, paleontology, mineralogy, petrology, and engineering, petroleum, and mining geology.

Although there were no organized scientific endeavors in geology until the late eighteenth century, early speculations about the earth can be traced to the times of the Greeks and Romans. The Greeks made important observations about the composition and structure of the earth, yet had little understanding of both internal and external earth processes. Aristotle was aware that the sea covers land that was previously dry, and he even predicted the emergence of land from the sea (McKeon, 1941). Even before Aristotle, the Greek philosophers Thales and Anaxinander understood the role of rivers (see STREAM) in shaping the landscape, and Herodotus interpreted marine fossils far inland as evidence that the sea once covered the land (Adams, 1954).

The Roman writings tended to follow those of the Greeks. In his far-reaching travels, Strabo observed much about earth processes. He observed alluvial deposits (see ALLUVIAL FAN) as well as volcanic processes (see VOLCANO), and much of his work appears in his massive seventeen-volume *Geography* written around 7 B.C. Seven decades later another Roman philosopher, Lucius Seneca, wrote *Questiones Naturales*, which contained detailed discussions about volcanoes, EARTHQUAKES, and fluvial processes. Shortly after Seneca, Pliny the Elder wrote a 37-volume natural history in which he discussed ROCKS, MINERALS, earthquakes, and volcanic eruptions. The Romans seemed to have been more interested in compiling geologic observations and less interested in speculation than the Greeks. Both the Greeks and Romans, however, made a major step forward by attributing geologic features to natural forces rather than the superhuman powers of the gods (Mears, 1970: 78).

With the downfall of the western Roman Empire, little advancement in the earth sciences took place for about 600 years. Then, in the early 1020s, Ibn-Sina, widely known as Avicenna, wrote an important book on EROSION, mountain building, and the properties of rocks and minerals. "Though Ibn-Sina admitted the existence of God, his checking by experience and practice of theoretical deductions regarding nature made him disagree with Islamic dogmas, which caused the reactionary Moslem clergy to persecute him and burn his papers" (Tikhomirov, 1969: 361). This is just one example of the persecution of geologists by religious fundamentalists.

The Renaissance was a period of renewed interests in many areas of learning, including the study of the earth. Among his many monumental accomplishments, Leonardo da Vinci made important contributions to geology. According to Brainerd Mears, "He clearly believed that fossils were once-living organisms, and not relics of the Biblical flood. He appreciated the shifting relations of sea and land and the nature of erosion, transport and deposition. An original thinker whose mind was not shackled by traditional concepts, Leonardo relied on his own observations and clear reasoning (1970: 79).

Shortly after the death of Leonardo da Vinci, a physician from Saxony, Georgius Agricola, published several books on geology. Of his published works, *De Re Metallica*, in which he laid the foundations for modern mineralogy and metallurgy, was the most far reaching (Hoover and Hoover, 1950).

In the seventeenth century, Nicolaus Steno made a very important geological discovery. In 1669 he demonstrated that rock layers are deposited in a definite sequence with the oldest layers on the bottom and the youngest on top. This "law of superposition" is critical in determining the order in which geologic events take place.

The beginnings of modern geology can probably be traced to the eighteenth-century pioneering work of Jean Etienne Guettard. He emphasized field work with a minimum of speculative theoretical ideas. Among his many contributions were the development of geologic maps, detailed studies of the origin of fossils, and the presentation of a very complete picture of the geologic forces of erosion (Rappaport, 1969).

Major disputes among geologists developed in the latter part of the eighteenth and early nineteenth centuries (Hallam, 1983). The way in which rocks were formed became the center of a major controversy. On one side were those who believed crustal rocks precipitated from sea water. These scientists were called Neptunists, after Neptune, the Roman god of the sea, and were led by Abraham Werner, a German professor of mineralogy. On the other side of this dispute were those who believed that rocks formed by the cooling and solidification of molten material. Those who believed in the cooling of molten materials as the origin of rocks were referred to as Plutonists, after Pluto, the Greek god of the lower world. They were led by the French geologist Nicolas Desmarest and in England by James Hutton. Desmarest (1774) had shown that the rocks of the Auvergne region of south-central France were of volcanic origin. James Hutton (1795) further extended the Plutonists' ideas with his work on extrusive and intrusive rocks in Scotland. The Neptunist-Plutonist controversy faded in the early nineteenth century because two of Werner's most famous students, Leopold von Buch and Alexander von Humboldt, joined the Plutonists, and William Smith (Eyles, 1969) developed the new research technique of correlation of strata by fossils. This work of Smith was very important, and according to Hallam, it "laid the basis for modern geology by allowing the establishment of a valid relative time-scale which could be used across the world" (1983: 25).

A second major geological dispute was that between the catastrophists and the uniformitarians. The catastrophists believed that geologic events occurred as a result of rapid catastrophic changes whereas the uniformitarian concept developed by Hutton (1795) and Charles Lyell (1830) said that rocks, changes in the fossil record, and geological features like landscapes were produced by the same natural processes and laws that operate at present to modify the earth. This principle of "the present is the key to the past" is a cornerstone of much of modern geology, although certain modifications have occurred since its inception by Hutton (Hallam, 1983).

Lyell's writings on gradual physical change had a major influence on the work of Charles Darwin (1859). Darwin's idea of organic evolution stemmed from geological and biological observations during his five-year voyage to the South Pacific. Darwin's work had a major impact on geological ideas, and according to Brainerd Mears, "Today the scientific evidence, much of it from the fossil record, overwhelmingly supports the past occurrence of organic evolution. Geologically we now recognize that organic evolution accounts for the faunal succession that made fossils an indispensable tool in unravelling and assembling a proper geologic time sequence" (1970: 86).

During the nineteenth century geology became an organized science, and perhaps most significant was the idea that geology was no longer forced to fit into dogmatic theological ideas. The latter half of the nineteenth century saw much general interest in geology as well as the introduction of geology into the science curriculum of major universities.

Geology also changed from being primarily a part-time venture conducted by many, like Hutton, who were independently wealthy and trained in other fields, to a full-time venture with adequate financial backing and the establishment of governmental agencies like the U.S. Geological Survey.

The twentieth century has seen many major advances in the geological sciences. Many allied sciences have also made advances that were instrumental in supporting geologic ideas. For example, the dating of radioactive elements in minerals has been responsible for major advancements in our concepts of geologic time (see GEOLOGIC TIME SCALE). The development of X-ray techniques following the work of Max Von Laue in 1912 is primarily responsible for important advances in the study of minerals.

The most significant advancement in the twentieth century, however, has probably been the development of the theory of PLATE TECTONICS. According to the theory of plate tectonics, the surface of the earth is composed of relatively thin and rigid plates that fit together like a jigsaw puzzle or patchwork quilt. The plates float on hot, soft rock of the earth's interior. In a recent discussion of plate-tectonics theory, Irene Kiefer introduced the concept by stating that "the theory of plate tectonics has become the key to understanding the earth's features—past, present, and future. . . . This new view of the earth ranks with other great scientific ideas that have revolutionized our understanding of the natural world in which we live" (1978: 4).

Many new areas of geological research are continuing to develop. The frontier of geology has reached the moon and beyond, yet geologists continue to try to understand earth processes.

References

Adams, Frank D. 1954. *The Birth and Development of the Geological Sciences*. New York: Dover Publications. A good discussion of the historical development of geology.

Darwin, Charles. 1859. *On the Origin of Species*. London: J. Murray. The classic study of organic evolution.

Desmarest, Nicolas. 1774. "Memoire sur L'Origine and La Nature du Basalte," *Memoires de l'Academie Royale des Sciences*, pp. 706–708. An early study on the origin of extrusive rocks.

Eyles, Joan M. 1969. "William Smith: Some Aspects of his Life and Work." In Cecil

J. Schneer, ed., *Toward a History of Geology*. Cambridge, Mass.: The M.I.T. Press, pp. 142–159. A good discussion of the contributions of William Smith to geological science.

Hallam, A. 1983. *Great Geological Controversies*. Oxford: Oxford University Press. A very good book that outlines the major geological controversies.

Hoover, H. C., and Hoover, L. H. 1950. *De Re Metallica*. English Translation. New York: Dover Publications. The first English translation of Agricola's classic book on mineralogy and metallurgy.

Hutton, James. 1795. *Theory of the Earth with Proofs and Illustrations*. Edinburgh: William Creech. A classic work, difficult to read, by the father of modern geology.

Kiefer, Irene. 1978. *Global Jigsaw Puzzle*. New York: Atheneum. An excellent, easy-to-read book on plate tectonics.

Lyell, Charles. 1830. *Principles of Geology*. 3 vols. London: J. Murray.

McKeon, Richard, ed. 1941. *The Basic Works of Aristotle*. New York: Random House.

Mears, Brainerd, Jr. 1970. "The History and Development of Geology." In Mears, Brainerd, Jr., *The Nature of Geology: Contemporary Readings*. New York: Van Nostrand-Reinhold Co., pp. 5–18.

Rappaport, Rhoda. 1969. "The Geological Atlas of Guettard, Lavoisier, and Monnet: Conflicting Views of the Nature of Geology." In Cecil J. Schneer, ed., *Toward a History of Geology*. Cambridge: The M.I.T. Press, pp. 272–288.

Steno, Nicolaus. 1669. *The Prodromus of Nicolaus Steno's Dissertation*. English Translation J. Winter (1916). New York: Macmillan.

Tikhomirov, V. V. 1969. "The Development of the Geological Sciences in the U.S.S.R. from Ancient Times to the Middle of the Nineteenth Century." In Cecil J. Schneer, ed., *Toward a History of Geology*. Cambridge: The M.I.T. Press, pp. 357–386.

Sources of Additional Information

Short definitions of geology can be found in Antony Wyatt, ed., *Challinor's Dictionary of Geology* (New York: Oxford University Press, 1986); Stella Steigler, ed., *A Dictionary of Earth Sciences* (New York: Pica Press); and Andrew Gaudie, ed., *The Encyclopaedic Dictionary of Physical Geography* (Oxford: Blackwell Reference, 1985). For good and thorough discussions of the historical evolution of geology see Roy Porter, *The Making of Geology* (Cambridge: Cambridge University Press, 1977); C. J. Schneer, "The Rise of Historical Geology in the Seventeenth Century," *Isis* 45 (1954): 266–67; K. V. Zittel, *History of Geology* (New York: Hafner Publishing Co., 1962); Archibald Gelkie, *The Founders of Geology*, 2d ed. (New York: Dover Publications, 1962); and H. B. Woodward, *The History of the Geological Society of London* (London: Longmans, Green and Co., 1908). For good discussions of the idea of plate tectonics and its historical development see A. Hallam, *A Revolution in the Earth Sciences* (Oxford: Clarendon Press, 1973); Ursula B. Marvin, *Continental Drift: The Evolution of a Concept* (Washington, D.C.: Smithsonian Institution Press, 1973); A. Wegener, *The Origin of Continents and Oceans* (London: Methuen & Co., 1966; translated from 4th German ed., 1929).

GEOMORPHOLOGY. 1. The science that studies the origin, arrangement, and changes in landforms on the earth's surface. 2. The study of surface and submarine relief features including the relationship between these features and their underlying geological structures.

Although the science of geomorphology grew out of, and is still considered part of, GEOLOGY, it is also a subdiscipline within PHYSICAL GEOGRAPHY, and many geographers have made important contributions to its development. The root of the word *geomorphology* gives some indication of its nature. It deals with *geo* (the earth), *morph* (shape), and *logos* (the science of). Thus we have the science of the earth's shape, geomorphology. According to Keith Tinkler, "The word 'geomorphology' probably was used first by Lauman . . . in the German language . . . , but first appeared in English in two papers by McGee" (1985: 4).

Although the word *geomorphology* first appeared in the latter half of the nineteenth century, interest in and the study of the earth's surface features can be traced to before the days of Herodotus (fifth century, B.C.). "The visual impact of scenery is the most obvious facet of geology and must have been with man since the earliest days of his consciousness" (Tinkler, 1985: 4).

The ancient Greeks were interested in LANDFORM study, and although modern ideas are not directly founded upon Greek studies, the works of early Greek writers have had an important impact on later geomorphological ideas. Tinkler believes the Greeks developed three major principles as part of their natural philosophy or holistic view of the world. "The first is the concept of infinite time—a universal eternal. The second is a belief in the reality of denudation—mass loss from the landscape. The third is an acceptance of the conservation of mass—nothing is lost and nothing is gained" (Tinkler, 1985: 19).

The idea of EROSION, or denudation, and its relationship to landform evolution can be traced to Homer's "Iliad" around the ninth century B.C. Further refinements of the idea of erosion can be traced through the works of Strabo (64 B.C.– c. A.D. 25), Erastothenes (c. 276–c. 196 B.C.), and Plato (427–347 B.C.). Plato's remarks are still being used as a basis for modern stratigraphical studies of erosion (Bintliff, 1976).

Roman and Arab scholars also contributed to early geomorphological thinking. Seneca had an interest in the causes of STREAM flow and speculated "how the earth supplies the continuous flow of rivers, and where such great quantities of water come from" (Book III–4). Among the important Arab scholars were Ibn-Sena (980–1037), also known as Avicenna, who discussed the effects of running water and wind on the hollowing out of valleys in soft ROCKS, and in the same period Omar el Aalem wrote a treatise on coastline changes in Asia during the previous 2,000 years (Geikie, 1905).

The revival of learning in Europe that took place after 1250 brought about an increased interest in understanding landforms and the processes that bring them about. Of particular importance was the work on the erosive powers of rivers by the Renaissance scholars Leonardo da Vinci (1452–1519), Agricola (Georg Bauer, 1494–1555), and Bernard Palissy (1510–1590). These early process-oriented ideas, however, were overwhelmed by the biblical interpretation of the earth's evolution by Bishop Ussher and others (Hallam, 1983). Ussher concluded that "heaven and Earth, center and circumference were made in the same instance of time, and clouds full of water, and man were created on the 26th October 4004 B.C. at 9 o'clock in the morning'' (Beckinsale and Chorley, 1968: 410). This idea of the catastrophic formation of the earth strongly influenced the study of landforms for many centuries to follow (Hallam, 1983: Chapter 2).

The birth of modern geomorphology can be traced to the latter part of the eighteenth and the early nineteenth centuries. The Frenchman Jean Baptists de Lamarck (1802), in his classic work *Hydrogeologie*, determined that running water was responsible for carving mountains, and in a similar fashion in Switzerland, Horace de Saussure (1799) looked at stream-eroded landscapes as well as glacial caused features.

It was, however, James Hutton, the Scottish physician, who made the most significant contributions. Hutton is considered by many as the father of modern geomorphology. In his monumental book *The Theory of the Earth* (1795), Hutton proposed his theory of uniformitarianism, or the belief that understanding the present is the key to the past. He believed that the most important element in landscape development was stream erosion and deposition. After his death, Hutton's friend John Playfair expanded Hutton's ideas and wrote them in a prose more lucid than that of Hutton. "Playfair stressed that moving water (rivers, waves, and glaciers) were the great agent of landscape formation; that rivers cut their own valleys and that in river systems trunks and tributaries were perfectly adjusted'' (Beckinsale and Chorley, 1968: 410).

Further support for Hutton came a few decades later from Charles Lyell (1830), who, using quantitative measurements, looked in detail at erosional and depositional processes. By the middle of the nineteenth century uniformitarianism was the dominant idea in both Britain and the United States.

About the same time, European geomorphologists became interested in mountain glaciers (see VALLEY GLACIER) and continental (see CONTINENTAL GLACIER) ice sheets. Foremost among the early glaciologists was Jean Louis Agissiz (1940), who emphasized the idea of ice sheets and an ice age. Two decades later A. C. Ramsey (1862) suggested that there had been more than one period of glacial activity, and Otto Troll looked at the movement of glacial ice across Scandinavia to the North German Plain (von Zittel, 1901).

The later decades of the nineteenth century saw further confirmation of Lyell's and Hutton's ideas. Quantitative data on stream erosive capacity was provided by A. A. Humphrey and H. L. Abbott (1861) in their study of the Mississippi River. John Wesley Powell (1875) made many contributions to geomorphology as a result of his exploration of the Colorado River, and Grove Karl Gilbert (1877) in his study of the Henry Mountains made major contributions in the analysis of fluvial processes and sediment formation.

Geomorphology at the end of the nineteenth century, "like an uncompleted jigsaw, awaited a visionary thinker who could stand back and see the whole. William Morris Davis was that man'' (Tinkler, 1985: 144). Davis developed a theoretical model called the "Geographical Cycle.'' This model stated that there would be a sequential and predictable set of landforms that would evolve under the influence of BASE LEVEL, unless interruptions or accidents intervened (Davis, 1899). This theory has become the most influential and perhaps the most misunderstood in modern geomorphology. Davis's ideas were prominent for almost half a century, and according to William Thornbury, "Geomorphology will probably retain his stamp longer than that of any other single person" (1954: 11).

By the time Davis had died in 1934, many scientists were critical of aspects of his geographical cycle. Foremost among these critics was N. M. Fenneman (1936) and Walter Penck (1924), who pointed out some of the theoretical weaknesses in Davis's work.

New developments in technology brought about by World War II gave several new tools or techniques to the geomorphologist. Sophisticated techniques of aerial photography (Parry, 1967), land-systems mapping, and REMOTE SENSING (McKee, 1979) added new dimensions to landform studies.

The past few decades have included studies that were process oriented as well as regionally oriented. Climatic geomorphology has received much attention with important contributions by J. Budel (1948, 1982) and P. Birot (1960). The influence of internal earth forces on landforms, called TECTONIC geomorphology, has received recent attention with much work done on the relationship between PLATE TECTONICS and landform development (Morisawa and Hack, 1985; Summerfield, 1986). Another area of geomorphology in which major strides have been made in the past two decades has been theoretical and quantitative studies. Numerical modeling studies (King and McCullagh, 1971; Vanderpool, 1982) have been greatly enhanced with modern computational and graphical facilities. Practical applications of geomorphology have also received increasing emphasis during the 1970s and 1980s (Cooke and Doornkamp, 1974; Hails, 1977).

The science of geomorphology continues to be a dynamic and exciting area of physical geography because, according to Tinkler, "geomorphology is, and always has been, the most accessible earth science to the ordinary person: we see scenery as we sit, walk, ride, or fly. It is a part of our daily visual imagery, and we do not even have to stop or stoop to examine it, although our perceptions are usually better if we do" (1985: 239).

References

Beckinsale, R. P., and Chorley, R. J. 1968. "History of Geomorphology." In Rhodes W. Fairbridge, ed., *The Encyclopedia of Geomorphology. Vol. 3, Encyclopedia of Earth Sciences.* New York: Reinhold Book Corp. An excellent short discussion of the historical evolution of geomorphology.

Bintliff, J. L. 1976. "The Plain of Western Macedonia and the Neolithic Site of Nea Nikomedeia." In *Proceedings, Prehistoric Society* 42: 241–262.

Birot, P. 1960. *Le Cycle d'Erosion Sous les Differents Climats.* Paris. English translation I. J. Jackson and K. M. Clayton (1968). London: Batsford.

Budel, J. 1948. "Das System der Climatischen Geomorphologie." *Verhandlungen Deutscher Geographische* 27: 65–100.

———. 1982. *Climatic Geomorphology.* Princeton, N.J.: Princeton University Press.

Cooke, R. U., and Doornkamp, J. 1974. *Geomorphology and Environmental Management*. London: E. Arnold.
Davis, W. M. 1899. "The Geographical Cycle." *Geographical Journal* 14: 481–504.
Fenneman, N. M. 1936. "Cyclic and Non-Cyclic Erosion." *Science* 83: 87–94. One of the major criticisms of the Davisian geographical cycle.
Geikie, A. 1905. *The Founders of Geology*. London: Macmillan.
Gilbert, G. K. 1877. *Report on the Geology of the Henry Mountains*. Rocky Mountain Region Report. Washington, D.C.: U.S. Geological Survey. A pioneering work by an American geomorphologist.
Hails, J. R., ed. 1977. *Applied Geomorphology*. Amsterdam: Elsevier.
Hallam, A. 1983. *Great Geological Controversies*. New York: Oxford University Press. An interesting account of the major controversies in modern geology.
Humphreys, A. A., and Abbott, H. L. 1861. "Report on the Physics and Hydraulics of the Mississippi River." *Corps of Topographical Engineers, Professional Paper*, no. 4, Washington, D.C.: Corps of Topographical Engineers. An early study of erosion and stream dynamics.
Hutton, James. 1795. *The Theory of the Earth*. 2 vols. Edinburgh: William Creech. A monumental work by the father of modern geology and geomorphology.
Isachenko, A. G. 1973. *Principles of Landscape Science and Physical-Geographic Regionalization*. Melbourne, Aust.: Melbourne University Press. A discussion of land system mapping and terrain analysis.
King, C.A.M., and McCullagh, M. J. 1971. "A Simulation Model of a Complex Recurved Spit." *Journal of Geology* 79: 22–37.
Lamarck, J. B. 1802. *Hydrogeologie*. Paris. English translation A. V. Carozzi. (1964). Urbana: University of Illinois Press.
Lyell, Charles. 1830. *The Principles of Geology*. London: J. Murray. A classic work by one of the founders of modern geology and geomorphology.
McGee, W. J. 1888a. "The Classification of Geographic Form by Genesis." *National Geographic Magazine* 1: 27–36.
———. 1888b. *The Geology of the Head of Chesapeake Bay*. Seventh Annual Report of the U.S. Geological Survey for 1885–86. Washington, D.C.: U.S. Geological Survey, pp. 545–646.
McKee, E. D., ed. 1979. "Global Sand Seas." U.S. Geological Survey Professional Paper, no. 1052, Washington, D.C.: U.S. Government Printing Office. A good discussion of the use of remote sensing techniques.
Morisawa, M., and Hack, J. T., eds. 1985. *Tectonic Geomorphology*. Boston: Allen and Unwin.
Parry, J. T. 1967. "Geomorphology in Canada." *Canadian Geographer* 9: 280–311.
Penck, W. 1924. *Die Morphologische Analyse: Ein Kapitel der Physicalischen Geologie*. Stuttgart. English translation H. Czech and K. Boswell (1953). *Morphological Analysis of Landforms*. London: Macmillan.
Playfair, John. 1802. *Illustrations of the Huttonian Theory of the Earth*. Edinburgh: William Creech. A lucid discussion of many of James Hutton's ideas.
Powell, John W. 1875. *Exploration of the Colorado River of the West (1869–72)*. Chicago: University of Chicago Press.
Ramsey, A. C. 1862. "On the Glacial Origin of Certain Lakes in Switzerland, The Black Forest, Great Britain, Sweden, North America, and Elsewhere." *Quarterly Journal of the Geologic Society* 18: 185–204.

Roglic, J. 1972. "Historical Review of Morphologic Concepts." In M. Herak and V. T. Springfield eds., *Karst: Important Karst Regions of the Northern Hemisphere.* Amsterdam: Elsevier, pp. 1–18.

Saussure, H. B. de. 1799. "Agenda, or a Collection of Observations and Researches." *Philosophical Magazine* 3: 33–41, 147–156, 194–199; 4: 68–71; 5: 24–29, 135–140, 217–221.

Seneca, L. A. 1971. *Natural Questions.* (trans. T. H. Corcoran). Cambridge: Harvard University Press.

Summerfield, M. A. 1986. "Tectonic Geomorphology: Macroscale Perspectives." *Progress in Physical Geography* 10, no. 2: 227–238. A recent review of work in tectonic geomorphology.

Thornbury, W. D. 1954. *Principles of Geomorphology.* New York: John Wiley and Sons, 2d ed., 1969. Although somewhat outdated, it contains an easy-to-read discussion of the major ideas in geomorphology.

Tinkler, Keith J. 1985. *A Short History of Geomorphology.* Totowa, N.J.: Barnes and Noble Books. An in-depth study of the development of the science of geomorphology.

Vanderpool, N. L. 1982. "Erode—A Computer Model of Drainage Basin Development under Changing Baselevel Conditions." In R. Craig and R. L. Craft, eds., *Applied Geomorphology.* London: Allen and Unwin, pp. 73–84.

von Zittel, F. A. 1901. *History of Geology and Paleontology.* London: Walter Scott.

Sources of Additional Information

Short discussions on the science of geomorphology can be found in William D. Thornbury, *The Encyclopedia Americana, International Edition*, vol. 12 (Danbury, Conn.: Grolier, 1984), p. 503, and Lester King, "Geomorphology," in Rhodes W. Fairbridge, *The Encyclopedia of Geomorphology. Vol. 3, Encyclopedia of Earth Sciences* (New York: Reinhold Book Corp., 1968), pp. 403–404. Much work has been done on the historical evolution of geomorphology, including K. J. Gregory, *The Nature of Physical Geography* (London: E. Arnold, 1985), and R. J. Chorley, A. J. Dunn, and R. P. Beckinsale, *The History of the Study of Landforms. Vol. 1, Geomorphology before Davis* (London: Methuen & Co., 1964).

GEOSYNCLINE. A large depression or linear trough generally hundreds of miles long and tens of miles wide in which thousands of feet of stratified sediment or extrusive volcanic rock has accumulated.

The geosynclinal concept originated in the United States in the middle of the nineteenth century (see GEOLOGY). In 1859 James Hall loosely formulated his theory of geosynclinal development and related it to mountain building. Hall speculated that ocean currents laid down sediments in shallow areas, and this in turn would depress the sea floor until the original water depth was regained. This subsidence would then cause lateral movement away from the trough and in turn uplift the surrounding areas. According to Hall, mountains would form where sediment accumulation was the greatest and therefore would correspond with the original line of greatest sediment accumulation (Hall, 1859).

Several decades after Hall's work, another American geologist, J. D. Dana

(1873), first coined the term *geosyncline* and offered one of the first public challenges to Hall's theory. Although Dana accepted much of Hall's work, he argued that subsidence provided the opportunity for sedimentation, rather than sedimentation causing subsidence, as proposed by Hall.

Although they had differing views on the subsidence process, by 1895 the Hall-Dana theory of geosynclinal development became widely known. It stated simply that mountains originate from geosynclines through a very long process of sedimentation, some manner of subsidence, and a relatively short period of uplift.

Most European geologists at this time agreed with the thick nature of sedimentary strata in geosynclines (Suess, 1875; Neumayr, 18 5), but since many of their observations were based on the Alps, they believed geosynclines developed between continental masses and not marginal to continents as proposed by Hall and Dana (see PLATE TECTONICS). Although attempts were made to reconcile the American-European schism (Walther, 1897) about the nature of geosynclines, the schism continued, exemplified by E. Haug's (1900) classic study at the turn of the century.

To clear up some of the confusion over the nature of geosynclinal development and to define geosynclines more accurately, several classification schemes were developed (Schuchert, 1923; Stille, 1940; Kay, 1951). These traditional approaches culminated in the theory of Jean Aubouin (1965), who attempted to place geosynclinal structures in concrete time-space models.

The most significant developments in geosynclinal research since 1970 have been the relating of geosynclinal types to modern ideas of plate and global tectonics (Dickinson, 1971; Dewey and Bird, 1971). According to Kenneth Hsu, in a review of the development of geosynclinal theory:

> In the view of plate tectonicians, geosynclines characterize tectonic settings of sedimentary sequences and their associated rocks existing at the present continental or plate margins: the Pacific type of geosyncline represents the setting at converging plate margins, the Atlantic type describes stable continental margins within plates, and the Mediterranean type denotes intercontinental realms of sedimentation. (1978: 352)

A discussion of classical theory as well as modern interpretations can be found in the papers presented in a symposium on geosynclinal sedimentation (Dott and Shaver, 1974). A recent book edited by F. L. Schwab (1982) traces the development of the geosynclinal concept, discusses the concept in relation to early mountain-building theories, and relates geosynclinal to plate-tectonic concepts developed in the 1960s and 1970s.

Although the basic ideas behind geosynclinal development have changed little in the past 100 years, earth scientists, armed with new tectonic insights, have been able to advance the study of geosynclines from the descriptive phase to a more genetic one.

References

Aubouin, Jean. 1965. *Geosynclines*. Amsterdam: Elsevier. A good discussion of the historical development of the concept.

Dewey, J. F., and Bird, J. 1971. "Geosynclines in Terms of Plate-Tectonics." *Tectonophysics* 40: 625–635.

Dana, J. D. 1873. "On the Origins of Mountains." *American Journal of Science* 5: 347–350. One of the early discussions of geosynclinal theory.

Dickinson, W. R. 1971. "Plate-Tectonic Models of Geosynclines." *Earth and Planetary Science* 10: 165–174.

Dott, R. H., Jr., and Shaver, Robert H., eds. 1974. *Modern and Ancient Geosynclinal Sedimentation*. Tulsa, Okla.: Society of Economic Paleontologists and Mineralogists. A good collection of papers looking at many aspects of geosynclines.

Hall, James. 1859. "Paleontology." *New York Geological Survey* 3, no. 1: 66–96. The first statement of the geosynclinal theory.

Haug, E. 1900. "Les Geosynclinaux et les Aires Continentales." *Societe Geologique de France Bulletin* 3(28): 617–711.

Hsu, Kenneth J. 1978. "Geosynclinal Sedimentation." In Fairbridge, Rhodes W., and Bourgeois, Joanne, eds., *The Encyclopedia of Sedimentology. Vol. 6, Encyclopedia of Earth Sciences*. Stroudsburg, Pa.: Dowden, Hutchinson and Ross. An excellent review of the historical development of geosynclinal research.

Kay, Marshall. 1951. *North American Geosynclines*. New York: The Geological Society of America.

Neumayr, M. 1875. *Erdgeschichte*. Leipzig: Bibliographisches Institute. One of the first papers on geosynclinal structures in Europe.

Schuchert, C. 1923. "Sites and Natures of the North American Geosyncline." *Geological Society of America Bulletin* 34: 151–260.

Schwab, F. L., ed. 1982. *Geosynclines: Concept and Place within Plate Tectonics*. Stroudsburg, Pa.: Hutchinson Ross Publishing Co. A recent discussion of geosynclines.

Seuss, E. 1875. *Die Entstehung der Alpen*. Vienna: W. Braunmuller.

Stille, H. 1940. *Einfuhrung in Den Bau Amerikas*. Berlin: Borntraeger.

Walther, J. 1897. "Uber Lebensweise Fossiler Meeresthiere." *Z. Deutsch, Geol. Ges.* 49: 209–273.

Sources of Additional Information

For good definitions of the concept see Andrew Goudie, ed., *The Encyclopeadic Dictionary of Physical Geography* (Oxford: Blackwell Reference, 1985); Daniel N. Lapedes, ed., *McGraw-Hill Encyclopedia of the Geological Sciences* (New York: McGraw-Hill Book Co., 1978); Stella E. Stiegeler, ed., *A Dictionary of Earth Sciences* (New York: Pica Press, 1976); Anthony Wyatt, ed., *Challinor's Dictionary of Geology* (New York: Oxford University Press, 1986).

GRATICULE. See MAP PROJECTION.

GREAT CIRCLE. See LATITUDE/LONGITUDE.

GROUNDWATER. 1. Water present below the surface of the earth in the soil, subsoil, and underlying rocks. 2. Water that occurs in the permanently saturated zone beneath the water table.

An all-encompassing definition of groundwater would include all of the water below the earth's surface whether in its solid, liquid, or gaseous states. Most earth scientists, however, usually include only subsurface water that is part of the HYDROLOGIC CYCLE, and some earth scientists define *groundwater* as only that water found below the water table. The *water table* is the upper level of the zone of groundwater saturation in permeable rocks and is the surface at which the groundwater pressure is equal to the atmospheric pressure (Whittow, 1984: 576).

Groundwater has been classified by its origin as either connate, meteoric, or juvenile. *Connate water* is water trapped in sediments at their time of deposition. *Meteoric water* is groundwater derived from the atmosphere and is part of the hydrologic cycle. *Juvenile water* comes from the earth's interior and, according to C. L. McGuinness, "is emitted in quantities that are insignificant in comparison to those involved in the hydrologic cycle" (1963: 18).

Attempts to understand groundwater can be traced to ancient times. Archaeological discoveries as well as documentary evidence points to the importance of groundwater in the lives of, for example, the Old Testament peoples, the ancient Egyptians and Persians, and later the Greeks and Romans (Ward, 1975: 3).

The Bible recounts many instances illustrating the importance of groundwater to the tribes of Israel. As O. E. Meinzer stated, "the twenty-sixth chapter of Genesis . . . reads like a water-supply paper" (1934: 6). The pools of Solomon and Shiloah, as well as Jacob's well, are a few examples of groundwater sources mentioned in the Bible.

Perhaps the most extraordinary achievement of ancient man was the building of water-supply tunnels, or *kanats*, by the ancient Egyptians and Persians. Although the exact origin of kanat building is lost in antiquity, it is believed to have started in Persia around 800 B.C. An excellent discussion of kanat development is found in the work of C. F. Tolman (1937).

Greek and Roman philosophers like Homer, Plato, Seneca, and Pliny provide a variety of theories on groundwater movement, and although much of it was based on guesswork and mythology, some of their ideas were remarkably close to the truth as we presently understand it (Baker and Horton, 1936). Thus Aristotle (384–322 B.C.) explained the mechanics of PRECIPITATION, and three centuries later Marcus Vitruvius derived the meteoric origin of groundwater. Vitruvius's arguments were further substantiated in the middle of the sixteenth century by Bernard Palissy (Biswas, 1969).

In the later part of the seventeenth century pioneering quantitative studies were conducted by Pierre Perrault and Edme Mariotte on the Seine drainage, and the English astronomer Edmund Halley showed that the total flow from springs and rivers could be accounted for by oceanic evaporation. Many scientists believe these early scientific works by Perrault, Mariotte, and Halley qualify these men to be regarded as the founders of modern scientific groundwater study (Jones et al., 1963). Meinzer believes that "Mariotte . . . probably deserves more than any other man the distinction of being regarded as the founder of ground-water hydrology'' (1934: p. 7).

After a period of modest consolidation during the eighteenth century the first half of the nineteenth century saw substantial increases in knowledge. Important work was done by the French hydraulic engineer Henry Darcy (1856), who studied the movement of water through sand. Other European contributions in the nineteenth century focused on the hydraulics of ground-water development (Todd, 1959: 4). The nineteenth century also saw the publication of the first groundwater textbook, *Manual of Hydrology* by Nathaniel Beardmore. In 1904 Daniel Mead published his *Notes on Hydrology* as the first American text.

The twentieth century has seen great advances in groundwater studies, particularly in the United States. The U.S. government provided large sums of money in the 1930s for projects and studies primarily dealing with irrigation, conservation, and flood (see FLOODPLAIN) control. Much of the progress in the United States can be attributed to Meinzer, who, after the turn of the century, pioneered groundwater studies for the U.S. Geological Survey. According to O. M. Hackett, Meinzer,

> recognized that aquifers are functional components of the hydrologic cycle and insisted that groundwater should be studied with this fact in mind. As a consequence, he drew together the two traditional approaches to groundwater investigations—that of the geologist, concerned principally with describing the earth medium, and that of the engineer or physicist, concerned principally with fluid mechanics. (1972: 477)

Another contribution of major significance was the mathematical model developed by Theis (1935). Theis's study was the first quantitative analysis of the transient nature of groundwater hydraulics.

Many of the early studies were descriptive and primarily concerned with exploring for groundwater resources. Although much research still continues in this area, many modern studies are concerned with "an understanding of the behavior of ground water as it passes into, through, and out of the aquifer system" (Hackett, 1972: 476).

The advent of the modern computer has greatly aided the study of groundwater. The use of simulation models in analyzing groundwater has shown great promise in recent years (Lai, 1986; Rushton and Redshaw, 1979). Another area of much concern in recent years has been groundwater contamination. In 1974 the Safe Drinking Water Act was passed by Congress, which gave the Environmental Protection Agency regulatory controls on water where it could threaten public health. Much work has been done on trying to analyze and understand groundwater POLLUTION (Mosher, 1980; Burmaster, 1982).

References

Baker, M. N., and Horton, R. E. 1936. "Historical Development of Ideas Regarding the Origin of Springs and Ground-Water." *Transactions of the American Geophysical Union* 17: 395–400.

Biswas, A. K. 1969. "A Short History of Hydrology." *The Progress of Hydrology.* Urbana, Ill.: University of Illinois, 2: 914–936.
Burmaster, David E. 1982. "The New Pollution: Groundwater Contamination." *Environment* 24: 7–13. A good discussion of groundwater pollution.
Darcy, H. 1856. *Les Fontaines Publiques de la Villa de Dijon.* Paris. A classic early work on groundwater.
Hackett, O. M. 1972. "Groundwater." In Rhodes Fairbridge, ed., *The Encyclopedia of Geochemistry and Environmental Sciences.* Encyclopedia of Earth Science Series. Vol. 4. New York: Van Nostrand-Reinhold Co. A general discussion and definition of groundwater.
Jones, P. B.; Walker, G. D.; Harden, R. W.; and McDaniels, L. L. 1963. "The Development of the Science of Hydrology." *Texas Water Commission Circular*, no. 63–03. Austin: Texas Water Commission.
Lai, Chintu. 1986. "Numerical Modeling of Unsteady Open-Channel Flow." In Ven Te Chow, ed., *Advances in Hydroscience.* Vol. 14. Orlando, Fla.: Academic Press. A new book concerned with recent quantitative trends in groundwater study.
McGuinness, C. L. 1963. "The Role of Ground Water in the National Water Situation." U.S. Geological Survey Water Supply Paper, no. 1800. Washington, D.C.: U.S. Geological Survey.
Meinzer, O. E. 1934. "The History and Development of Ground-Water Hydrology." *Journal of the Washington Academy of Sciences* 24: 6–32. A discussion of the history of groundwater research by one of America's foremost authorities.
Mosher, Lawrence, 1980. "A Host of Pollutants Threaten Drinking Water from Underground." *National Journal* 12: 1353–1356. A discussion of groundwater pollution.
Rushton, K. R., and Redshaw, S. C. 1979. *Seepage and Groundwater Flow.* New York: John Wiley and Sons. A quantitative analysis of groundwater movement.
Theis, C. U. 1935. "Relation between the Lowering of the Piezometric Surface and the Rate and Duration of Discharge of a Well Using Groundwater Storage." *American Geophysical Union Transactions* 16: 519–524. A classic paper using quantitative techniques.
Todd, David Keith. 1959. *Ground Water Hydrology.* New York: John Wiley and Sons.
Tolman, C. F. 1937. *Ground Water.* New York: McGraw-Hill Book Co. An early textbook on groundwater with an excellent discussion of the history of research activities.
Ward, R. C. 1975. *Principles of Hydrology.* London: McGraw-Hill Book Co. (UK).
Whittow, John. 1984. *Dictionary of Physical Geography.* New York: Penguin Books.

Sources of Additional Information

Short definitions and discussions of groundwater can be found in David G. Smith, ed., *The Cambridge Encyclopedia of Earth Sciences* (Cambridge: Cambridge University Press, 1981), p. 472; W. G. Moore, *A Dictionary of Geography: Definitions and Explanations of Terms Used in Physical Geography* (New York: Barnes and Noble, 1975), p. 102; and John Challinor, *Challinor's Dictionary of Geology*, 6th ed. (New York: Oxford University Press, 1986), p. 147. An interesting collection of papers on groundwater and geomorphology is R. G. LaFleur, ed., *Groundwater as a Geomorphic Agent* (Boston: Allen and Unwin, 1984). A recent book on geography and groundwater is Brian Knapp, *Elements of Geographical Hydrology* (London: Allen and Unwin, 1979). Recent books that deal with quantitative analysis and modelling groundwater systems are Arved J. Raudkivi, *Hydrology: An Advanced Introduction to Hydrological Processes and Modelling* (New York: Pergamon Press, 1979), and T. N. Narasimhan, ed., *Recent Trends in Hydrology* (Boulder, Colo.: Geological Society of America).

H

HADLEY CELL. See ATMOSPHERIC CIRCULATION.

HAIL. See THUNDERSTORM.

HEAT ISLAND. A zone of increased air temperature usually associated with a large urban area.

It has been recognized for more than a century and a half that urban centers are warmer than surrounding rural areas given the same set of synoptic weather conditions. Luke Howard (1818) discovered that London was consistently warmer than adjacent outlying areas. As H. E. Landsberg wrote of Howard's discovery, "This in a nutshell is the recognition of the urban heat island.... Howard as many after him, attributed the greater temperature of the city to the extensive use of fuel" (1981: 5). Further confirmation came from Paris in 1855 when E. Renou (1855) described the temperature increment while traversing the city. Renou's primary concern seemed to be the positioning of thermometers in the city for achieving accurate readings. Renou wrote of the temperature differences between Paris and surrounding areas and concluded that the differences average about 1°C. Even with discoveries of this nature, the nineteenth-century research on heat islands was sporadic, and systematic research was not begun until the early twentieth century.

Wilhelm Schmidt (1917) started a detailed program of sophisticated monitoring of temperature variations using instrumented motor vehicles. This was a real innovation in mesoclimatic investigations. According to Landsberg, "This [instrumentation procedure] permitted the plotting of isolines of meteorological elements, essentially on a synoptic basis.... Since that time instruments on other vehicles, including bicycles and streetcars, have served such survey purposes" (1981: 8).

M. I. Budyko (1982: 196) listed the chief factors affecting urban climates (see CLIMATOLOGY). The first three of them are the primary factors affecting the heat element of urban climate:

1. Variations in albedo of the earth surface
2. Changes in evaporation rates due to urban structures and materials
3. Increased heat release from urban sources
4. Increased roughness of urban landscapes
5. Atmospheric POLLUTION concentrations in urban areas

According to the heat-island concept, each section of a city contributes and modifies heat at different rates, and other meteorological phenomena modify the effects of this heat. A very real example of the additive influence of progressive city development was cited in the work of H. E. Landsberg and T. N. Maisel (1972). These writers looked at the progressive development of the city of Columbia near Washington, D.C. In 1968 only a small portion of the city was complete, and the average temperature difference between the urban and rural areas was 0.5°F. In 1970, after significant development of the city, the temperature difference reached 4.5°F between the city center and the surrounding rural area.

The variation of urban heat regimes due to seasonal and locational changes was studied by M. A. Atwater (1977), who calculated mean temperature variations for cities in four climatic zones: tropics, desert, middle latitudes, and tundra. He found that the greatest differences between urban and rural temperatures are found when anthropogenic heat generation is greatest as at the tundra sites in his study.

Other advanced studies of this phenomenon include the research of urban heating using energetics as a basis. T. R. Oke (1982), for example, showed that the urban-rural thermal anomaly is three dimensional, caused by the intrinsic design nature of cities including city size, building density, and land-use pattern. Theoretical modeling techniques have been used to predict the reaction of the thermal regime to varying urban designs (e.g., Landsberg, 1979; Oke, 1980).

The urban-heat island has also been studied on a wide regional basis using satellite imagery. M. Matson and associates (1978) detected more than 50 urban-heat islands in the eastern United States in the middle of the summer. This imagery gave information on the quantitative nature of the heat islands. Urban-rural differences ranged from 2.6°C for Petersburg, Virginia, to 6.5°C for Louisville, Kentucky.

As a summary of the reasons for urban-heat islands, one can refer to the thorough description in Jäger. She summarized the changes between rural and urban landscapes thus:

> The spongy, often moist, soil cover of rural areas, which has low heat conductivity, is converted into an impermeable surface layer, with a high capacity for conducting heat and, because of generally low reflectivity, high absorptivity for radiation. The

changes in surface conditions also lead to more rapid run-off of precipitation and therefore to a reduction in local evaporation, which is equivalent to a further heat gain. In addition, heat release from energy use contributes to the heat island. (1983: 120)

References

Atwater, M. A. 1977. "Urbanization and Pollution Effects on the Thermal Structure in Four Climatic Regimes." *Journal of Applied Meteorology* 16: 888–895. Discussed the variations in heat islands according to city location and season.

Budyko, M. I. 1982. *The Earth's Climate: Past and Future*. New York: Academic Press. Provides an excellent compilation of climate changes including anthropogenic heat generation.

Howard, L. 1818. *Climate of London Deduced from Meteorological Observations*. 1st ed. London: Harvey and Darton. The first known work on urban-heat-island observation.

Jäger, J. 1983. *Climate and Energy Systems*. New York: John Wiley and Sons. Has an excellent chapter on waste heat and climate interactions.

Landsberg, H. E. 1979. "The Effects of Man's Activities on Climate." In M. R. Biswas and A. K. Biswas, eds., *Food, Climate, and Man*. New York: John Wiley and Sons, pp. 187–236.

———. 1981. *The Urban Climate*. New York: Academic Press. A very useful compendium on all aspects of urban climate phenomena.

Landsberg, H. F., and Maisel, T. N. 1972. "Micrometeorological Observations in an Area of Urban Growth." *Boundary Layer Meteorology* 2: 365–370.

Matson, M.; McClain, E. P.; McGinnis, D. F., Jr.; and Pritchard, J. A. 1978. "Satellite Detection of Urban Heat Islands." *Monthly Weather Review* 106: 1725–1734.

Oke, T. R. 1980. "Climatic Impacts of Urbanization." In Bach, W.; Pankrath, J.; and Williams, J., eds., *Interactions of Energy and Climate*. Dordrecht, Holland: Reidel, pp. 339–356.

———. 1982. "The Energetic Basis of the Urban Heat Island." *Quarterly Journal of the Royal Meteorological Society* 108: 1–24. Looks at the urban heat island in light of theoretical and practical energetics.

Renou, E. 1855. "Instructions Meteorologiques." *Annuaire Societe Meterologique de France* 3: 73–160. Confirmed Howard's work on London with the situation in Paris on increased urban temperatures.

Schmidt, W. 1971. "Zum Einfluss Grosser Stadte auf das Klima." *Naturwissenschaften* 5: 494–495. This started the systematic study of urban climate changes.

Sources of Additional Information

Good review works on heat islands and urban climatology include T. J. Chandler, "Urban Climatology and Its Relevance to Urban Design," *Technical note*, no. 149 (Geneva: World Meteorological Organization, 1976); T. R. Oke, "Review of Urban Climatology, 1968–1973," *Technical note*, no. 134 (Geneva: World Meteorological Organization, 1974); T. R. Oke, "Review of Urban Climatology, 1974–76," *Technical note*, no. 169 (Geneva: *World Meteorotological Organization*, 1979); W. Bach, "Waste Heat and Climatic Change," in L. Theodore, A. J. Buonicore, and E. J. Rolinski, eds., *Perspectives on Energy and the Environment*, vol. 1 (West Palm Beach, Fla.: CRC Press, 1979). For good general works integrating world climate and urban climate see R. G.

Barry and A. H. Perry, *Synoptic Climatology: Methods and Applications* (London: Methuen & Co., 1973); P. R. Crowe, *Concepts in Climatology* (New York: St. Martin's Press, 1971); and J. E. Oliver, *Climate and Man's Environment: An Introduction to Applied Climatology* (New York: John Wiley and Sons, 1973). For definitions and short explanations of heat island and urban climate see R. W. Fairbridge, ed., *The Encyclopedia of Geochemistry and Environmental Sciences*, Encyclopedia of Earth Science Series, vol. 4 (New York: Van Nostrand-Reinhold Co., 1972); S. P. Parker, *McGraw-Hill Dictionary of Scientific and Technical Terms*, 3d ed. (New York: McGraw-Hill Book Co., 1984); and J. T. Peterson, *McGraw-Hill Encyclopedia of Science and Technology*, 5th ed. (New York: McGraw-Hill Book Co., 1982). For two very good recent review articles on all aspects of urban climates see Y. Goldreich, "Urban Topoclimatology," *Progress in Physical Geography* 8 (1984): 336–364, and D. O. Lee, "Urban Climates," *Progress in Physical Geography* 8 (1984): 1–31.

HORSE LATITUDES. See ATMOSPHERIC CIRCULATION.

HORTON ANALYSIS. See DRAINAGE BASIN.

HUMIDITY. See PRECIPITATION.

HUMUS. 1. A complex and resistant mixture of dark colored, amorphous, and colloidal size substances within the soil that were modified from organic tissue. 2. These same substances synthesized by a variety of soil organisms. 3. That stable portion of soil organic matter remaining after the major portion of plant and animal residues have decomposed.

The idea that plants feed upon the humus in soils was initially proposed by Warllerius in 1761 (Simonson, 1968: 7). Humus was considered the only primary constituent of the SOIL, and all others, such as the MINERAL matter, were merely ancillary to the plant-growing process. This idea was given added emphasis in the early nineteenth century by von Wullfen and A. D. Thaer (Usher, 1923). Von Wullfen and Thaer proposed that the level of soil fertility was dependent upon the level of organic matter added to the soil each year. The standard computation of gains and losses of humus was made the essential idea in soil-fertility maintenance.

From these early attempts at linking organic matter (humus) with soil productivity evolved a very complex science of organic-matter decomposition and nutrient recycling. It is now realized that humus is really a near-end product in the decomposition of organic matter in the soil. N. Flaig, H. Beutelspacker, and E. Rietz (1975) have done extensive work on reviewing the research on the formation of humic substances in soils. Essentially, polysaccharides, proteins, lignin, mercapto-compounds, nucleic acid derivatives, and other phospho-acids are partially decomposed by fast-acting enzymes in the soils to release energy, monomers, and single polymers. When these monomers and single polymers recopolymerize, they create very stable humic substances.

These humic substances, although extremely complex and not yet completely known, have several properties in common. They all appear to have a very similar carbon to nitrogen ratio (C/N) of about 10:1. They also act like clays with very large cation-exchange capacity (CEC). This CEC comes from the formation of carboxyl groups with the readily replaceable hydrogen ion. Soil cations such as calcium, magnesium, and potassium can be absorbed by these carboxyl groups and saved for future plant use. Humus is also an excellent water absorber and exhibits shrinking and swelling characteristics. The final property is the humus substance use in creating good soil structure: a valuable physical soil property for good plant production (Foth, 1984: 151–154).

Most of the recent research has been concentrated in the areas of organic matter decomposition and translocation within soil profiles. This work necessarily deals with the soil-fraction involvement in the formation of the various soil orders. Examples of these works include that of L. Peterson (1976), B. K. Daly (1982), and J. J. Dawson et al. (1978).

References

Daly, B. K. 1982. "Identification of Podzols and Podzolized Soils in New Zealand by Relative Absorbence of Oxalate extracts of A and B Horizons." *Geoderma* 28: 29–38.

Dawson, J. J.; Ugolini, F. C.; Hrutfiord, B. F.; and Zachara, J. 1978. "Role of Soluble Organics in the Soil Processes of a Podzol, Central Cascades, Washington," *Soil Science* 126:290–296.

Flaig, W.; Beutelspacher, H.; and Rietz, E. 1975. "Chemical Composition and Physical Properties of Humic Substances." In J. E. Gieseking, ed., *Soil Components, Vol. 1, Organic Components*. New York: Springer-Verlag, pp. 1–211.

Foth, Henry D. 1984. *Fundamentals of Soil Science*. 7th ed. New York: John Wiley and Sons.

Peterson, L. 1976. *Podzols and Podzolization*. Copenhagen: DSR Forlag.

Simonson, Roy W. 1968. "Concept of Soil." *Advances in Agronomy* 20: 1–47.

Usher, A. P. 1923. "Soil Fertility, Soil Exhaustion, and Their Historical Significance." *Quarterly Journal of Economics* 37: 385–411.

Sources of Additional Information

For a good earlier work on humus see S. A. Waksman, *Humus* (Baltimore: Williams and Wilkins, 1948). A good source for information on organic matter and crop production is F. E. Allison, *Soil Organic Matter and Its Role in Crop Production* (New York: Elsevier, 1973). Several examples of works dealing with organic matter alteration of soil include D. Challinor, "Alteration of Surface Soil Characteristics by Four Tree Species," *Ecology* 49 (1968): 286–290; P. L. Gersper and N. Holowaychuk, "Some Effects of Stem Flow from Forest Canopy Trees on Chemical Properties of Soils," *Ecology* 52 (1971): 691–702; H. Jenny, "Role of the Plant Factor in the Pedogenic Functions," *Ecology* 39 (1958): 5–16; R. E. Lucas, and M. C. Vitosh, *Soil Organic Matter Dynamics*, Michigan Agricultural Experimental Station Research Report, no. 358 (Michigan Agricultural Experimental Station, 1978); and F. C. Ugolini, R. E. Reanier, G. H. Rau, and J. I. Hedges, "Pedological, Isotopic, and Geochemical Investigations of the Soils at the

Boreal Forest and Alpine Tundra Transition in Northern Alaska," *Soil Science* 131 (1981): 359–374. Much work has been done concerning soil organic matter and soil classification including W. W. Pettapiece, "The Forest Grassland Transition," in S. Pawluk, ed., *Pedology and Quaternary Research* (Edmonton: University of Alberta Printing Department, 1969), pp. 103–113, and C. D. Sawyer and S. Pawluk, "Characteristics of Organic Matter in Degrading Chernozenic Surface Soils," *Canada Journal of Soil Science* 43 (1963): 275–286. Also some examples of work on soil organic matter and the development of paleosols include R. V. Ruhe, "Soils, Paleosols, and Environment," in W. Dort, Jr., and J. K. Jones, Jr., eds., *Pleistocene and Recent Environments of the Central Great Plains* (Lawrence: University Press of Kansas, 1969), pp. 37–52; C. J. Sorenson, J. C. Knox, J. A. Larsen, and R. A. Bryson, "Paleosols and the Forest Border in Keewatin, N. W. T.," *Quarternary Research* 1 (1971): 468–473; and C. J. Sorenson, R. D. Mandel, and J. C. Wallis, "Changes in Bioclimate Inferred from Paleosols and Paleohydrologic Evidence in East-Central Texas," *Journal of Biogeography* 3 (1976): 141–149.

HURRICANE/TYPHOON. See TROPICAL CYCLONE.

HYDROLOGIC CYCLE. The continuous movement or circulation of water vapor and water through the earth-atmosphere system. Sometimes called the water cycle, this continual circulation includes water found in the atmosphere, the hydrosphere (surface water), and the lithosphere or solid-rock part of the earth system.

The most abundant single substance found in the BIOSPHERE is the unusual inorganic compound we call water. If we were to add up all of the water found in the earth's oceans, lakes, rivers (see STREAM), ice caps, glaciers, SOILS, and atmosphere (see ATMOSPHERIC CIRCULATION), we would find approximately 1.5 billion cubic kilometers of water in one form or another (Penman, 1970: 39).

Because water is so ubiquitous and necessary to life, throughout history humankind has been interested in its properties and distribution. Since water was the only substance known to ancient philosophers in its solid, liquid, and gaseous states, many earth theories about earth substances focused on water. Pre-Socratic philosophers, like Thales (c. 624–546 B.C.) speculated that water was the basic element from which all things were made (Brock, 1981: 92).

The idea of the cyclical nature of water vapor and water can be traced at least to Xenophanes, who believed that the source of rainfall was the sea, which is "the begetter of clouds and winds and rivers." Aristotle was even more specific when he wrote in his *Meteorological* that by the Sun's head "the finest and sweetest water is every day carried up and is dissolved in vapor and rises to the upper region, where it is condenses again by the cold and so returns to the earth. . . . So the sea will never dry up, for before that can happen the water that has gone up beforehand will return it."

It was therefore primarily the Greek philosophers who were the first serious scholars concerned with the hydrologic cycle, and thereafter, scholars continued to study various aspects of the cycle. An important milestone in the study of the hydrologic cycle was the work in the seventeenth century by the Frenchman Pierre Perrault (1674). Perrault measured streamflow and rainfall in the catchment of the upper Seine River and proved that sufficient quantities of rainfall were present to sustain river (see STREAM) flow. Another important work at almost the same time was done by the English astronomer Edmund Halley (1694). Halley used a measurement of evaporation from a small pan to estimate evaporation from the Mediterranean Sea.

The measurement of PRECIPITATION in an organized and scientific manner started in the United States in 1819 under the auspices of the Surgeon General of the U.S. Army. This work was continued primarily by the U.S. Weather Bureau, which was established in 1891 and was renamed the National Weather Service in 1970. Some streamflow measurements were made as early as 1848 on the Mississippi River, but a systematic program was not started until the U.S. Geological Survey undertook this work in 1848 (Linsley et al., 1982: 4).

One aspect of the hydrologic cycle that made many advances in the nineteenth century was the theory of GROUNDWATER movement. Important work was done by the French hydraulic engineer Henry Darcy (1856), who studied the movement of water through sand. Other contributions in the nineteenth century focused on the hydraulics of groundwater development (Todd, 1959: 4).

Most of the studies in the nineteenth century were more qualitative than quantitative. The early part of the twentieth century saw the advent of many quantitative studies of all aspects of the hydrologic cycle. Modern advances in scientific instrumentation provide the hydrologist with excellent tools to measure the various parameters of the hydrologic cycle (Shaw, 1983).

The past several decades have seen a great deal of interest, particularly among geographers, in the role humans play in altering the hydrologic cycle. According to Richard Chorley:

> The hydrologic cycle is a great natural system, but it should become apparent that it is increasingly a technological and social system as well. It has been estimated that 10% of the national wealth of the United States is found in capital structures designed to alter the hydrologic cycle: to collect, divert, and store about a quarter of the available surface water, distribute it where needed, cleanse it, carry it away, and return it to the natural system. (1969: 3)

Geographers have looked at many aspects of the hydrologic cycle including weather modification (Sewell, 1966), river-basin development (White, 1963), climatic geomorphology (Stoddart, 1969), and water-quality protection (Knapp, 1979).

The emphasis on environmental POLLUTION problems that started in the decades of the 1960s and 1970s has continued, and much work in recent years has focused on water-pollution problems and the hydrologic cycle (LeBlanc, 1985; Moore, 1985).

References

Brock, W. H. 1981. "Cycles." in W. F. Bynum, E. J. Browne, and Ray Porter, eds., *Dictionary of the History of Science*. Princeton, N.J.: Princeton University Press. A short discussion of various cycles including the water cycle.

Chorley, Richard J., 1969. *Water, Earth, and Man*. London: Methuen & Co. An excellent geographical analysis of the hydrologic cycle.

Darcy, H. 1856. *Les Fontaines Publiques de la Villa de Dijon*. Paris. A classic early work on groundwater movement.

Halley, E. 1694. "An Account of the Evaporation of Water." *Philosophical Transaction* Vol. 18. London: Royal Society, pp. 183–190.

Knapp, Brian. 1979. *Elements of Geographical Hydrology*. London: Allen and Unwin. A good analysis of the geographical aspects of the hydrologic cycle.

LeBlanc, Dennis R. 1985. "Sewage Plume in a Sand and Gravel Aquifer, Cape Cod, Massachusetts." *U.S. Geological Survey Water Supply Paper*, no. 2218. Washington, D.C.; U.S. Geological Survey. A good discussion, with maps and diagrams, of what happens to various pollutants once they enter an aquifer.

Linsley, Ray K.; Kohler, Max A.; and Paulhus, Joseph H. 1982. *Hydrology for Engineers*. 3d ed. New York: McGraw-Hill Book Co. A recent textbook that covers many aspects of the hydrologic cycle.

Moore, Taylor. 1985. "Groundwater: Examining a Resource at Risk." *Journal of the Electric Power Research Institute*. October, pp. 6–19. An easy to understand analysis of groundwater pollution problems. A balance of economic and ecological impacts is emphasized.

Penman, H. L. 1970. "The Water Cycle." In *The Biosphere*. San Francisco: W. H. Freeman and Co., pp. 39–45. An excellent article on the hydrologic cycle containing excellent graphics.

Perrault, P. 1674. *De l'origine des fontanes*. Paris. English Translation A. LaRoque. New York: Hafner Publishing Co., 1967.

Sewell, W.R.D., ed. 1966. "Human Dimensions of Weather Modification." *Department of Geography Research Paper*, no. 105. Chicago: University of Chicago.

Shaw, Elizabeth, 1983. *Hydrology in Practice*. England: Van Nostrand-Reinhold (UK) Co. A very technical analysis of hydrologic problems.

Stoddart, D. R. 1969. "Climate Geomorphology." In Chorley, Richard J., *Water, Earth, and Man*. London: Methuen & Co.

Todd, David K., 1959. *Ground Water Hydrology*. New York: John Wiley and Sons.

White, Gilbert F. 1963. "Contribution of Geographical Analysis to River Basin Development." *Geographical Journal* 129: 412–436. A good discussion of the geographer's role in river basin development.

Sources of Additional Information

Short discussions of the hydrologic cycle can be found in John Challinor, *Challinor's Dictionary of Geology*, 6th ed. (New York: Oxford University Press, 1986), pp. 157–158; Robert W. Durrenberger, *Dictionary of the Environmental Sciences* (Palo Alto, Calif.: National Press Books, 1977), p. 116; John Wittow, *Dictionary of Physical Geography* (New York: Penguin Books, 1984), pp. 257–258; Douglas M. Considine, ed., *Van Nostrand's Scientific Encyclopedia*, 6th ed., (New York: Van Nostrand-Reinhold Co., 1983), pp. 1559–1560; and Arthur N. Strahler and Alan H. Strahler, *Modern Physical*

Geography, 2d ed., (New York: John Wiley and Sons, 1983), pp. 154–155. Longer, more detailed descriptions of the hydrologic cycle can be found in the following chapters of hydrology and water resources books: C. F. Tolman, *Ground Water* (New York: McGraw-Hill Book Co., 1937), Chapter 2; Arved J. Raudkivi, *Hydrology: An Advanced Introduction to Hydrological Processes and Modelling* (New York: Pergamon Press, 1979), Chapter 1; R. C. Ward, *Principles of Hydrology*, 2d ed., (London: McGraw-Hill Book Co. [UK], 1975), Chapter 1; and Roger Barry, "The World Hydrological Cycle," in Richard J. Chorley, ed., *Water, Earth, and Man* (London: Methuen & Co., 1969), Chapter 1. A good general discussion of the global significance of water resources is found in Alan O. Tweedie, *Water and the World* (London: Thomas Nelson and Sons, 1966); Ralph E. Olson, *A Geography of Water* (Dubuque, Ia.: W. C. Brown, 1970); and David H. Miller, *Water at the Surface of the Earth: An Introduction to Ecosystem Hydrodynamics* (New York: Academic Press, 1977).

HYGROSCOPIC. See CONDENSATION NUCLEI.

I

INFRARED. 1. The energy in the electromagnetic spectrum with wavelengths from 0.7 to 100 micrometers (1 μm = $10^{-6}m$). 2. As used in remote sensing, it is subdivided into near infrared (IR) (7.0–15.0μm), middle IR (1.3–3.0μm), and far IR (7.0–15.0μm). 3. Far IR is also referred to as thermal IR.

In 1800 William Herschel broadened the knowledge of the ELECTROMAGNETIC RADIATION beyond the range of visible light. He discovered "heat" or thermal radiation in an area of the spectrum that was of longer wavelength than visible energy. Today we realize that the thermal portion of IR is only part of the IR range; the near and middle IR wavelengths also belong to this range. Thermal IR is referred to as emissive IR since objects detected by thermal measuring instruments actually emit this energy. Any object that has a temperature above 0° Kelvin emits energy within this wavelength range. The near and middle IR ranges are reflected energy areas. Objects at normal earth temperatures can be sensed in this range only if the energy (usually the sun's energy) is reflected from them.

The concept of infrared energy is pertinent to geographers because of its use in REMOTE SENSING. One of the limiting characteristics in remote sensing is atmospheric absorption by which certain gases and small particles in the atmosphere attenuate incoming solar energy in certain wavelengths. Little of the near IR energy is attenuated, so it can be used throughout the range. The middle and far IR ranges are seriously affected, however, and much care needs to be used in the selection of wavelengths for sensing purposes (Wiesnet and Matson, 1983: 1333). At the earth's surface, anything beyond 15 μm is virtually useless as a sensing range.

There are several instruments used for obtaining remotely sensed information in the IR range. Cameras can be used with near IR film (either color or black and white) up to 0.9 μm. Beyond this range film is very unstable and unreliable. Radiometers can be used to measure radiation throughout the IR range. These instruments quantitatively measure radiation in wide bands using scanning techniques. Spectrometers are used to obtain readings throughout the IR range in very narrow, selected bands. Spectrometers use dispersive prisms, grating, or circular interference filters to separate the desired band selectively. Lasers have now been developed for use in the IR range. They are generally used to measure backscattered radiation, especially from atmospheric particulates. Solid state detectors are also being used for scanners and radiometers (Suits, 1983: 41).

The application for uses of IR energy are as varied as the imagination. M. T. Chahine stated that "any of the physical interactions between the radiation field and the environment offer potential means for probing that part of the environment which is involved in the interaction" (1983: p. 166). However, this depends upon the strength of the interactions and the energy available for detection. Applications for the use of IR (near, middle, and far) include but are not limited to the following: snow and ice monitoring, water-resource detection, volcanic-hazard research, geologic resource evaluation, geobotanic investigation, agricultural disease monitoring, crop estimation, forest-fire detection, forest- resource inventories, and interplanetary exploration (Colwell, 1983).

Since 1972 the most used sensing system in the near IR range has been the multispectral scanner (MSS) on the LANDSAT satellites launched by NASA in 1972, 1975, 1978, and the more advanced systems, in 1982 and 1985. The 1982 and 1985 satellites included several additional bands incorporated in the Thematic Mapper (TM) including the thermal range. Other advanced systems include the Heat Capacity Mapping Mission (HCMM), the first of which was launched in 1978 with subsequent satellites being considered, and the Shuttle Multispectral Infrared Radiometer (SMIRR). NASA has also developed the Thermal Infrared Multispectral Scanner (TIMS) for airborne use. Instruments being developed for the future are sensors using the Solid State Multispectral Linear Arrays (MLA), which require no moving parts and have higher spectral sensitivity and better geometric precision than current instruments (Elachi, 1983).

References

Chahine, M. T. 1983. "Interaction Mechanisms within the Atmosphere." In R. N. Colwell, ed., *Manual of Remote Sensing*. Falls Church, Va.: pp. 165–230.

Colwell, R. N., ed. 1983. *Manual of Remote Sensing*. Falls Church, Va.: The single best resource for work in the infrared in particular and remote sensing in general.

Elachi, C. 1983. "Microwave and Infrared Satellite Remote Sensors." In R. N. Colwell, ed., *Manual of Remote Sensing*. Falls Church, Va.: American Society of Photogrammetry, pp. 571–650.

Suits, G. H. 1983. "The Nature of Electromagnetic Radiation." In R. N. Colwell, ed., *Manual of Remote Sensing*. Falls Church, Va.: pp. 37–60.

Wiesnet, D. R., and Matson, M. 1983. "Remote Sensing of Weather and Climate." In R. N. Colwell, ed., *Manual of Remote Sensing*. Falls Church, Va.: pp. 1305–1369.

Sources of Additional Information

There is a plethora of works in many application areas using infrared energy including: (Archaeology and Anthropology) T. E. Avery and T. R. Lyons, *Remote Sensing: Practical Exercises on Remote Sensing in Archeology* (Washington D.C.: National Park Service, 1978), a handbook for archeologists and cultural resource managers; F. P. Conant, "The Use of LANDSAT Data in Studies of Human Ecology," *Current Anthropology* 19 (1978): 382–384; and T. R. Lyons and F. J. Mathien, eds., *Cultural Resources Remote Sensing* (Washington, D.C.: Cultural Resources Management Division, National Park Service, 1980). (Weather and Climate) D. S. Johnson and I. P. Vetlov, *The Role of Satellites in WMO Programs in the 1980's* (Geneva: World Meteorological Organization, 1977); *World Weather Watch Planning Report* no. 36, (World Weather Watch, 1977); and W. M. Washington and R. M. Chervin, "Regional Climatic Effects of Large-Scale Thermal Pollution," *Journal of Applied Meteorology* 18 (1979): 3–16. (Water Resources) V. Carter, M. K. Garrett, L. Shima, and P. Gannon, "The Great Dismal Swamp: Management of a Hydrologic Resource with the Aid of Remote Sensing," *Bulletin of Water Resources* 13 (1977): 1–12; R. J. Cermak, A. D. Feldman, and R. P. Webb, "Hydrologic Land Use Classification Using LANDSAT," *Hydrologic Engineering Center Technical Paper*, no. 67 (Hydrologic Engineering Center, 1979); and M. Deutsch, D. R. Wiesnet, and A. Rango, eds., *Satellite Hydrology* (Minneapolis: American Water Resources Association, 1981). (Geology) F. P. Agterberg, "Application of Image Analysis and Multivariate Analysis to Mineral Resource Appraisal," *Economic Geology* 76 (1981): 1016–1031; A.F.H. Goetz and L. C. Rowan, "Geologic Remote Sensing," *Science* 211 (1981): 781–791; and F. F. Sabins, Jr., *Remote Sensing, Principles and Techniques* (San Francisco: W. H. Freeman and Co., 1978). (Agriculture) J. Cihlar, T. Sommerfeldt, and B. Paterson, "Soil Water Content Estimation in Fallow Fields from Airborne Thermal Scanner Measurements," *Canadian Journal of Remote Sensing* 5 (1979): 18–32; C. J. Tucker, "Red and Photographic Infrared Linear Combinations for Monitoring Vegetation," *Remote Sensing of the Environment* 8 (1979): 127–150; and R. A. Weismiller, F. R. Kirschner, S. A. Kaminsky, and E. J. Hinzel, *Spectral Classification of Soil Characteristics to Aid the Soil Survey of Jasper County, Indiana*, LARS Technical Report, no. 040179 (West Lafayette, Ind.: Purdue University, 1979). (Forest Resources) E. S. Bryant, A. G. Dodge, Jr., and S. D. Warren, *Satellites for Practical Natural Resource Mapping? A Forestry Test Case* (Fort Collins, Colo.: Rocky Mountain Forest and Range Experiment Station, General Technical Department RM–55, 1978); R. C. Heller, "Case Applications of Remote Sensing for Vegetation Damage Assessments," *Photogrammetric Engineering* 44 (1978): 1159–1166; and A. H. Strahler, T. L. Logan, and C. E. Woodcock, *Forest Classification and Inventory System using LANDSAT, Digital Terrain, and Ground Sample Data*, Proceedings of the Thirteenth International Symposium on Remote Sensing of Environment (Ann Arbor, Mich.: 1979).

INTERNATIONAL DATE LINE. In a strict sense it is the 180th meridian of longitude, the other half of the great circle that begins with the prime meridian, although in reality the line deviates both east and west of the 180th meridian somewhat to avoid local land areas and island groups.

The International Meridian Congress met in Washington, D.C., in 1884 to establish an international basis for time zones. The result was the establishment of 24 time zones, one for each hour in the day. Each time zone would be 15 degrees of longitude. As Arthur Strahler and Alan Strahler (1987: 37) noted, "In all global time calculations the prime meridian of Greenwich, England, is taken as the reference meridian." The time zone around Greenwich would extend for 7.5 degrees in either direction from the prime meridian, and each additional time zone, then, would extend 7.5 degrees on either side of the next standard meridian. Each time zone can be described in terms of how much it differs in time from the time at Greenwich.

Twelve hours away from Greenwich, either east or west, is the 180th meridian, which is the International Date Line. Only precisely at midnight on the International Date Line do we find all parts of the globe to have the same date. Otherwise, there are always two dates. On the west side of the International Date Line it is always, with the exception mentioned already, a day ahead of what exists on the east side of the International Date Line.

Thus if you travel westward across the International Date Line you lose a day, because you must advance your calendar by one day. Conversely, if you cross the International Date Line from west to east you gain a day.

References

Moore, W. G. 1975. *A Dictionary of Geography: Definitions and Explanations of Terms Used in Physical Geography*. London: Adam and Charles Black. A useful reference work for geographers.

Strahler, Arthur N., and Strahler, Alan H. 1987. *Modern Physical Geography*. 3d ed. New York: John Wiley and Sons. Probably the best of the current physical geography texts that are available.

Sources of Additional Information

Discussions of time zones and meridians can be found in virtually any of the physical geography texts that are currently available. These discussions are usually supplemented with maps that illustrate the location of time zones, including the International Date Line. Either of the following provides a good section on these matters: William M. Marsh, *Earthscape: A Physical Geography* (New York: John Wiley and Sons, 1987), or Tom L. McKnight, *Physical Geography: A Landscape Appreciation* (Englewood Cliffs, N.J.: Prentice-Hall, 1984). Those interested in matters of time may also want to look at I. R. Bartky and E. Harrison, "Standard and Daylight-Saving Time," *Scientific American* 240, no. 5 (1979): 46–53.

INTERTROPICAL CONVERGENCE ZONE. A trough of almost continuous low pressure in the equatorial zone, where the tropical easterlies converge.

The intertropical convergence zone is one feature of ATMOSPHERIC CIRCULATION. Within the equatorial zone air rises and ultimately is carried poleward. The void left by this rising air is filled by air flowing in at the surface from the subtropics, forming the trade-wind belts in both hemispheres. These winds come together, or converge, in the intertropical convergence zone, as the name implies.

The intertropical convergence zone exists primarily over the ocean, and it moves back and forth seasonally, reaching its northernmost point during the northern hemisphere summer and its southernmost point during the southern hemisphere summer. It is most closely associated with the equator around the time of the equinoxes. These seasonal changes in the location of the intertropical convergence zone turn out to be extremely important for weather in many equatorial and tropical regions, where its presence usually brings cloudiness and precipitation.

Convergence does not occur continuously, either geographically or chronologically, throughout all of the equatorial zone. As R. G. Barry and R. J. Chorley commented, "Equatorward of the main *root zones* of the trades over the eastern Pacific and eastern Atlantic are regions of light, variable winds, known traditionally as the *doldrums* and much feared in past centuries by the crews of sailing ships" (1970: 107). For sailors the doldrums were similar to the horse latitudes (see ATMOSPHERIC CIRCULATION).

The importance of the intertropical convergence zone in the overall atmospheric circulation pattern was suggested by L. Hasse and F. Dobson, who stated that "hence in the ITCZ the latent heat collected in the trade wind belt is released and drives the Hadley circulation" (1983: 17).

References

Barry, R. G., and Chorley, R. J. 1970. *Atmosphere, Weather, and Climate.* New York: Holt, Rinehart and Winston. Although somewhat dated in many ways, this remains a very useful reference on many meteorological and climatological topics.

Hasse, L., and Dobson, F. 1983. *Introductory Physics of the Atmosphere and Ocean.* Boston: D. Reidel Publishing Co. A concise introduction to atmospheric circulation, including the intertropical convergence zone and its importance.

Oliver, John E., and Hidore, John J. 1984. *Climatology: An Introduction.* Columbus, Ohio: Charles E. Merrill Publishing Co. A very good introduction to the study of climatology by two well-known geographers.

Sources of Additional Information

Any discussion of the earth's atmospheric circulation system includes a section on the intertropical convergence zone. A good introduction can be found in the following: Arthur N. Strahler and Alan H. Strahler, *Modern Physical Geography,* 3d ed. (New York: John Wiley and Sons, 1987); Tom L. McKnight, *Physical Geography: A Landscape*

Appreciation (Englewood Cliffs, N.J.: Prentice-Hall, 1984); and E. Willard Miller, *Physical Geography: Earth Systems and Human Interactions* (Columbus, Ohio: Charles E. Merrill Publishing Co., 1984). Somewhat more detailed discussions of the intertropical convergence zone and its importance in regional climatic patterns can be found in Paul E. Lydolph, *The Climate of the Earth* (Totowa, N.J.: Rowman and Allanheld, Publishers, 1985), or Glenn T. Trewartha and Lyle H. Horn, *An Introduction to Climate*, 5th ed. (New York: McGraw-Hill Book Co., 1980).

IONOSPHERE. A region in the upper atmosphere that is characterized by the presence of relatively large numbers of free electrons.

According to S. K. Mitra (1952), the existence of a conducting layer in the upper atmosphere was first hypothesized as an explanation of ground-level variations in the earth's magnetic field and later confirmed to exist. Because of its ability to trap radio waves, the ionosphere rapidly became important in long-distance communications. Kenneth Davies (1965) summarized our knowledge of the ionosphere and its impact on radio waves.

Free electrons in the ionosphere are accompanied by positive ions, and according to S. A. Bowhill, "These electrons, with their accompanying positive ions, form an ionized plasma which is substantially electrically neutral" (1967: 498).

The ionosphere actually exists in several layers within the atmosphere at elevations between 50 and 600 kilometers above the earth's surface. It can be subdivided into three regions, D, E, and F, as they are known, on the basis of variations in atomic ionization, as noted by Bowhill (1967), who also presented a detailed discussion of these three regions.

Although the ionosphere is important for radio-wave propagation and communications, geographers have not found it of much interest in their dealing with climate and other phenomena on or near the earth's surface; thus the concept is seldom found today in introductory texts and receives scant attention in more advanced texts.

References

Bowhill, S. A. 1967. "Ionosphere." In Rhodes W. Fairbridge, ed., *The Encyclopedia of Atmospheric Sciences and Astrogeology*. Encyclopedia of Earth Sciences Series. Vol. 2. New York: Reinhold Publishing Co.

Davies, K., ed. 1965. *Ionospheric Radio Propagation*. Washington, D.C.: U.S. Government Printing Office. Still represents an excellent summary of ionospheric effects on radio waves.

Mitra, S. K. 1952. *The Upper Atmosphere*. 2d ed. Calcutta: The Asiatic Society. Contains a summary of early research dealing with the ionosphere and its properties.

Sources of Additional Information

The ionosphere is seldom mentioned in introductory physical geography texts anymore, although it did receive mention in E. Willard Miller, *Physical Geography: Earth Systems and Human Interactions* (Columbus, Ohio: Charles E. Merrill Publishing

Co., 1985). The ionosphere is discussed, though only briefly, in most climatology books, including the following: John E. Oliver and John J. Hidore, *Climatology: An Introduction* (Columbus, Ohio: Charles E. Merrill Publishing Co., 1984); Paul E. Lydolph, *The Climate of the Earth* (Totowa, N.J.: Rowman and Allanheld, Publishers, 1985); and Glenn T. Trewartha and Lyle H. Horn, *An Introduction to Climate* (New York: McGraw-Hill Book Co., 1980). A brief discussion of the ionosphere can also be found in Herbert J. Spiegel and Arnold Gruber, *From Weather Vanes to Satellites: An Introduction to Meteorology* (New York: John Wiley and Sons, 1983). The physics of chemical activity in the ionosphere is thoroughly discussed in C. O. Hines et al., eds., *Physics of the Earth's Upper Atmosphere* (Englewood Cliffs, N.J.: Prentice-Hall, 1965).

ISOBAR. See ATMOSPHERIC PRESSURE.

J

JET STREAM. A narrow, high-altitude (around 9 to 12 kilometers), high-velocity stream of air, generally westerly.

The presence of high-altitude winds was recognized no later than the early part of the twentieth century, as suggested by cloud studies such as that of W. H. Dines (1911) and the balloon observations reported by G. M. B. Dobson (1920). However, most of these studies were not pursued very far, and the first use of the term *jet stream* (*Strahlstrom*) in German was introduced some years later by H. Seilkopf (1939).

Early research on the jet stream was largely a product of the expansion of modern aviation during World War II. Among the first major works on this new research was that of C. G. Rossby and H. C. Willet (1948). During and after that war, research progressed rapidly, and the first summaries of literature on the jet stream soon appeared, including those by Thomas F. Malone (1951), C. J. Van der Ham (1954), and F. P. Rodriguez (1955).

Almost a decade later Elmer Reiter (1963) published his classic book on jet streams, and it became clear that research to that date had led to a much more complex picture of the jet-stream phenomena. Reiter noted that "the phenomenon of jet streams is by no means confined to the upper troposphere or lower stratosphere" (1963: 2). Rather, he argued, jet streams seem "to appear wherever large-scale motions exist under the influence of quasi-stationary fields of forces and of the deflecting force of the earth's rotation" (1963: 2).

For geographers the most important of the jet streams is the polar-front jet stream, which forms at the level of the tropopause (see TROPOSPHERE). This jet stream develops along the polar front, as its name implies, where contact occurs between cold polar air and warm tropical air. Arthur Strahler and Alan Strahler described the situation for the polar front jet stream as follows:

> The tropopause drops sharply in altitude at the polar front, being much lower over the cold troposphere layer than over the warm troposphere. Atmospheric pressure also changes abruptly at the polar front. Isobaric surfaces . . . drop steeply at the polar front, the steepest drop being at the level of the jet-stream core. The very steep pressure gradient at this point causes the high-speed flow of air. (1978: 86)

The northern hemisphere polar-front jet stream has been studied considerably, although a polar-front jet stream exists in the southern hemisphere as well. In the northern hemisphere the polar-front jet stream is located on the average at the 300 millibar level at a typical elevation of around 10 kilometers. Although located in the temperate latitudes, its actual latitudinal location varies considerably, both on a day-to-day and seasonal basis.

The polar-front jet stream is closely tied to Rossby wave patterns, the circumpolar vortex, and overall weather patterns in the northern hemisphere, and its location and velocity are also important considerations in the flight of jet aircraft that are flying at high altitudes.

Behavior of the jet stream, along with standing waves in the upper troposphere, can be used to predict the paths that storms are likely to take and the trend toward strengthening or weakening of a particular storm along its probable path. According to James Hanwell:

> Northbound divergent streams exert such strong controls over the formation and subsequent location of cold-core lows that they are referred to as being *cyclogenic*, meaning simply "cyclone-generating." Likewise, southbound convergent streams appear to be a major reason for warm-core highs and may be regarded as *anticyclogenic*. (1980: 29)

The other important jet stream is known as the subtropical jet stream, usually located near 30 degrees north latitude at or near the 200 millibar level, which is generally at an altitude of nearly 12 kilometers. Its location is at the tropopause, just above the Hadley Cell. There is also a subtropical jet stream in the southern hemisphere, although as with the polar front jet stream, much less is known about the southern hemisphere version.

Both the polar-front jet stream and the subtropical jet stream are westerly, but another jet stream, the tropical easterly jet stream, flows in the other direction, as its name implies. It is primarily a summer phenomena, located at a height of nearly 15 kilometers, and is thought to be important as a factor in Asia's summer monsoon.

Although of less interest to geographers, there are other jet streams, with locations at much higher elevations, namely, at the stratopause and mesopause. These jet streams are described in the work of Rhodes Fairbridge (1967).

References

Dines, W. H. 1911. "The Vertical Temperature Distribution in the Atmosphere over England, and Some Remarks on the General and Local Circulation." *Philosophical Transactions of the Royal Society*, Series A, 211: 277–300. One of the earliest studies of vertical temperature distribution.

Dobson, G.M.B. 1920. "Winds and Temperature Gradients in the Stratosphere." *Quarterly Journal of the Royal Meteorological Society* 46: 54–62. Somewhat outdated but still of some interest.
Fairbridge, Rhodes W. 1967. *The Encyclopedia of Atmospheric Sciences and Astrogeology.* Encyclopedia of Earth Sciences Series. Vol. 2. New York: Reinhold Publishing Corp.
Hanwell, James D. 1980. *Atmospheric Processes.* London: Allen and Unwin. A simple and useful introduction.
Malone, Thomas F. 1951. *Compendium of Meteorology.* Boston: American Meteorological Society. Mostly only of historical interest.
Rodriguez, F. P. 1955. "Notas Sobre las Corrientas de Chorro." *Revista de Geofisica* 14: 313–346.
Reiter, Elmar R. 1963. *Jet-Stream Meteorology.* Chicago: University of Chicago Press. (Translated from *Meteorologie der Strahlstrome.* Vienna: Springer-Verlag, 1961.) An excellent and detailed look at the jet stream.
Rossby, C. G., and Willet, H. C. 1948. "The Circulation of the Upper Troposphere and Lower Stratosphere." *Science* 108: 643–652. A landmark study in its day.
Seilkopf, H. 1939. *Maritime Meteorologie: Handbuch der Fliegerwetterkunde.* Bd. II. Berlin: Radetzki.
Strahler, Arthur N., and Strahler, Alan H. 1978. *Modern Physical Geography.* New York: John Wiley and Sons.
Van der Ham, C. J. 1954. "De Straalstrom," *Hemel en Dampkring* 52: 201–210.

Sources of Additional Information

The basic concept of a jet stream is introduced in most physical geography texts, although it is usually not presented in detail. Typical examples include the following: Tom L. McKnight, *Physical Geography: A Landscape Appreciation* (Englewood Cliffs, N.J.: Prentice-Hall, 1984); Arthur N. Strahler and Alan H. Strahler, *Elements of Physical Geography*, 3d ed. (New York: John Wiley and Sons, 1984); and E. Willard Miller, *Physical Geography: Earth Systems and Human Interactions* (Columbus, Ohio: Charles E. Merrill Publishing Co., 1985). Somewhat more detailed discussions of jet streams can be found in Joe R. Eagleman, *Meteorology: The Atmosphere in Action* (New York: D. Van Nostrand Co., 1980), or Paul E. Lydolph, *Weather and Climate* (Totowa, N.J.: Rowman and Allanheld, Publishers, 1985).

K

KARST. 1. A type of landscape or terrain found in areas composed primarily of carbonate rock. 2. A class of topography arising from a high degree of rock solubility (limestone or dolomite) with distinctive drainage and relief characteristics such as sinkholes, caverns, and dry valleys.

The study of karst topography can be traced to antiquity. The first known attempt to understand karst phenomena was an expedition sent by Assyrian King Salmanassar III to the caves and springs at the source of the Tigris River (Jennings, 1971: 6). The distribution of karst phenomena and their association with early Mediterranean civilizations helps to explain the interest by Greek and Roman poets, philosophers, and natural historians. Thales (640–547 B.C.) and Aristotle (384–322 B.C.) were interested in the movement of GROUNDWATER as was Lucretius (98–55 B.C.), who described the circulation of water between the earth and the sea (Herak and Stringfield, 1972: 19).

Throughout the Middle Ages and the Renaissance, little detailed investigation was undertaken except for the thirteenth century inscriptions in the Postojna Cave in Slovenia and some direct observations of caves, the result of the search for valuable minerals. The seventeenth century saw the emergence of the first book devoted to karst written by Jacques Gaffarel of Paris in 1654 (Jennings, 1971: 6). Unfortunately, little survives of Gaffarel's work; however in 1689 J. W. Valvasor, a master of drawings and descriptions, published his impressive work on the Slovenian karst. Useful data and observations of karst features in Slovenia continued into the eighteenth century with the works of B. Hacquet (1778) and T. Gruber (1781). According to J. Roglic (1972: 2), however, "these observations were too early for the general scientific climate, and the country was outside the main flow of cultural aspirations and decisive events." Hacquet's work, however, was a forerunner to the great English geologist Charles Lyell's (1839) explanation of the chalk of southern England. Except for this work by

Lyell, most of the early karst investigations focused on the limestone regions of present-day Yugoslavia. The word *karst* is the German form of the Slovene word *kras* or *kas*, meaning crag or stone. The current scientific international terminology still includes many words like *uvala*, *polje*, and *datine* that have Yugoslavian origins (Sweeting, 1973: 1).

The railway constructed through the karst region from middle Europe to Trieste in the middle of the nineteenth century (1850–1857) brought about an increasing concern for karst landscapes, especially since these karst features were related to the civil engineering problems of railroad construction. This time of renewed scientific interest in karst in the second half of the nineteenth and early twentieth centuries is usually referred to as the "classical period" of the karst investigation.

The study of caves took on increasing importance with the pioneering work on the karst of Trieste by Schmidl (1854). Other studies in England (Prestwich, 1854), the United States (Owen, 1856; Cox, 1874), and Jamaica (Sawkins, 1869) added to the increasing scientific concern with karst.

An impetus to further geomorphological research (see GEOMORPHOLOGY) on karst was given when the noted geomorphologist Alfred Penck became interested in the development of plains in the Yugoslavian karst region. Penck (1894) wrote the first textbook on karst, and many of his students went on to do further research on karst processes. Together with the famous American geomorphologist William Morris Davis, Penck led students on field trips in Bosnia and Hercegovina. Penck (1900) and Davis (1901) looked at "karstification" as a cyclic evolution process. Penck's most important contribution, however, was his suggestion to his talented student Jovan Cvijic that he should prepare a doctoral dissertation on the subject of karst.

Cvijic is considered by many to be the "father" of karst geomorphology research. His pioneering work, "Das Karstphanomen," published in 1893, was a turning point and also the beginning of an intensified study of karst topography. This original study in 1893 was followed by many other important research efforts (Cvijic, 1909, 1918, 1924).

After World War I karst research continued with the development of institutes in Austria and Romania, and after World War II karst studies became part of the global scientific exchange of ideas. International exchanges of ideas were greatly enhanced by the Commission of Karst Phenomena of the International Geographical Union.

Because the field of karst LANDFORMS had its roots in central and eastern Europe, there were few research contributions by English-speaking scientists during the early years of its development. In recent years, however, these scientists have made many significant contributions especially in the understanding of karst processes (Palmer, 1984).

The past two decades have seen a trend toward describing karst processes and landforms in quantitative terms (Williams, 1972; 1983; LaValle, 1967; Baker, 1973). The results of these studies have shown karst drainage to be more organized and less chaotic than thought by early investigators.

The belief that climate was the most significant factor in controlling karst landforms (Lehmann, 1936; Corbel, 1959) is now being challenged by those who have found a variety of landforms in the same local region where climatic regimes were identical (Verstappen, 1964). Much of the recent work has concentrated on the spatial or geographic characteristics of landforms as they relate to hydrologic and geologic parameters (Kemmerly, 1976, 1982; Palmquist, 1979; Mills and Starnes, 1983).

Research in karst processes and landforms is now undertaken throughout the world. Geographers and other earth scientists are continually trying to understand the mechanisms associated with karst topography.

References

Baker, V. R. 1973. "Geomorphology and Hydrology of Karst Drainage Basins and Cave Channel Networks in East-Central New York." *Water Resources Research* 9: 695–706.

Corbel, J. 1959. "Vitesse de L'Erosion." *Zeitschrift für Geomorphologie* 3: 1–28.

Cox, E. J. 1874. *Fifth Annual Report of the Geological Survey of Indiana.* Indianapolis: Geological Survey of Indiana, pp. 280–305. An early karst study in the United States.

Cvijic, Jovan 1893. "Das Karstphanomen." *Geograph. Abhandl.* 5: 218–329. Cvijic is considered by many geomorphologists to be the "father" of karst studies; this is his classical and seminal paper on karst development.

———. 1909. "Bildung und Dislozierung der Dinarischen Rumpfflache." *Petermanns Geographische Mittheilungen* 6: 121–127; 7: 156–163; 8: 177–181.

———. 1918. "Hydrographie Souterraine et Evolution Morphologique du Karst." *Recùeil des Travaux Institut Geograph. Alpine* 6: 376–420.

———. 1924. "The Evolution of Lapies." *Geographical Review*, pp. 26–49. One of the few English-language articles by the "father" of karst studies.

Davis, William Morris. 1901. "An Excursion in Bosnia, Hercegovina, and Dalmatia." *Geographical Society Bulletin* 3: 47–50. An early American analysis of the Yugoslavian karst region by one of America's pioneering geographers.

Gruber, T. 1781. *Briefe Hydrographischen und Physikalischen Inhalts aus Krain.* Wien: Krauss.

Hacquet, B. 1778. *Oryctographia Carniolica oder Physicalische Erd-Beschreibung des Herzogthums Krain, Istrien und zum Teil der Benachbarten Lander.* Leipzig: Gottlob und Breitkopf. A pioneering work in karst analysis.

Herak, M., and Stringfield, V. T. 1972. "Historical Review of Hydrogeologic Concepts." In Herak, M., and Stringfield, V. T., eds., *Karst: Important Karst Regions of the Northern Hemisphere.* New York: Elsevier, chapter 2, pp. 19–24. An excellent review of the historical development of karst.

Jennings, J. N. 1971. *Karst.* Cambridge, Mass.: The M.I.T. Press. Part of a series on systematic geomorphology.

Kemmerly, P. R. 1976. "Definitive Doline Characteristics in the Clarksville Quadrangle." *Geological Society of America Bulletin* 87: 42–46.

———. 1982. "Spatial Analysis of a Karst Depression Population—Clues to Genesis." *Geological Society of America Bulletin*, 93, no. 7: 1078–1086. A detailed analysis of depression spatial distribution on parts of the Pennyroyal Plain between Tennessee and Kentucky.

LaValle, P. 1967. "Some Aspects of Linear Karst Depression Development in South-Central Kentucky." *Annals of the Association of American Geographers* 57: 49–71. A quantitative analysis of karst development.

Lehmann, H. 1936. "Morphologische Studien auf Java." *Geogr. Abh.* 3: 9. Primarily focused on the role of climate in karst development.

Lyell, Charles. 1839. "On the Tubular Cavities Filled with Gravel and Sand Called 'Sand Pipes' in the Chalk Near Norwich." *London Edinburg Phil. Mag. J. Sci.* 15: 257–266. An early work dealing with karst by one of the fathers of geology.

Mills, H., and Starnes, D. 1983. "Sinkhole Morphometry in a Fluviokarst Region: Eastern Highland Rim, Tennessee." *Zeitschrift für Geomorphologie* 27: 39–54.

Owen, D. D. 1856. *Annual Report of the Geological Survey in Kentucky.* Frankfort, Ky.: Geological Survey of Kentucky, pp. 169–172. An early karst study in the United States.

Palmer, A. N. 1984. "Recent Trends in Karst Geomorphology." *Journal of Geological Education* 32: 247–253. An excellent review article on recent trends in karst geomorphology.

Palmquist, R. C. 1979. "Geological Controls on Doline Characteristics in Mantled Karst." *Zeitschrift Für Geomorphologie* Supplement, 32: 90–106.

Penck, A. 1894. *Morphologie der Erdoberflache.* Stuttgart: Englehorn. The first textbook in karst studies.

———. 1900. "Geomorphologische Studien aus der Hercegovina." *Z. Deut. Osterreich Alpenver.* 31: 25–41.

Prestwich, J. 1854. "On Some Swallow Holes on the Chalk Hills Near Canterbury." *Quarterly Journal*, pp. 222–224, 241. An early English-language study in England.

Roglic, J. 1972. "Historical Review of Morphologic Concepts." In Herak, M., and Stringfield, V. T., eds., *Karst: Important Karst Regions of the Northern Hemisphere.* New York: Elsevier, Chapter 1, pp. 1–18. A good review of the historical development of karst studies.

Sawkins, J. Q. 1869. *Reports on the Geology of Jamaica.* London: Geological Survey.

Schmidl, A. 1854. *Die Grotten und Hohlen von Adelsberg, Leug, Planina und Lass.* Wien: Braumuller. An important early work on the study of caves in karst regions. Primarily focused on caves near Trieste and how cave development related to civil engineering.

Sweeting, Marjorie M. 1973. *Karst Landforms.* New York: Columbia University Press. A most helpful book dealing with aims and applications of karst landform study. Many good photographs and figures.

Valvasor, J. W. 1689. *Die Ehre des Herzogsthums krain.* Nurenberg: Endter.

Verstappen, H. 1964. "Karst Morphology of the Star Mountains (Central New Guinea) and Its Relation to Lithology and Climate." *Zeitschrift für Geomorphologie* 8: 40–49.

Williams, P. W. 1972. "The Analysis of Spatial Characteristics of Karst Terrains." In R. J. Chorley, ed., *Spatial Analysis in Geomorphology*, London: Methuen Co. A quantitative analysis of karst primarily dealing with spatial or geographic parameters.

———. 1983. "The Role of the Subcutaneous Zone in Karst Hydrology." *Journal of Hydrology* 61: 45–67.

Sources of Additional Information

Brief discussions of karst processes and landforms can be found in Karl W. Butzer, *Geomorphology from the Earth* (New York: Harper and Row, Publisher, 1976); Arthur M. Strahler and Alan H. Strahler, *Modern Physical Geography*, 2d ed. (New York: John Wiley and Sons, 1983); and William M. Marsh and Jeff Dozier, *Landscape: An Introduction to Physical Geography* (Reading, Mass.: Addison-Wesley Publishing Co., 1981). A thorough discussion of karst processes and a review of current research are found in Dale F. Ritter, *Process Geomorphology*, 2d ed. (Dubuque, Ia.: William C. Brown, 1986), pp. 445–481. A very technical book with excellent plates and figures is Laslo Jakues, *Morphogenetics of Karst Regions—Variants of Karst Evolution* (New York: John Wiley and Sons, a Halstead Press Book, 1977). The popular press has recently included short articles on karst features and the hazards associated with their development. These articles include "Florida Sinkhole: Still Nibbling Away," *Newsweek* 99 (1982): 16; Richard Phelan, "The Little Critters Leaving Devil's Sinkhole Are Like Bats Out of Hell," *Sports Illustrated* 59 (1983): 23; and "That Sinking Feeling," *Time* 117 (1981): 30. For a recent textbook see J. N. Jennings, *Karst Geomorphology* (Oxford: Basil Blackwell).

KÖPPEN SYSTEM. An empirical classification system of climate types based on easily obtainable climatic data.

The Greeks made the first attempt at climate classification (see CLIMATOLOGY) when they divided the world into three zones: torrid, temperate, and frigid (Lutgens and Tarbuck, 1986: 389). Few, if any, other attempts at classification were made until the end of the nineteenth and the beginning of the twentieth centuries. The person most responsible for ending this hiatus was Wladimir Köppen (1846–1940).

Köppen's training in natural science in general, and plant growth in particular, is seen in his first attempt at creating a climate-classification system (Köppen, 1884). In this publication he was mainly concerned with heat zones as related to temperature and latitude (see LATITUDE/LONGITUDE) and the effect of temperature on organic matter. The shortened and translated title of this work is illuminating: "The Heat Zones of the Earth According to the Effect of Heat on the Organic World." He expanded this first attempt in 1900 when he included not just temperature but climatic factors as a whole (Köppen, 1900). These first two attempts have little in common with Köppen's classification system that is used today.

The 1918 system developed by Köppen is the true precursor to the present system. Using observable meteorological data, Köppen claimed that this new system was much more clearly a climate-classification system rather than a plant/geography/climate system (Köppen, 1918). Because this system is global, many critics see it as oversimplified and less than useful at the regional level. Köppen would reply to these critics that the system was meant to be simple and global, not regional. Most of the succeeding publications by Köppen that dealt with climate classification were revisions of this work.

The next iteration of Köppen's system was published in 1923 as a textbook

(Köppen, 1923). In it he stated the case for a general system very articulately: "The classification has, here as elsewhere, the purpose of arranging in review a mass of facts overlarge and difficult to master. It is in many cases only in this way [simply] that they can be brought out clearly. The briefer the presentation is the more valuable is this aid [Köppen's system]" (p. iv).

In 1928 Köppen collaborated with another eminent climatologist, R. Geiger, to create a wall map of the climates of the earth (Köppen and Geiger, 1928). Revisions that took place between 1923 and 1928 caused confusion. Differences existed between the 1923 text and the 1928 map. Therefore, in 1931 Köppen issued new roles in textual form to coincide with the 1928 map (Köppen, 1931).

Köppen's final contribution to the classification was written in 1936 and included in the 1939 handbook edited by him and Geiger (Köppen, 1936). In this publication he reviewed his revisions and his claims for the system. He also looked at other authors' amendments to his work. He restated his purpose for the system of simple, global climate classification.

An excellent article by Arthur Wilcock reviews in detail the many publications by other authors concerned with the Köppen system (Wilcock, 1968). Wilcock classified the writings about Köppen and the system into five categories: applications, clarifications, modifications, amplifications, and rejections of the system.

The works that fell into the applications area were done by authors who may have had criticisms of Köppen's work but used Köppen's rules anyway. Even relatively recently we can see the use of these rules with little modification (Doerr and Sutherland, 1964). The clarifications came about due primarily to Köppen himself. As stated by Wilcock (1968: 14), "successive versions published by Köppen himself were revised, sometimes incompletely. Thus, even if we decide to accept his own last publication as the final authority we encounter inconsistency." Many of the clarifications involved the definition of Köppen's B or dry climates (Bailey, 1948). Dozens of modifications of the system took place, especially for regional studies, to give more precise limits to climate types. Several of these modified systems were created for general use also (Trewartha, 1943). There is always a desire to map in more detail, and thus the original system has been amplified greatly by some authors. H. P. Bailey (1962) and J. A. Shear (1964) have added categories to Köppen's system and split some original categories into several classes. The articles that promoted total rejection of Köppen's system were less widespread but more vehement. Most who rejected the system simply did not use it; some openly attacked it. C. W. Thornthwaite (1943), for example, undertook to expose the weaknesses of the Köppen system. His paper was well documented and precise, but it did not take into account the total of Köppen's own aims and criticisms of the scheme.

Köppen's system is still in wide pedagogical use in geography today. It is simple, understandable, and useful. As a teaching tool for climate classification, it will be used for years to come.

References

Bailey, H. P. 1948. "Proposal for a Modification of Köppen's Definitions of the Dry Climates." *Yearbook, Association of Pacific Coast Geographers* 10: 33–38.

———. 1962. "Some Remarks on Köppen's Definitions of Climatic Types and Their Mapped Representations." *Geographical Review* 52: 444–447.

Doerr, A. H., and Sutherland, S. M. 1964. "Variations in Oklahoma's Climate as Depicted by the Köppen and Early Thornthwaite Classifications." *Journal of Geography* 63: 62–66.

Köppen, W. 1884. "Die Warmezonen der Erde, nach der Dauer der Heissen, Gemassigten und Kalten Zeit und Nach der Wirkung der Warme auf die Organische Welt Betrachtet." *Meteorologische Zeitschrift* 1: 215–226. This is the first attempt by Köppen at climate classification.

———. 1900. "Versuch einer Klassifikation der Klimate, Vorzugweise nach ihren Beziehungen zur Pflanzenwelt." *Geographische Zeitschrift* 6: 593–611, 657–679.

———. 1918. "Klassifikation der Klimate nach Temperatur, Niederschlag und Jahreslauf." *Petermann's Geographische Mitteilungen* 64: 193–203, 243–248. This was Köppen's precursor to the current system now widely in use.

———. 1923. *Die Klimate der Erde. Grundriss der Klimakunde*. Berlin and Leipzig: W. de Gruyter. The first large text dealing with his climate-classification system.

———. 1931. *Grundriss der Klimakunde*. Berlin and Leipzig: W. de Gruyter. This is the second edition of the 1923 book with numerous revisions to accompany the 1928 map.

———. 1936. "Das Geographischen System der Klimate." In Köppen, W., and Geiger, R., eds. (1939), *Handbuch der Klimatologie*. Berlin: Borntraeger, pp. 1–44. This was Köppen's final work on his classification system.

Köppen, W., and Geiger, R. 1928. *Klimakarte der Erde*. Gotha: Justus Perthes. This is the world climate-classification map.

Lutgens, F. K., and Tarbuck, E. J. 1986. *The Atmosphere: An Introduction to Meteorology*. 3d ed. Englewood Cliffs, N.J.: Prentice-Hall.

Shear, J. A. 1964. "The Polar Marine Climate." *Annals of the Association of American Geographers* 54: 310–317.

Thornthwaite, C. W. 1943. "Problems in the Classification of Climates." *Geographical Review* 33: 232–255. This is a very comprehensive critique of the Köppen system.

Trewartha, G. T. 1943. *An Introduction to Weather and Climate*. 2d ed. New York: McGraw-Hill Book Co.

Sources of Additional Information

For good discussions of the mechanics of Köppen's classification system see A. N. Strahler and A. H. Strahler, *Modern Physical Geography*, 3d ed. (New York: John Wiley and Sons, 1983); J. F. Griffiths and D. M. Driscoll, *Survey of Climatology* (Columbus, Ohio: Charles E. Merrill Publishing Co., 1982; J. E. Oliver, *Climate and Man's Environment: An Introduction to Applied Climatology* (New York: John Wiley and Sons, 1973) and Arthur A. Wilcock,"Köppen After Fifty Years," *Annals of the Association of American Geographers* 58(1): 12–28. For several publications advocating Köppen's approach see P. E. James, *A Geography of Man* (Boston: Ginn and Co., 1949); F. M. Exner and R. Suring, "W. Köppen-zum 80," *Meteorologische Zeitschrift* 43 (1926):

321; and R. Geiger and W. Pohl, *Revision of the Köppen-Geiger Klimakarte de Erde* (Darmstadt: Justus Perthes, 1953), Map. Thornthwaite's final assault on Köppen's system is seen in C. W. Thornthwaite, "The Task Ahead," *Annals, Association of American Geographers* 51 (1961): 345–356.

L

LAND AND SEA BREEZES. Diurnal winds set in motion by differences in atmospheric pressures that occur between land and water as the land heats up more rapidly than water during the daytime and cools more rapidly than the water at night.

The phenomena of land and sea breezes have been recognized for centuries, and use of the concept in English can be traced at least to the seventeenth century, according to Murray et al. (1933). During the twentieth century, meteorologists and climatologists have developed a considerable interest in land and sea breezes. The work of E. G. Bilham (1934) was among the early studies of sea breezes as climatic factors, and conceptual and theoretical ideas about land and sea breezes were suggested by H. Arakawa and M. Utsugi (1937). Following World War II several new studies of sea breezes appeared, including a lengthy discussion in W. Donn (1951).

The sea breeze develops during the daytime as the land heats up and pressure over the land drops, thus drawing air in from the nearby ocean surface. At night the land cools faster than the ocean, thus setting up higher pressure over the land than over the water, with a consequent flow of air seaward. As with other winds, these winds are named according to the direction from which they are coming. Thus the sea breeze is moving onshore and the land breeze is moving offshore.

Sea breezes and land breezes are important in local weather conditions in many localities. R. G. Barry and R. J. Chorley (1970: 245) discussed the influence of land and sea breezes in the distribution and timing of rainfall in the tropics, noting that "the effects of land and sea breezes . . . greatly complicate the rainfall pattern by their interactions with the low-level southwesterly monsoon current." E. Willard Miller (1985: 66) commented that "Brighton in southern England is noted for its cool sea breezes late in the afternoon and its warm land breezes late in the evening."

References

Arakawa, H., and Utsugi, M. 1937. "Theoretical Investigation on Land and Sea Breezes." *Geophysical Magazine* 11: 97–104. One of the first conceptual and theoretical studies of land and sea breezes.

Barry, R. G., and Chorley, R. J. 1970. *Atmosphere, Weather, and Climate.* New York: Holt, Rinehart and Winston. Although now somewhat dated, this book still offers a good exposition of the basics of climate and weather.

Bilham, E. G. 1934. "The Sea Breeze as a Climatic Factor." *Journal of Medicine* (London) 42: 40–50. An early study of how the sea breeze is related to other aspects of climate.

Donn, W. 1951. *Meteorology with Marine Applications.* New York: McGraw-Hill Book Co. An early text that included a good discussion of land and sea breezes.

Fairbridge, Rhodes W. 1967. "Sea and Land Breezes." In Rhodes W. Fairbridge, ed., *The Encyclopedia of Atmospheric Sciences and Astrogeology.* Encyclopedia of Earth Sciences Series. Vol. 2, New York: Reinhold Publishing Co., pp. 857–858.

Miller, E. Willard. 1985. *Physical Geography: Earth Systems and Human Interactions.* Columbus, Ohio: Charles E. Merrill Publishing Co. A sound introduction to physical geography.

Murray, James A. H., et al. eds. 1933. *The Oxford English Dictionary.* Oxford: Clarendon Press.

Sources of Additional Information

Sea and land breezes are discussed briefly in most introductory physical geography texts including the following: William M. Marsh, *Earthscape: A Physical Geography* (New York: John Wiley and Sons, 1987), and Arthur N. Strahler and Alan H. Strahler, *Modern Physical Geography* (New York: John Wiley and Sons, 1987). Meteorology texts such as Joe R. Eagleman, *Meteorology: The Atmosphere in Action* (New York: D. Van Nostrand Co., 1980), also discuss land and sea breezes, as do climatology texts such as Paul E. Lydolph, *The Climate of the Earth* (Totowa, N.J.: Rowman and Allanheld, Publishers, 1985). The development of land and sea breezes is also discussed in L. Hasse and F. Dobson, *Introductory Physics of the Atmosphere and Ocean* (Boston: D. Reidel Publishing Co., 1983).

LANDFORM. The configuration and nature of a specific portion of the earth's land surface.

Landforms are the surface expressions of a portion of the earth at a given time, and what we see represents only part of what is actually there, because landforms evolve and change with time. The ancient Greeks were well aware of landforms, as is apparent in the writings of Herodotus and Strabo, although their scientific study did not progress much until the fifteenth and sixteenth centuries with the work of Leonardo da Vinci and Agricola (Mather and Mason, 1939).

During the nineteenth century the science of geology emerged as a subject separate from "natural philosophy" and gradually geomorphology, the study of landforms, surfaced as a distinct field within geology. Distinctions developed between landscapes and landforms, and as Arthur Bloom stated, "Each element of the landscape that can be observed in its entirety, and has consistence of form or regular change of form, is defined as a *landform*" (1978: 3). Earlier A. D. Howard and L. E. Spock (1940) noted that decisions about landform definitions were somewhat subjective, a problem that remains but does not hinder progress in geomorphology.

The bases for landform studies were laid during the nineteenth century by John Playfair (1802) and Charles Lyell (1830) among others, and the work of William Morris Davis (1899, 1902, 1905) led to much of the study of landform evolution that occurred during the earlier part of the twentieth century. J. T. Hack (1960) provided an alternative way of interpreting landforms by focusing on equilibrium conditions as opposed to the more direct idea of a cycle of erosion. More about the evolution of these two views of landform interpretation can be found under STREAM.

Howard and Spock (1940) were the first to review landform-classification methods and problems, and they discussed six methods that had been used. They then offered a genetic landform-classification schema that considered the geomorphic process and resulting reduction and residual forms. Arthur Strahler (1946) provided a somewhat different classification system, and it is still used widely, sometimes with minor modifications. Rhodes Fairbridge (1968) provided an overview of other landform-classification systems.

Arthur Strahler and Alan Strahler (1987: 243) suggested that "the configuration of continental surfaces reflects the balance of power, so to speak, between internal earth forces, acting through volcanic and tectonic processes, and external forces, acting through the agents of denudation." They subsequently suggested that landforms fall into two broad categories, initial and sequential. *Initial landforms* are those produced by volcanic and tectonic activity, whereas *sequential landforms* are those produced over a period of geologic time as the initial landforms are shaped by weathering, mass wasting, running water, glaciers, wind, and waves. "Any landscape," wrote Strahler and Strahler, "is really nothing more than the existing stage in a great contest" (1987: 243). That contest is between those internal forces that force the crust upward and those external forces that wear it down once it is exposed to erosion.

References

Bloom, Arthur L. 1978. *Geomorphology: A Systematic Analysis of Late Cenozoic Landforms*. Englewood Cliffs, N.J.: Prentice-Hall. One of several very good texts on geomorphology.

Davis, William Morris. 1899. "The Geographical Cycle." *Geographical Journal* 14: 481–504. The classic statement on the cycle of erosion.

———. 1902. "Base Level, Grade, and Peneplain." *Journal of Geology* 10: 77–111. Extends and clarifies his earlier work on the cycle of erosion.

———. 1905. "The Geographical Cycle in an Arid Climate." *Journal of Geology* 13: 381–407. Carries the idea of an erosion cycle into arid environments.

Fairbridge, Rhodes W., ed. 1968. *The Encyclopedia of Geomorphology*. Encyclopedia of Earth Sciences Series. Vol. 3. New York: Reinhold Book Corp.

Hack, J. T. 1960. "Interpretation of Erosional Topography in Humid Temperate Regions." *American Journal of Science* 258-A: 80–97. Focused attention on a view of landform interpretation that differed from that offered by William Morris Davis.

Howard, A. D., and Spock, L. E. 1940. "Classification of Landforms." *Journal of Geomorphology* 3: 332–345. A thoughtful presentation of landform classification and associated problems.

Lyell, Charles. 1830. *Principles of Geology: Being an Attempt to Explain the Former Changes of the Earth's Surface by Reference to Causes now in Operation*. Vol. I. London: J. Murray. A classic of nineteenth-century geological literature.

Mather, Kirtley F., and Mason, Shirley L. 1939. *A Source Book in Geology*. New York: McGraw-Hill Book Co. An excellent selection of writings from early geological studies.

Playfair, John. 1802. *Illustrations of the Huttonian Theory of the Earth*. London: Cadell and Davies. A major early work in geology that presented the ideas of James Hutton, an eighteenth-century naturalist who first recognized what is now called the principle of uniformitarianism, the idea that all geological phenomena can be explained as a result of existing forces operating over long periods of geologic time.

Strahler, Arthur N. 1946. "Geomorphic Terminology and Classification of Land Masses." *Journal of Geology* 54: 32–42. A useful classification of landforms that is still widely used today in physical geography.

Strahler, Arthur N., and Strahler, Alan H. 1987. *Modern Physical Geography*. New York: John Wiley and Sons. Perhaps the best of the physical geography books currently on the market and an excellent source for anyone interested in physical geography.

Sources of Additional Information

Landforms comprise a major portion of most physical geography texts, and the following provide an excellent discussion of landforms: Tom L. McKnight, *Physical Geography: A Landscape Appreciation* (Englewood Cliffs, N.J.: Prentice-Hall, 1984); William M. Marsh, *Earthscape: A Physical Geography* (New York: John Wiley and Sons, 1987); and E. Willard Miller, *Physical Geography: Earth Systems and Human Interactions* (Columbus, Ohio: Charles E. Merrill Publishing Co., 1985). Thorough discussions of landforms can be found in introductory geology texts and in some earth science texts as well, and the following would be useful for those interested in landforms: Edgar W. Spencer, *Physical Geology* (Reading, Mass.: Addison-Wesley Publishing Co., 1983); Richard Foster Flint, *The Earth and Its History* (New York: W. W. Norton and Co., 1973); and Arthur N. Strahler, *Physical Geology* (New York: Harper and Row, Publishers, 1981). Those who seek to pursue landforms in more detail should go to an excellent geomorphology text, such as the following: R. J. Chorley, S. A. Schumm, and

D. E. Sugden, *Geomorphology*, (New York: Methuen & Co., 1985); W. D. Thornburg, *Principles of Geomorphology*, 2d ed. (New York: John Wiley and Sons, 1969); H. F. Garner, *The Origin of Landscapes: A Synthesis of Geomorphology* (New York: Oxford University Press, 1974); Karl W. Butzer, *Geomorphology from the Earth* (New York: Harper and Row, Publishers, 1976); and R. J. Small, *The Study of Landforms: A Textbook of Geomorphology* (Cambridge: Cambridge University Press, 1972). Studies of specific landforms, and of landforms in specific environments, can be found in the following: M. F. Thomas, *Tropical Geomorphology* (London: Macmillan, 1974); E. Douglas, *Humid Landforms* (Cambridge, Mass.: The M.I.T. Press, 1977); J. A. Mabbutt, *Desert Landforms* (Cambridge, Mass.: The M.I.T. Press, 1977); S. A. Schumm, *The Fluvial System* (New York: John Wiley and Sons, 1977); C.A.M. King, *Beaches and Coasts*, 2d ed. (New York: St. Martin's Press, 1972); Richard Foster Flint, *Glacial and Quaternary Geology* (New York: John Wiley and Sons, 1971; and D. E. Sugden and B. S. John, *Glaciers and Landscape: A Geomorphological Approach* (New York: Halsted Press, John Wiley and Sons, 1976).

LANDSAT. See REMOTE SENSING.

LANDSLIDE. See MASS WASTING.

LAPSE RATE. A measure of the rate of change in air temperature with respect to elevation.

There is not one lapse rate but, rather, a family of lapse rates. Each of these different lapse rates, however, measures the rate of change of temperature with elevation. Differences relate to what an air mass is doing.

The easiest lapse rate to understand, and the one that is usually meant when the term *lapse rate* is used without any further qualification, is the *environmental lapse rate*. Within the TROPOSPHERE air temperature decreases regularly with elevation at a rate of 3.5°F per 1,000 feet (6.4°C per 1,000 meters), providing that the air itself is not moving upward or downward.

If a parcel of air is moving, the applicable lapse rate must take into consideration ADIABATIC HEATING AND COOLING. If a parcel of air is moving, it will experience not only the change in temperature measured by the environmental lapse rate but also a further change in temperature created by expansion or compression of the parcel of air itself, as it rises or descends, respectively. Thus two adiabatic lapse rates exist, the dry adiabatic lapse rate and the wet adiabatic lapse rate, although their use is not simply one for rising air and the other for descending air. The picture is somewhat more complicated.

The *dry adiabatic lapse rate*, which applies to all descending air masses and to rising air masses in which condensation is not taking place, is 5.5°F per 1,000 feet (10°C per 1,000 meters). A mathematical means of calculating the dry adiabatic lapse rate for any atmosphere can be found in the work of H. R. Byers (1959). Richard Rommer and Rhodes Fairbridge (1967) presented a mathematical discussion of lapse rates in general.

The *wet adiabatic lapse rate* applies to rising air masses when condensation is occurring. *Condensation* is the process in which water vapor in the atmosphere is converted to a liquid, a process first defined as early as 1614 according to Murray et al. (1933). When condensation occurs, heat is given off, thus slowing the rate of cooling in a rising air mass below what it would be otherwise. The source of the heat is latent heat, stored during evaporation. *Evaporation* is a process in which liquid water is converted to water vapor, and this process requires heat. That heat is then stored within the water vapor in the form of latent heat. According to Murray et al. (1933), the term *evaporation* was in use no later than 1718. Condensation begins to occur when the air mass has been cooled to the *dew point*, the temperature at which an air mass is saturated, or contains all of the water vapor that it can hold. Any further cooling results in condensation. Unlike the environmental lapse rate and the dry adiabatic lapse rate, the wet adiabatic lapse rate varies, depending on the amount of moisture that is present in an air mass when it cools below the dew point. The wet adiabatic rate normally falls within the range of 2°–3°F per 1,000 feet (3°–6°C per 1,000 meters).

As Tom McKnight noted, "One of the most significant facts in the study of physical geography is that the only way in which large masses of air can be cooled to the dew point is by expansion of rising air" (1984: 102). Thus an understanding of adiabatic processes and their related lapse rates is essential for understanding precipitation and the mechanisms that cause it. One last addition is necessary, the *dew point lapse rate*. The dew point temperature falls with increasing elevation at a rate of 1°F per 1,000 feet (2°C per 1,000 meters). As Arthur Strahler and Alan Strahler noted, once a rising air mass reaches the dew point temperature, "further rise results in condensation of water vapor into minute liquid particles, and so a cloud is produced" (1984: 95).

All of these lapse rates are important in order to gain an understanding of atmospheric dynamics and weather. Under certain circumstances, usually for short periods, still other variations of these lapse rates exist. Joe Eagleman (1980), for example, used the term *negative lapse rate* to describe conditions that exist when there is a temperature inversion and the term *isothermal lapse rate* to describe a situation in which temperature remains constant with changes in elevation. Furthermore, there is a *superadiabatic lapse rate*, as noted by R. G. Barry and R. J. Chorley, who stated that, "close to the surface the vertical temperature gradient sometimes greatly exceeds the dry adiabatic lapse rate, that is, it is superadiabatic" (1970: 65).

References

Barry, R. G., and Chorley, R. J. 1970. *Atmosphere, Climate, and Weather*. New York: Holt, Rinehart and Winston. Contains an excellent discussion of lapse rates and related phenomena such as stability and instability.

Byers, H. R. 1959. *General Meteorology*. 3d ed. New York: McGraw-Hill Book Co. One of several texts of that era.

Eagleman, Joe R. 1980. *Meteorology: The Atmosphere in Action*. New York: D. Van

Nostrand Co. One of the best current meteorology texts and an excellent reference on most facets of meteorology.

McKnight, Tom L. 1984. *Physical Geography: A Landscape Appreciation*. Englewood Cliffs, N.J.: Prentice-Hall. A well-accepted new physical geography text.

Murray, James A. H., et al., eds. 1933. *The Oxford English Dictionary*. Oxford: Clarendon Press.

Rommer, Richard J., and Fairbridge, Rhodes W. 1967. "Lapse Rate." In Rhodes W. Fairbridge, ed., *The Encyclopedia of Atmospheric Sciences and Astrogeology*. Encyclopedia of Earth Sciences Series. Vol. 2, New York: Reinhold Publishing Co., pp. 527–531. Provides an excellent summary of the mathematics of lapse rates.

Strahler, Arthur N., and Strahler, Alan H. 1984. *Elements of Physical Geography*. 3d ed. New York: John Wiley and Sons. Remains one of the major texts in physical geography.

Sources of Additional Information

All of the terms mentioned in this discussion of lapse rates are discussed in most introductory physical geography texts, so in addition to those listed under the references above one could look at E. Willard Miller, *Physical Geography: Earth Systems and Human Interactions* (Columbus, Ohio: Charles E. Merrill Publishing Co., 1985). More detailed discussions can be found in any of the following: John E. Oliver and John J. Hidore, *Climatology: An Introduction* (Columbus, Ohio: Charles E. Merrill Publishing Co., 1984); Paul E. Lydolph, *The Climate of the Earth* (Totowa, N.J.: Rowman and Allanheld, Publishers, 1985); Glenn T. Trewartha and Lyle H. Horn, *An Introduction to Climate* (New York: McGraw-Hill Book Co., 1980); and A. S. Monin, *An Introduction to the Theory of Climate* (Boston: D. Reidel Publishing Co., 1986).

LATENT HEAT. The amount of energy used or released when a body changes phase. During this phase change the body's temperature remains constant.

Latent heat is a concept critical to the mechanisms involved in weather phenomena. It deals with the changes of state of water from gas to liquid to solid and vice versa. The explanations in H. Riehl (1978: 77) and R. J. Rommer (1967: 530) provide a good basic understanding of latent heat. At one atmosphere of air pressure and 15°C, it takes 79.71 calories/gram to melt ice. This energy is not turned into sensible heat but is added to the molecular structure of the ice to overcome the holding forces of the crystal form. This is called the latent heat of fusion. Similarly, it takes 595.9°C/gram to vaporize water at this same pressure and temperature. This energy is required to give the water molecules enough energy to vaporize without a temperature change and is called the latent heat of vaporization. It is also possible to go directly from the solid to the vapor, which requires the total energy in the latent heat of fusion and the latent heat of vaporization. This is the latent heat of sublimation. When water goes to the solid state from the liquid and to the liquid state from the vapor, the respective latent heats are released back into the atmosphere as sensible heat.

The importance of latent heat in meteorology was known but little understood until the twentieth century. Pettersson (1904, 1907) realized the importance of the latent heat of fusion for oceanic circulation and, therefore, for ATMOSPHERIC CIRCULATION. Physical climatologists began to look in earnest at the physical characteristics of latent heat in terms of overall climatic influence in the 1930s and 1940s (e.g., Brunt, 1941). J. London (1957) calculated that latent heat transfer represented 18.5 percent of the solar radiation impact of the earth and that this latent heat was eventually released as sensible heat during PRECIPITATION events.

Two scales of research have been ongoing concerning latent heat in meteorology and CLIMATOLOGY. The first scale is localized storm events, particularly intense convective storms as single entities and as grouped phenomena such as hurricanes (see TROPICAL CYCLONE). As early as 1914 Sir Napier Shaw realized the importance of the latent heat of vaporization and heat transfers in THUNDERSTORM development. When a large storm system like a hurricane is produced, tremendous amounts of energy are stored, moved, and released due to latent heat processes. J. Simpson (1967: 95) estimated that the average amount of water vapor moving through a hurricane is 15 billion tons a day with half of this amount falling as rain. The latent heat released as sensible heat from this is equivalent to 400 twenty megaton hydrogen bombs a day.

The second scale of study concerning latent heat is the global heat balance. Because poleward of approximately 40 degrees north and south latitude there is a radiation deficit and equatorward of these latitudes a radiation surplus, a mechanism for moving large amounts of energy is needed. According to W. D. Sellers, it takes both movements of sensible and latent heat.

> In order to keep the poles from getting colder and the tropics from getting warmer, energy, therefore, must be transferred meridionally from lower to higher latitude. This horizontal heat exchange, which is induced partly by differential heating of continents and oceans, is carried out mainly by the poleward transfer of sensible heat by the atmospheric circulation and ocean currents and by the release of latent heat through condensation of water vapor carried poleward in the atmosphere. (1965: 67)

Most of the research on global scale latent heat effects deals with the patterns of circulation involved in heat transfers from oceans and seas. Examples of this work are seen in studies of J. S. Winston (1955) for the Gulf of Alaska, S. Manabe (1957, 1958) for the Sea of Japan, and T. Laevastu (1965) for the North Pacific. The development of cyclones as studied using synoptic weather charts indicates that release of latent heat is pronounced in these areas (Sawyer, 1965).

The tremendous contribution of latent heat to local, regional, and global climates is undeniable. Future work with latent heat will undoubtedly be concerned with climate anomalies that alter generally accepted synoptic weather patterns on a global scale. This work will be intertwined with such phenomena as El Niño. An excellent review of this global, cyclical event is seen in the work of B. Yarnal (1985).

References

Brunt, Sir David. 1941. *Physical and Dynamical Meteorology*. 2d ed. Cambridge: Cambridge University Press.

Laevastu, T. 1965. "Daily Heat Exchange in the North Pacific." *Societas Scientiarum Fennica, Commission on Physics and Mathematics* 31: 3–53. Showed the value of latent heat transfers over cold ocean waters.

London, J. 1957. *A Study of the Atmospheric Heat Balance*. Final Report, no. AF19(122)–165. New York: New York University. Calculated the contribution of latent heat to global heat transfers.

Manabe, S. 1957. "On the Modification of Air Mass over the Japan Sea When the Outburst of Cold Air Predominates." *Journal of the Meteorological Society of Japan* 35: 311–326.

———. 1958. "On the Estimation of Energy Exchange between the Japan Sea and the Atmosphere during Winter Based upon the Energy Budget of Both the Atmosphere and the Sea." *Journal of the Meteorological Society of Japan* 36: 123–133. Deals with latent heat transfers over cold water areas.

Pettersson, O. 1904. "On the Influence of Ice Melting upon Ocean Circulation, Part I." *Geographical Journal* 24: 285–333. Showed the importance of latent heat to ocean characteristics.

———. 1907. "On the Influence of Ice-Melting upon Ocean Circulation, Part II." *Geographical Journal* 30: 273–303.

Riehl, H. 1978. *Introduction to the Atmosphere*. 3d ed. New York: McGraw-Hill Book Co. Gives a good explanation of the process of latent heat.

Rommer, R. J. 1967. "Latent Heat." In R. W. Fairbridge, ed., *The Encyclopedia of Atmospheric Sciences and Astrogeology*. Encyclopedia of Earth Science Series. Vol. 2. New York: Reinhold Publishing Co., pp. 530–531. A good explanation of the latent heat phenomenon.

Sawyer, J. S. 1965. "Notes on Possible Physical Causes of Long-Term Weather Anomalies." *World Meteorological Organization Technical Note*, no. 66: 227–248.

Sellers, W. D. 1965. *Physical Climatology*. Chicago: University of Chicago Press. Discusses the physical science aspects of latent heat.

Shaw, N. 1914. "Principia Atmospherica: A Study of the Circulation of the Atmosphere." In *Proceedings of the Royal Society, Edinburgh* 34: 166–186. An early discussion of atmospheric circulation including latent heat.

Simpson, J. 1967. "An Experimental Approach to Cumulus Clouds and Hurricanes." *Weather* 22: 95–114. Calculated latent heat flux in hurricanes.

Winston, J. S. 1955. "Physical Aspects of Rapid Cyclogenesis in the Gulf of Alaska." *Tellus* 7: 481–500. Studied latent heat processes in the Gulf of Alaska.

Yarnal, B. 1985. "Extratropical Teleconnections with El Niño/Southern Oscillation (ENSO) Events." *Progress in Physical Geography* 9: 315–352. A good review article on global circulation patterns including latent heat processes.

Sources of Additional Information

For more extensive definitions of latent heat see R. W. Fairbridge, *The Encyclopedia of Atmospheric Sciences and Astrogeology*, Encyclopedia of Earth Science Series, vol. 2 (New York: Reinhold Publishing Co., 1967); and J. Whitlow, *Dictionary of*

Physical Geography (New York: Penguin Books, 1984). For good explanations of the role of latent heat see F. K. Lutgens and E. J. Tarbuck, *The Atmosphere: An Introduction to Meteorology*, 3d ed. (Englewood Cliffs, N.J.: Prentice-Hall, 1986).

LATITUDE/LONGITUDE. 1. The angle (measured in degrees, minutes, and seconds) between the imaginary line to the center of the earth from any point on the earth and the equator (latitude). 2. The angle measured from any point on the earth to the Greenwich meridian (longitude).

Latitude is the standard measure north and south from the equator. It ranges from 0° at the equator to 90° at the poles (90°N at the North Pole, 90°S at the South Pole). The lines defining latitude are the parallels that run east-west and consist of concentric circles decreasing in size toward the poles. The only great circle parallel is the equator. A great circle is the arc formed on the surface of the earth by a plane passing through the center of the earth. The path along a great circle that passes through two points on the earth's surface is the shortest distance between these two points (McDonnell, 1979: 10).

Longitude is the standard measure east and west from the Greenwich (prime) meridian located at 0° longitude. The maximum measure of longitude is 180°E/180°W from Greenwich. The lines defining longitude are the meridians that run north-south and meet at the poles. Each meridian is one-half of a great circle.

The Greeks are credited with creating the coordinate system for a sphere that we use today. Taking the work done by the Greeks, Claudius Ptolemy (A.D. 90–168) created a framework upon which the world grid system was defined and coined the terms *latitude* and *longitude* (Boorstin, 1983: 98). During the European Dark Ages, Muslim scientists revised and perfected the Ptolemaic figures for latitude and longitude because they needed precise measurements of their position to site mosques facing Mecca (Boorstin, 1983: 219).

After the Great Interruption, Christian Europe began the age of exploration in which latitude and longitude were vital. The Portuguese, under the influence of Prince Henry the Navigator (1394–1460), developed a relatively easy method of latitude determination but failed to create a usable longitude measuring system. Spain and England also began their explorations without a sound longitude measuring method. The theory behind latitude measurement was well established, and the only new ideas were in instrumentation. L. A. Brown (1979: 180–207) described in detail the progress of instrumentation from astrolabe through the sextant. His detailed description covers the years from the early Middle Ages to the nineteenth century.

The real key to exploration, however, lay in the development of a useful method for determining longitude. There were many proposed methods of which some were theoretically possible and others were pure fancy. The one accepted method for longitude determination was through comparison of local solar noontime to the time at the prime meridian. Therefore, the instrument most essential for longitude calculation was an accurate and dependable clock. Throughout the seventeenth and eighteenth centuries, pendulum clocks were the standard, but they could not be used on ships. The spring-powered clock was developed over many years and was finally perfected between 1728 and 1770 by John Harrison, an Englishman who created five models of very accurate, dependable, and large spring-powered clocks (Brown, 1979: 230–240).

At this time there was no one recognized reference meridian established to systematize longitude readings throughout the world. In October 1884 the International Meridian Conference was convened in Greenwich, England, and produced a generally accepted "prime" meridian. This meridian of 0° east or west ran north and south through the Greenwich Observatory and is, even today, the standard 0° meridian for most of the world (Tehan, 1979: 3–4).

Until recently, the measurement of latitude and longitude had not changed significantly from the late nineteenth-century methods. Today, with the study of geodesy as the exacting science of precise earth measurements, we see more precise and varied kinds of measurements. With sophisticated equipment we have determined at least three sets of geographic coordinates: (1) the astronautical coordinates, which use the direction of gravity for the reference direction; (2) the geodetic coordinates, which use the sea level geoid as the reference surface; and (3) the geocentric coordinates, which use the reference direction that actually passes through the center of mass of the earth (Richardus and Adler, 1972).

Future work will undoubtedly concentrate on geographic measurements from satellites. We can also expect a growth in the science and technology of measuring extraterrestrial bodies for the purposes of creating coordinate systems.

References

Boorstin, D. J. 1983. *The Discoverers*. New York: Random House. An articulate and extensive volume on discoveries throughout history that details much of the information on latitude and longitude.

Brown, L. A. 1979. *The Story of Maps*. New York: Dover Publications. A thorough history of maps and mapping that contains two extensive chapters on latitude and longitude.

McDonnell, P. W., Jr. 1979. *Introduction to Map Projections*. New York: Marcel Dekker.

Richardus, P., and Adler, R. K. 1972. *Map Projections*. London: North-Holland Publishing Co. A very technical reference source that details precise methods of latitude/longitude determination.

Tehan, R. 1979. "Greenwich Mean Time." *Oceans* 12: 3–4.

Sources of Additional Information

Two classic works on latitude and longitude related to navigation include G. Forbes, *The Birth of Scientific Navigation* (Greenwich, Eng.: National Maritime Museum, 1974); and David Watkins Waters, *The Art of Navigation in England in Elizabethan and Early Stuart Times* (London: Hollis and Carter, 1958). An example of studies of coordinate systems and extraterrestrial bodies is M. E. Davies and R. M. Batson, "Surface Coordinates and Cartography of Mercury," *Journal of Geophysical Research* 80 (1975): 2417–

2430. References on coordinate systems and geodetic research include the following: O. S. Adams, *Latitude Developments Connected with Geodesy and Cartography with Tables*, Special Publication, no. 130. (Washington, D.C.: U.S. Coast and Geodetic Survey, 1927); G. Bamford, *Geodesy* (Oxford: Clarendon Press, 1971); J. Dozier, *Improved Algorithm for Calculation of UTM and Geodetic Coordinates*, Technical Report NESS, no. 81 (Washington, D.C.: National Oceanic and Atmospheric Administration, 1980); and D. P. Rubincam, "Latitude and Longitude from Van der Grinten Grid Coordinates," *American Cartographer* 8 (1981): 177–180.

LAW OF SUPERPOSITION. See GEOLOGIC TIME SCALE.

LEACHING. 1. The downward movement of dissolved substances in percolating water within soil. 2. The removal by percolating water of dissolved substances from soil. 3. The depletion of dissolved substances from certain soil horizons (e.g., E, A2, A3).

Leaching is a complex concept with no universally accepted meaning. T. R. Paton (1978: 17), for example, believed that leaching is only one of three major processes in his concept of epimorphism, or the adjustment by the SOIL to surface environmental conditions. He thought that the unearthing of primary MINERALS, the leaching of simpler materials acted on by surface chemical forces, and the creation of new minerals in equilibrium with surface conditions are all part of a continuum process of epimorphism. Whereas S. W. Buol, F. D. Hole, and R. J. McCracken (1980: 96) defined *leaching* as the process of removal of substances from the entire solum, C. W. Finkl (1979: 260) defined it as merely the downward movement of substances dissolved in percolating water with no mention of complete removal of these substances from the solum. Others such as P. W. Birkeland (1984: 6–7) referred to leached-soil horizons where removal of substances has occurred without specific reference to where the substances have been transported.

Historically, the study of leaching has concentrated on the determination of the relative mobility of colloids. F. W. Clarke (1908) and B. B. Polynov (1937) determined the mobility of soil constituents (ions) as compared to the abundance of these ions in the earth's crust. In his seminal work on soil-forming factors, Hans Jenny (1941) devised an index to compare leaching in various soils. This "leaching factor" compares the ratios of given oxides with silicon dioxide in weathered horizons to the ratios of the same oxide in the parent material. Another area of research has been on ion bonding in soils according to ion size and valency ratios (Wickmann, 1944). Others have looked at ion solubility and pH/Eh levels (Hem and Cropper, 1959; Okamoto et al., 1957; Davis, 1964). More recent work has focused on leaching and environmental conditions (Simonson, 1970) and the factors that determine leaching rates.

The specific mechanisms of leaching are very complex. As C. W. Finkl stated:

[The leaching mechanism's] complexity can be indicated by a partial list of important factors: (1) nature of the soil medium, (2) presence and effects of microorganisms, (3) formation of complex ions, (4) membrane effects and exchange phenomena associated with silicate clay minerals, (5) carbon dioxide pressure, and (6) porosity and hydraulic conductivity of the soil. These variables and others govern the relative mobilities of elements in soils, i.e., whether they will move, how far, and in what direction (downward or laterally). (1979: 262)

A soil characteristic that is particularly deterministic for leaching is the soil acidity. D. A. Andrews-Jones (1968) showed that leaching is affected greatly by whether the pH is acidic, neutral, and alkaline.

The measurement of leaching losses has also been a rich research area during the past several decades. Measures of leaching from the analysis of drainage waters is an indirect method used by N. M. Johnson et al. (1968). Direct measures using controlled lysimeter studies have been done by J. S. Joffe (1940) and R. S. Stauffer and R. H. Rust (1954). Of critical concern in current research is the amount of leached ions present in groundwater due to pollution from waste-disposal sites and the effects of acid rain on soils and groundwater (e.g., Burns et al., 1981; Meiwes and Khanna, 1981; Strayer et al., 1981; Kollig and Hall, 1982). With these more recent works, an ecological approach is being pursued to link all factors in the environment, including leaching rates, to environmental degradation. (See also WEATHERING.)

References

Andrews-Jones, D. A. 1968. "The Application of Geochemical Techniques to Mineral Exploration." *Colorado School of Mines Mineral Industrial Bulletin* 11: 1–31.
Birkeland, P. W. 1984. *Soils and Geomorphology.* New York: Oxford University Press. A reference that contains a tremendous amount of soils information for use in the study of several subareas of soil research including leaching.
Buol, S. W.; Hole, F. D.; and McCracken, R. J. 1980. *Soil Genesis and Classification.* 2d ed. Ames, Ia.: Iowa State University Press. An invaluable sourcebook for use in soil research.
Burns, D. A.; Galloway, J. N.; and Hendrey, G. R. 1981. "Acidification of Surface Waters in Two Areas of the Eastern United States." *Water, Air, and Soil Pollution* 16: 277–285.
Clarke, F. W. 1908. "The Data of Geochemistry." U.S. Geological Survey Bulletin, no. 330. Washington, D.C.: U.S. Geological Survey.
Davis, S. N. 1964. "Silica in Streams and Groundwater." *American Journal of Science* 262: 870–891.
Finkl, C. W., Jr. 1979. "Leaching." In R. W. Fairbridge and C. W. Finkl, Jr., *The Encyclopedia of Soil Science, Part I.* Stroudsburg, Pa.: Dowden, Hutchinson and Ross, 260–266. This is an excellent reference for the historical perspective of the study of leaching.
Hem, J. D., and Cropper, W. H. 1959. "Survey of Ferrous-Ferric Chemical Equilibria and Redox Potentials." U.S. Geological Survey Water Supply Paper, no. 1459A. Washington, D.C.: U.S. Geological Survey.

Jenny, H. 1941. *Factors of Soil Formation.* New York: McGraw-Hill Book Co. The classic work on soil formation and the environmental factors that affect it.

Joffe, J. S. 1940. "Lysimeter Studies: IV. Movement of Anions through the Profile of a Gray-Brown Podzolic Soil." *Soil Science* 50: 57–63.

Johnson, N. M.; Likens, G. E.; Bormann, F. H.; and Pierce, R. S. 1968. "Rate of Chemical Weathering of Silicate Minerals in New Hampshire." *Geochimica et Cosmochimica Acta* 32: 531–541.

Kollig, H. P., and Hall, T. L. 1982. "The Effects of Acid Perturbation on a Controlled Ecosystem." *Water, Air, and Soil Pollution* 17: 225–233.

Meiwes, K. J., and Khanna, P. K 1981. "Distribution and Cycling of Sulphur in the Vegetation of Two Forest Ecosystems in an Acid Rain Environment." *Plant and Soil* 60: 369–378.

Okamato, G.; Okura, T.; and Goto, K. 1957. "Properties of Silica in Water." *Geochimica et Cosmochimica Acta* 12: 123–132.

Paton, T. R. 1978. *The Formation of Soil Material.* London: Allen and Unwin. A very useful text with a highly chemical explanation of leaching processes.

Polynov, B. B. 1937. *The Cycle of Weathering* (trans. A. Muir). London: Murby and Co.

Simonson, R. W. 1970. "Loss of Nutrient Elements during Soil Formation." In Engelstad, O. P. ed., *Nutrient Mobility in Soils: Accumulation and Losses.* Madison, Wis.: Soil Science Society of America, pp. 21–45.

Stauffer, R. S., and Rust, R. H. 1954. "Leaching Losses, Runoff, and Percolate from Eight Illinois Soils." *Agron. Journal* 46: 207–211.

Strayer, R. F.; Lin, C. J.; and Alexander, M. O. 1981. "Effect of Simulated Acid Rain on Nitrification and Nitrogen Mineralization in Forest Soils." *Journal of Environmental Quality* 10: 547–552.

Wickmann, F. E. 1944. "Some Notes on the Geochemistry of Elements in Sedimentary Rocks." *Arkiv für Kemi.* 19B: 1–7.

Sources of Additional Information

For additional work on soil chemistry, the geochemistry of water, and weathering see H. J. Atkinson and J. R. Wright, "Chelation and the Vertical Movement of Soil Constituents," *Soil Science* 84 (1957): 1–11; R. M. Garrels and C. L. Christ, *Solutions, Minerals, and Equilibria* (New York: Harper and Row, Publishers, 1965); C. D. Ollier, *Weathering* (Edinburgh: Oliver and Boyd, 1969); H. I. Bohn, B. L. McNeal, and G. A. O'Connor, *Soil Chemistry* (New York: John Wiley and Sons, 1979); and J. I. Drever, *The Geochemistry of Natural Waters* (Englewood Cliffs, N.J.: Prentice-Hall, 1982). For a useful reference on soil classification and leaching characteristics see Soil Survey Staff, *Soil Taxonomy*, Handbook no. 436 (Washington, D.C.: U.S. Department of Agriculture, 1975). Examples of classic works in leaching of soils and nutrient mobility are J. Thorp, L. E. Strong, and E. Gamble, "Experiments in Soil Genesis—The Role of Leaching," In *Soil Science Society of America Proceedings* 21 (1957): 99–102; and O. P. Engelstad, ed., *Nutrient Mobility in Soils: Accumulation and Losses* (Madison, Wis.: Soil Science Society of America, 1970). An excellent short reference on leaching is R. W. Fairbridge and C. W. Finkl, Jr., *The Encyclopedia of Soil Science, Part I* (Stroudsburg, Pa.: Dowden, Hutchinson and Ross, 1979). Current research directions in soils and other physical geography areas can be learned in *Progress in Physical Geography*, vol. 1- , 1977-).

LIFE ZONE. See ECOSYSTEM.

LITHOSPHERE. 1. The solid or rock portion of the earth sphere. 2. The solid outer portion of the earth about 75 kilometers thick in which the tectonic plates are found. It includes continental and oceanic crust as well as the rigid uppermost part of the mantle.

Although the term *lithosphere* can refer to the solid portion of the earth excluding the liquid (hydrosphere) and gaseous (atmosphere), it is more commonly used today to describe the rigid, outer shell of the earth. This shell includes the entire crust and a portion of the upper mantle. The mantle is the region between the uppermost layer, the crust, and the earth's core. It begins somewhere between 5 and 50 kilometers below the surface of the earth and extends to about 2,900 kilometers below the surface. The upper part of the mantle, included in the lithosphere, is important because it plays an important rule in PLATE TECTONICS (Brush, 1984; Ringwood, 1975).

Another part of the upper mantle, the asthenosphere, is an easily deformed and soft layer. The asthenosphere lies directly under the lithosphere, and because it is softer and easily deformed, these rigid lithospheric plates glide or move horizontally across it (Officer and Drake, 1983).

The processes associated with lithospheric movement and development exert a dominant influence on many surface phenomena. The lithosphere is also the reservoir for almost all of the chemical elements and hence NATURAL RESOURCES. For these reasons the nature and composition of the lithosphere have been studied by the scientific community.

Although the term *lithosphere* was not used before the nineteenth century, mankind has taken an interest in the nature and origin of the earth's crust and internal properties throughout history. People in ancient times imagined that the earth's interior was a realm of darkness or filled with water or fire. Early Roman and Greek authors were fascinated by VOLCANOS and considered their function to be a natural safety valve designed to permit the escape of dangerous vapors from a volatile interior. Ancient Greeks hypothesized that marine fossils found far inland in layers of ROCK were left by retreating seas (Smith, 1981: 12–22).

Scientific, western European study of the lithosphere and its properties essentially began after 1500, apparently as a result of a need for MINERALS and precious metals for coinage. From the fifteenth to the eighteenth century, scientists studying the earth generally accepted the theory that the entire planet was hard and rigid, "holding the continents in place like rocks in cement" (*Planet We Live On*, 1976: 8).

Early in the nineteenth century geologists believed that the earth had a relatively thin crust surrounding a hot, liquid interior, therefore easily accounting for volcanos and EARTHQUAKES (Brush, 1980: 705–724). In September 1896 Emil Wiechert advanced a paper announcing a quantitative model for the earth's internal structure, assuming an iron core surrounded by a thick stony shell (Brush,

1980: 705). Debate concerning the structure of the earth continued, and at the end of the nineteenth century the weight of the scientific opinion favored a completely solid earth, with the solid earth having a gaseous center as the second choice (Brush, 1980: 707).

The theory of a solid earth remained during the early twentieth century. In 1911 Alfred Wegener proposed a theory of "wandering continents" or "continental drift," as it became known (Marvin, 1973). He claimed that a supercontinent broke up millions of years ago, and current continents drifted to where they are today. Wegener's theory also advanced the idea that the entire earth was not solid, that the crust could move. The theory was largely disregarded until after World War II.

Wegener's theory led to the current study of PLATE TECTONICS and the now accepted model for the earth's interior. The lithosphere was emphasized as a component of the earth as a result of these theories.

The basic hypothesis of plate tectonics was delivered by Jason Morgan in 1968 (Turcotte and Schubert, 1982: 6). In his hypothesis he referred to "lithospheric plates," thereby incorporating the term *lithosphere* with the theory of plate tectonics. This hypothesis also served to further define the lithosphere as an entity with its own characteristics and functions. Instead of referring to just the surface, crust, or mantle of the earth, reference to the lithosphere was now reference to the part of the earth that consisted of huge pieces that move over the asthenosphere.

Studies of the lithosphere today focus on its conductive versus convective qualities, its size, its chemical composition, and its thermal profile. A major component of the "Interunion Commission on Geodynamics Project" highlights several of these recent studies (Anderson, 1981). Of special interest is the study concerning thermal conductivity in the lithosphere. This study, done in 1977, relates the depth of the lithosphere to pressure and planet size as well as providing new information on the chemical composition of the lithosphere and axioms for an accurate dynamic model of the earth.

More recently, a study was done using heat-flow measurements to support a theory that the lithosphere thickens in proportion to its age (Brown, 1979). Another interesting study by Claude Allegre (1982) presented an analysis of the history of the continental lithosphere as recorded by ultramafic xenoliths (rock fragments foreign to the area in which they are found).

Research concerning the history and composition of the lithosphere will no doubt continue. Arie Poldervaart, a well-known earth scientist, said several years ago that "the vexing problem of the evolution of the lithosphere and its unequal distribution between continents and ocean bottoms is still far from solved." Certainly, tremendous progress has been made in this field, but many mysteries remain and many theories need further refinement and documentation.

Conclusions about the lithosphere have been largely based upon indirect evidence such as studies of rocks and minerals, seismic waves, heat flow, gravity, and magnetic fields. New technology may provide the ability actually to drill through the crust of the earth into the mantle and perhaps bring forth some well-kept secrets concerning the nature and history of the lithosphere and the interior of the earth.

References

Allegre, Claude, 1982. "History of the Continental Lithosphere Recorded by Ultramafic Xenoliths." *Nature* 296: 732–735. A recent study on the development of the continental lithosphere.

Anderson, Orson L. 1981. "A Decade of Progress in Earth's Internal Properties and Processes." *Science* 213: 76–82. A good summary of recent research.

Brown, Geoff. 1979. "Oceanic Heat Flow, Lithospheric Age, and Thickness." *Nature* 280: 198.

Brush, Stephen G. 1980. "Discovery of the Earth's Core." *American Journal of Physics* 48: 705–724.

———. 1984. "Inside the Earth." *Natural History*, February, pp. 26–34. A good discussion of the history of geologic study associated with the earth's interior.

Marvin, U. B. 1973. *Continental Drift: The Evolution of a Concept*. Washington, D.C.: Smithsonian Institute Press. A good discussion of Alfred Wegener and his continental drift theory.

Officer, C. B., and Drake, C. L. 1983. "Plate Dynamics and Isostacy in a Dynamic System." *Journal of Geophysics* 54: 1–19. A detailed discussion of the lithospheric plate-asthenosphere coupling processes.

Planet We Live On, The. 1976. New York: Harry N. Abrams. A good discussion of earth processes and lithospheric properties.

Ringwood, A. E. 1975. *Composition and Petrology of the Earth's Mantle*. New York: McGraw-Hill Book Co. A detailed analysis of the mantle is presented.

Smith, David G., ed. 1981. *The Cambridge Encyclopedia of the Earth Sciences*. Cambridge: Cambridge University Press, pp. 12–22. An excellent discussion of the history of the earth sciences.

Turcotte, Donald L., and Schubert, Gerald. 1982. *Geodynamics*. New York: John Wiley and Sons.

Sources of Additional Information

Short discussions of the concept can be found in Arthur M. Strahler, *Physical Geology* (New York: Harper and Row, Publishers, 1981); William M. Marsh and Jeff Dozier, *Landscape* (Menlo Park, Calif.: Addison-Wesley Publishing Co., 1981); and Richard Noster Flint, *The Earth and Its History* (New York: W. W. Norton and Co., 1973). For a thorough discussion of the earth's mantle see D. P. McKenzie, "The Earth's Mantle," *Scientific American*, September 1983), and Keith Bullen, "Interior of the Earth," in *McGraw-Hill Encyclopedia of Geological Sciences* (New York: McGraw-Hill Book Co., 1978). For a discussion of the asthenosphere see H. P. Bott, *The Interior of the Earth* (London: E. Arnold, 1982).

M

MAGMA. 1. Molten rock material from which igneous rocks are formed. 2. Naturally occurring rock material, primarily molten and mobile, that may contain suspended solids and/or gas phases.

Magma is the hot, molten liquid ROCK material that cools and solidifies into igneous rocks. The depth and rate at which this cooling takes place can be used to subdivide igneous rocks further into extrusive rocks that cool at or very near the surface and intrusive rocks that cool below the surface.

Attempts to define and understand the nature of magma has a long history starting with the writings of the ancient Greeks. The word *magma* is of Greek origin and means a plastic mass or a paste of solid and liquid matter (Shand, 1947: 4). People were wondering about volcanic eruptions, and in the early fourth century B.C Aristotle put forth his ideas about how magma was created. He recognized the importance of heat and believed the heat came from a fire that was produced underground (Fenton and Fenton, 1952: 8). Almost four centuries after Aristotle, further investigations into the magma associated with VOLCANOES were undertaken during Roman times by Strabo and Pliny the Elder.

It was not until the latter part of the nineteenth century that the scientific study of magma was undertaken. During this time there was much disagreement among geologists about the origin of rocks. A German mineralogist, Abraham Werner, believed that an ocean once covered the entire surface of the earth, and chemicals in the water slowly settled on the OCEAN floor where they formed granite and other types of igneous rocks (Ospovat, 1971). Werner explained the existence of volcanoes and their associated molten rock materials by arguing that the eruption of volcanic materials took place where basalt and other rocks were melted by the combustion of underlying coal seams (Hallam, 1983: 7). The scientists who believed these ideas were called Neptunists, after the Roman god of the sea Neptune.

A much different idea about the origin of rocks was held by the French geologist Nicolas Desmarest and his colleagues (von Zittel, 1901) and the Scottish physician James Hutton (1795). These men believed that igneous rocks were formed by the cooling of magma from volcanoes and were referred to as Plutonists, after Pluto, the Greek god of the lower world. This argument over rock origins and the nature of magma was settled in the early 1800s, when two of Werner's most famous students, Leopold von Buch and Alexander von Humboldt, switched sides after studying regions with volcanic rocks. Many former believers of the Neptunist school were later convinced of their misinterpretation of rock origins.

Prompted by the Neptunist-Plutonist controversy, the nineteenth century saw a great deal of research into the evolution and nature of magma. Sir John Hall (1805) undertook a series of experiments with both igneous and metamorphic rocks. Following these initial experiments by Hall, experimental studies in petrology developed rapidly in both Europe and the United States (Loewinson-Lessing, 1954). Important advances in equipment and the rapid development of physical chemistry in the latter half of the nineteenth century led to many important ideas about the nature of magma.

Foremost among the scientists studying magma at the time was James Dana. His early observations were based on an analysis of data collected from a scientific exploration expedition that he participated in from 1838 to 1842. As a result of that expedition to the southern Atlantic and Pacific oceans, he determined that textural differences in igneous rocks were due to the physical conditions at the time of magma consolidation. Later work by Dana, primarily concerning Hawaii's volcanoes, enabled him to establish the relationship between the liquidity or viscosity of magma and its crystallization properties (Fenton and Fenton, 1952: 232).

In the latter part of the nineteenth century, J. P. Iddings studied the formation of igneous rocks and their relationship to the depths of the earth. Iddings found the "recurrence of eruptions of similar magmas in one region indicate magmas of like composition may be erupting from the same general source at widely remote periods" (1914: 158). This disproved earlier theories that the earth's crust was solidified over a molten globe of concentric zones.

The first half of the twentieth century saw another increase in the scientific investigation of magmatic processes. Classification of magma by R. A. Daly (1914, 1933) as well as the experimental studies and meticulous field observations by Norman Bowen (1913, 1915) were responsible for major advances in the understanding of magma. With the publication of *The Evolution of Igneous Rocks* (1928), Bowen firmly established the importance of experiments in rock analysis as well as emphasizing the significance of combining laboratory and field methods.

The past two decades have brought new developments in technology as well as further understanding of the complexity of magmatic processes. Many advances have been the result of in-depth analysis of the role of PLATE TECTONICS in rock formation (Ehlers and Blatt, 1982). Many recent advances in magma studies include information about the role of magma formation in exploration for economically valuable minerals (Alabaster, 1985; Pearce, 1985), information gathered from recent volcanic eruptions at Mt. Saint Helens (Merzbacher and Eggler, 1984) and the Hawaiian volcanoes (Gerlach and Graeber, 1985; Thurber, 1984), and new information about magma processes in outer space (Hill, 1984; Kerr, 1983).

In a recent book commemorating the fiftieth anniversary of the publication of Bowen's book on igneous rocks, H. S. Yoder (1979: 515) concluded the book with a statement by Bowen 50 years earlier that can sum up the state of current research: "The early stages of the development of the human race are shrouded in mystery. So it is with the early stages of the development of the magmatic cycle."

References

Alabaster, T. 1985. "The Interrelationship between the Magmatic and Ore-Forming Hydrothermal Processes in the Oman Ophiolite." *Economic Geology and the Bulletin of the Society of Economic Geologists*, January-February, pp. 1–16.

Bowen, N. L. 1913. "The Melting Phenomena of the Plagioclase Feldspar." *American Journal of Science* 35: 577–599.

———. 1915. "The Crystallization of Haplobasaltic, Haplodioritic, and Related Magmas." *American Journal of Science* 40: 161–185. An early study on magmas that focused on laboratory experiments.

———. 1928. *The Evolution of Igneous Rocks*. Princeton, N.J.: Princeton University Press. A classic work combining laboratory experimentation and field methods.

Clarke, J. W., and Hughes, T. McK. 1890. *Life and Letters of Adam Sedgwick*. Vol. 1. Cambridge: Cambridge University Press.

Daly, R. A. 1914. *Igneous Rocks and Their Origins*. New York: McGraw-Hill Book Co. A classic work on the evolution of igneous rocks.

———. 1933. *Igneous Rocks and Depth of the Earth*. New York: McGraw-Hill Book Co.

Ehlers, Ernest G., and Blatt, Harvey. 1982. *Petrology: Igneous, Sedimentary, and Metamorphic*. San Francisco: W. H. Freeman and Co. A recent textbook on the study of rocks.

Fenton, Carroll, and Fenton, Mildred. 1952. *Giants of Geology*. Garden City, N.Y.: Doubleday and Co. A good discussion of the history of geologic thought.

Gerlach, Terrance M., and Graeber, Edward. 1985. "Volatile Budget of Kilauea Volcano." *Nature*. January, pp. 273–277. A recent study of magma processes.

Hall, Sir J. 1805. "Experiments on Whinstone and Lava." *Transactions, Royal Society of Edinburgh* 5: 43–75. A pioneering work in experimental petrology.

Hallam, A. 1983. *Great Geological Controversies*. Oxford: Oxford University Press. A good discussion of the vulcanist-Plutonist controversy.

Hill, M. 1984. "New Evidence Indicated Huge Volcanoes on Venus." *Earth Science*, Summer, pp. 8–10.

Hutton, James. 1795. *Theory of the Earth*. 2 vols. Edinburgh.

Iddings, J. P. 1914. *The Problem of Volcanism*. New Haven: Yale University Press.

Kerr, R. A. 1983. "Volcanism at 100° C Below." *Science* 221: 449–451.
Loewinson-Lessing, F. Y. 1954. *A Historical Survey of Petrology*. Edinburgh: Oliver and Boyd. A good discussion of the historical roots of rock studies.
Merzbacher, Celia, and Eggler, David H. 1984. "Amagmatic Geohygrometer: Application to Mt. St. Helens and Other Dacitic Magmas." *Geology*, October, pp. 587–590.
Ospouat, A. M. 1971. *Abraham Gottlob Werner, Short Classification and Description of the Various Rocks*. New York: Hafner Publishing Co.
Pearce, J. A. 1985. "Chronology of Magmatism, Skarn Formation, and Uranium Mineralization." *Economic Geology and the Bulletin of the Society of Economic Geologists*, March-April, pp. 513–518.
Shand, S. James. 1947. *Eruptive Rocks*. New York: Hafner Publishing Co.
Thurber, C. H. 1984. "Seismic Detection of the Summit Magma Complex of Kilauea Volcano, Hawaii." *Science*, January, 1B, pp. 165–167.
Yoder, H. S. 1979. *The Evolution of the Igneous Rocks: Fiftieth Anniversary Perspectives*. Princeton, N.J.: Princeton University Press.
von Zittel, K. A. 1901. *History of Geology and Palaeontology to the End of the Nineteenth Century*. London: Scott.

Sources of Additional Information

For short definitions of the concept see Anthony Wyatt, *Challinor's Dictionary of Geology* (New York: Oxford University Press, 1986), p. 188, and Stella E. Steigeler, ed., *Dictionary of Earth Sciences* (New York: Pica Press, 1976), p. 168. Some recent books that give a detailed discussion of magma evolution are R. B. Hargraves, *Physics of Magmatic Processes* (Princeton, N.J.: Princeton University Press, 1980); Robert E. Reicker, *Rio Grande Rift: Tectonics and Magmatism* (Washington, D.C.: American Geophysical Union, 1979); E. R. Oxburgh, *Heat Flow and Magma Genesis* (Cambridge: Cambridge University Press, 1980); and M. P. Atherton and J. Tarney, eds., *Origin of Granite Batholiths* (Cheshire, Eng.: Shiva Publishing Limited, 1981). Some recent journal articles on magma are Angelo Peccerillo, "Roman Comagmatic Province: Evidence for Subduction-Related Magma Genesis," *Geology*, February 1985, pp. 103–106; R. I. Hill, L. T. Silver, B. W. Chappel, and H. P. Taylor, Jr., "Solidification and Recharge of SiO_2-Rich Plutonic Magma Chambers," *Nature*, February 1985, pp. 643–646; and S. N. Williams, "Soil Radon and Elemental Mercury Distribution in Relation to Magmatic Resurgence at Long Valley Caldera," *Science*, August 9, 1985, pp. 551–553.

MANTLE. See LITHOSPHERE.

MAP PROJECTION. 1. A geometric algorithm for transforming spherical coordinates into plane coordinates. 2. A method for transferring three-dimensional surface features onto a two-dimensional medium. 3. A systematic method of drawing the earth's meridians and parallels onto a flat surface.

The need for map projections arises from the fact that the spherical coordinates of a globe cannot be transferred to flat maps without some distortion. A systematic and consistent transferal of points from a globe to a flat map is a map projection. There are an infinite number of such systematic methods of transformation. Each method deforms the original LATITUDE/LONGITUDE graticule in a particular way. Certain projections are better at maintaining areal relationships, others are better for direction and angular relationships, and still others are more appropriate for distance determination.

The now common term *projection* comes from one of the two general methods of graticule transformation. In this method a real or hypothetical light source is shown through the global graticule and projected onto a flat surface. The type of projection depends upon the relative positions of the light source, the graticule, and the flat developable surface. The second method of graticule transfer comes from mathematical manipulation of the spherical coordinates. The term *projection* now is commonly used for either method.

The earliest evidence for the use of mathematics in cartography comes from Dicaearchus (350–290 B.C.), who realized the need for an orienting line on a world map. Three centuries later Marinus of Tyre produced the first rigorous mathematical formulation of projections using meridians and parallels (Bagrow, 1966: 33). Although Marinus can be credited with devising the mathematical projection, it was Claudius Ptolemy, primarily in *Almagest* and *Geographia*, who is credited with providing a more workable and satisfactory projection that was used for the next 1,200 years without significant alteration (Brown, 1979: 58–59). Ptolemy created the first conical projection, which, though later modified to a more realistic looking spherical projection, retained at least a semblance of spherical proportions on the flat map. The main drawback to Ptolemy's work was his erroneous estimate of the circumference of the earth (Bunbury, 1879: 546–644).

The sixteenth century saw the rebirth of learning in all areas including the making and using of maps. Nicolaus Donus, a Benedictine monk, took the Ptolemaic texts and improved the projections. At this time navigation of the seas was also expanding. There was an acute need for charts where direction could be easily and accurately calculated. Gerardus Mercator devised a projection for such a chart. In 1569 this Dutch mathematician and toolmaker produced the now well-known Mercator projection, wherein rhumb lines between any two points (loxodromes) cut meridians at constant angles, and provided a simple, if inefficient, tool for sea navigation (Brown, 1979: 134–136).

From the sixteenth to the twentieth century many developments concerning projections occurred. They include, but are not limited to, the planisphere of Cassini, the polyconic projection used in the international map, and progress in geodetic surveying (Brown, 1979: 303). All of these advancements led to a plethora of projections and more complex transformations.

Modern cartographers have three basic projection surfaces from which to choose: the plane (azimuthal projections), the cone (conic projections), and the cylinder (cylindrical projections) (Richardus and Adler, 1972: 4). A multitude of variations of these basic surfaces exist; projections can be categorized into tangent, secant, or polysurficial projections and into normal, transverse, or oblique projections (Richardus and Adler, 1972: 5).

An excellent review of the most-used projections and of their special properties was produced by James Hilliard, Umit Basoglu, and Phillip Muehrcke (1978). A particularly good reference for projections used by the U.S. Geological Survey was written by John Snyder (1982). The most salient features of this volume are the sections on projections used in space imagery such as the Space Oblique Mercator projection used in the LANDSAT. An ideal book written for an introductory understanding of all aspects of projection construction is the *Introduction to Map Projections* (McDonnell, 1979).

Current research is generally concentrated in three areas: projections used in satellite surveying (e.g., Batson, 1973, 1976), the production of higher quality geodetic surveys for use with all projections (e.g., Sigl, 1984; Allman, 1982), and the influence of projections on interpretation of thematic data (e.g., Gilmartin, 1983; Hsu, 1981). The emphasis in most projection research today lies in the use of precise satellite information and global positioning. All of the above three areas rely heavily on satellite data.

References

Allman, J. S. 1982. "A Geoid for South-East Asia and the Pacific." *Australian Journal of Geodesy, Photogrammetry and Surveying* 36: 59–63.

Bagrow, Leo (revised and enlarged by Skelton, R. A.). 1966. *History of Cartography*. Cambridge: Harvard University Press. Relatively old but nonetheless excellent text on the history and nature of cartography.

Batson, R. M. 1973. "Cartographic Products from the Mariner 9 Mission." *Journal of Geophysical Research* 78: 4424–4435.

———. 1976. "Cartography of Mars: 1975." *American Cartographer* 3:57–63.

Batson, R. M.; Bridges, P. M.; Inge, J. L.; Isbell, C.; Masursky, H.; Strobell, M. E.; and Tyner, R. L. 1980. "Mapping the Galilean Satellites of Jupiter with Voyager Data." *Photogrammatric Engineering and Remote Sensing* 46: 1303–1312.

Brown, Lloyd A. 1979. *The Story of Maps*. New York: Dover Publications. Another very good overall book on the history of cartography.

Bunbury, E. H. 1879. *A History of Ancient Geography*. Vol. 1. London.

Gilmartin, P. P. 1983. "Aesthetic Preferences for the Proportions and Forms of Graticules." *Cartographic Journal* 20: 95–100.

Hilliard, James A.; Basoglu, Umit; and Muehrcke, Phillip C. 1978. *A Projection Handbook*. Madison: University of Wisconsin Cartographic Laboratory. A very good guide to the production, uses, and properties of the most used projections.

Hsu, Mei-Ling. 1981. "The Role of Projections in Modern Map Design." *Cartographica* 18: 151–186.

McDonnell, Porter W., Jr. 1979. *Introduction to Map Projections*. New York: Marcel Dekker. A good reference text for students in geography, surveying, and cartography.

Richardus, Peter, and Adler, Ron K. 1972. *Map Projections*. Amsterdam: North-Holland Publishing Co. A mathematically oriented primer to map projections.

Sigl, R. 1984. "The Contribution of Satellite Geodesy to the Geosciences." *GeoJournal* 8: 341–362.

Snyder, John P. 1981a. "Map Projections for Satellite Tracking." *Photogrammetric Engineering and Remote Sensing* 47: 205–213.

———. 1981b. "The Space Oblique Mercator Projection-Mathematical Development." U.S. Geological Survey Bulletin, no. 1518. Washington, D.C.: U.S. Geological Survey.

———. 1982. "Map Projections Used by the U.S. Geological Survey." U.S. Geological Survey Bulletin, no. 1532. Washington, D.C.: U.S. Geological Survey.

Sources of Additional Information

For more examples of the use of computers and projections see R. K. Adler, J. P. Reilly, and C. R. Schwartz, "A Generalized System for the Evaluation and Automatic Plotting of Map Projections," *The Canadian Surveyor* 22 (1968): 442–450; C. N. Claire, *State Plane Coordinates by Automatic Data Processing*, U.S. Coast and Geodetic Survey Publication, no. 62–4 (Washington, D.C.: U.S. Coast and Geodetic Survey, 1968); M. F. Hutchinson, "MAPROJ—A Computer Map Projection System," *Cartography* 12 (1982): 144–145; and W.B.P. Williams, "The Transverse Mercator Projection—Simple but Accurate Formulae for Small Computers," *Survey Review* 26 (1982): 307–320. For additional examples of the uses of specific projections see Tau Rho Alpha and Marybeth Gerin, *A Survey of the Properties and Uses of Selected Map Projections*, U.S. Geological Survey Miscellaneous Geologic Inventory Map, no. I–1096 (Washington, D.C.: U.S. Geological Survey, 1978); P. Roy and A. Sarkar, "Some Selected Map Projections for India: Their Relative Efficiencies," *Geographical Review of India* 43 (1981): 182–195; and J. P. Snyder, "The Modified Polyconic Projection for the IMW," *Cartographica* 19 (1982): 31–43. Other work in geodesy and projections include Defense Nuclear Agency, *Geodesy for the Layman*, U.S. Defense Nuclear Agency-TR–80–003 (Washington, D.C., 1983); and S. W. Henriksen, "Photogrammetric Geodesy over Large Regions," *Photogrammetric Engineering and Remote Sensing* 50 (1984): 557. Two excellent texts on the history of the use of projections and mapping are M. M. Thompson, *Maps for America*, 2d ed. (Washington, D.C.: U.S. Geological Survey, 1981), and J. N. Wilford, *The Map Makers* (New York: Vintage Books, 1981).

MASS WASTING. 1. The downhill movement under the force of gravity of surface materials including solid bedrock. 2. The variety of processes by which masses of earth material are moved at various speeds downhill.

Mass-wasting phenomena are subaerial or subaqueous processes of material movement that are very important to landscape evolution and development. Under the influence of gravity as a constant and environmental factors as variables, surficial materials may move by sliding, flowing, or falling. The main controlling factor is whether the shear strength of the material is greater than the shear stresses caused by mass and gravity. If the stress is greater, movement occurs. C. N. Savage stated this well:

> Changes in environmental conditions, including climate factors, vegetative cover, and artificial changes within or adjacent to potentially unstable areas of rock material may result in mass wasting.... A complex of forces is usually involved in mass wasting; one set may affect the shearing strength and the internal shearing stresses of the materials, and another set of forces may promote or increase the effects of the other forces. (1968: 697).

The main concern in mass-wasting research is how to predict and control mass-wasting events. Classification of mass-wasting processes is the key to prediction and control. The difficulty in classifying mass-wasting phenomena arises from the fact that there are many material types (e.g., rock, regolith, water, snow), many movement rates (e.g., very fast to very slow), and several types of movements (e.g., falls, slips, flows, and combinations of these) (Young, 1972: 75). During the past 100 years, there have been many classification schemes proposed including those by A. Hein (1882, 1932), W. H. Ward (1945), D. J. Varnes (1958), and J. N. Hutchinson (1968). However, the most widely accepted classification system over the past five decades has been the one proposed by C.F.S. Sharpe (1938). This system defines mass-wasting phenomena based on the properties of water, ROCK and debris, and ice. It also depends upon the characteristics of the movement itself in terms of speed and type of movement. A more recent system developed by Varnes (1978) is an extension and elaboration of Varnes's own work of 1958 and Sharpe's system. Type of movement is the primary concern in this scheme with six classes of movement including complex slope movements. Movement material is secondary with materials divided into three classes: rock, debris, and SOIL (Varnes, 1978: 11). This system has a very applied character, since it is the system adopted by the Transportation Research Board of the National Academy of Sciences, which emphasizes engineering aspects of mass wasting.

In recent years mass wasting has been studied in a more process-response manner. According to T. P. Burt (1984: 570), "Over the last two decades, the study of contemporary processes has been the main preoccupation of hillslope geomorphologists; traditional theoretical interests in landform [see LANDFORM] evolution, whilst continuing to occur, have provided an almost separate area of study." This process-oriented work has been integrated into more sophisticated process models such as those proposed by A. C. Armstrong (1982), M. J. Kirkby (1983), and F. Ahnert (1987). As development moves into more restrictive environments, these process-oriented studies will become more valuable in a very practical sense. Look for this emphasis to continue for years to come.

References

Ahnert, F. 1987. "Theoretical Geomorphology: Hillslopes and Rivers." *Earth Surface Processes and Landforms* 12: 3–15.

Armstrong, A. C. 1982. "A Comment on the Continuity Equation Model of Slope Profile Development and its Boundary Conditions." *Earth Surface Processes and Landforms* 7: 283–284.

Burt, T. P. 1984. "Slopes and Slope Processes." *Progress in Physical Geography* 8: 570–582. A good review of current research directions dealing with mass wasting and slope processes.

Heim, A. 1882. "Ueber Bergsturze." *Neujahrsblatt Naturforschenden Gesellschaft in Zurich* 84: 31.

———. 1932. *Bergsturz und Menschenleren*. Zurich: Fretz and Wasmuth. The two

references by Heim compose one of the earliest mass-wasting classification systems.
Hutchinson, J.N. 1968. "Mass Movement." In *Encyclopedia of Earth Sciences*. R. W. Fairbridge, ed. N.Y.: Reinhold, pp. 688–695.
Kirkby, M. J. 1983. "The Continuity Slope Model and Basal Boundary Conditions: A Further Comment." *Earth Surface Processes and Landforms* 8: 287–288.
Savage, C. N. 1968. "Mass Wasting." In R. W. Fairbridge, ed., *The Encyclopedia of Geomorphology*. Encyclopedia of Earth Sciences Series. Vol. 3. Stroudsburg, Pa.: Dowden, Hutchinson and Ross, pp. 696–700. This has an excellent description of mass-wasting processes and landforms.
Sharpe, C.F.S. 1938. *Landslides and Related Phenomena*. New York: Columbia University Press. This is the preeminent classification for mass-wasting phenomena.
Varnes, D. J. 1958. "Landslide Types and Processes." In E. B. Eckel, ed., *Landslides and Engineering Practical*. Highway Research Board Special Report, no. 29, Washington, D.C.: Highway Research Board, pp. 20–47.
———. 1978. "Slope Movement Types and Processes." In R. L. Schuster and R. J. Krizek, eds., *Landslides: Analysis and Control*. Special Report, no. 176. Washington, D.C.: Transportation Research Board, National Academy of Sciences, pp. 11–33. This is the standard engineering classification system for mass wasting.
Ward, W. H. 1945. "The Instability of Natural Slopes." *Geographical Journal* 105: 170–197.
Young, A. 1972. *Slopes*. Edinburgh: Oliver and Boyd.

Sources of Additional Information

For works on the history of research in mass wasting see R. W. Fairbridge, ed., *The Encyclopedia of Geomorphology*, Encyclopedia of Earth Science Series, vol. 3 (Stroudsburg, Pa.: Dowden, Hutchinson and Ross, 1968), and S. A. Schuman and M. P. Mosley, eds., *Slope Morphology* (Stroudsburg, Pa.: Dowden, Hutchinson and Ross, 1973). For a variety of works dealing with slope and mass-wasting processes see A. Rapp, "Recent Development of Mountain Slopes in Karkevagge and Surroundings, Northern Scandinavia," *Geografiska Annales* 42 (1960): 65–200; I. V. Popov and F. W. Kotlov, eds., *The Stability of Slopes* (New York: Consultants Bureau Enterprises, 1961); M. A. Carson and M. J. Kirkby, *Hillslope Form and Process* (Cambridge: Cambridge University Press, 1972); and D. Brunsden and D. B. Prior, *Slope Instability* (New York: John Wiley and Sons, 1984). Some examples of engineering concerns and mass wasting include E. B. Eckel, ed., *Landslides and Engineering Practice*, Highway Research Board Special Report, no. 29 (Washington, D.C.: Highway Research Board, 1958), and Q. Zaruba and V. Mencl, *Landslides and Their Control* (Amsterdam: Elsevier, 1969). A very good anatomy of a single mass-wasting disaster can be seen in J. B. Hadley, "Landslides and Related Phenomena Accompanying the Hebgen Lake Earthquake of August 17, 1959," *U.S. Geological Society Professional Paper*, no. 435K (Washington, D.C.: U.S. Geological Society, 1969), pp. 107–138.

MERIDIANS. See LATITUDE/LONGITUDE.

MILLIBAR. See ATMOSPHERIC PRESSURE.

MINERAL. A naturally occurring homogeneous solid that is formed inorganically and has a definite chemical composition and an ordered atomic arrangement.

Although there has not been a general agreement for the definition of a mineral (Berry and Mason, 1959: 3), most scientists believe certain characteristics are essential for a substance to be called a mineral. A mineral must be "naturally occurring"; that is, it must be formed by a natural process, not in a laboratory. It is common for industrial and research laboratories to produce synthetic equivalents of many naturally occurring minerals, but these synthetic productions are not considered minerals.

A second characteristic of a mineral is that it must be a homogeneous solid. This means that it must consist of a single solid phase that cannot be physically subdivided into simpler compounds. A further restriction to "inorganically formed" substances means that a mineral cannot be produced by plants and animals, although it does not eliminate the possibility of minerals being organic compounds. Having a "definite chemical composition" means that a mineral is a chemical compound and therefore can be expressed by a specific chemical formula. Finally, "an ordered atomic arrangement" indicates an internal structural framework of atoms (or ions) arranged in a regular geometric pattern.

The emergence of mineralogy as a science (see GEOLOGY) is relatively recent, but the attempt to understand and use minerals can be traced far into antiquity. Early peoples used mineral pigments from hematite and manganese oxides for cave paintings. The colors red and black were produced from these mineral forms (Berry and Mason, 1959: 5). The development of tools by Stone Age people led to the use of minerals in the toolmaking process. Of particular significance was the manufacture of the adze, an axlike tool made of a hard, tough material called fibrous actinolite or nephrite jade. These adzes were found in widely scattered areas indicating a significant amount of long-distance trade. The mining and smelting of metallic ores can also be traced more than 4,000 years. The discovery and use of metals in the Bronze and Iron ages brought about an increasing interest in minerals. According to John Sinkankas, "When metals found increasing use in cultures throughout the world, the usefulness of many other substances taken from the mineral kingdom was also explored and developed" (1966: 3).

The Greeks and Romans made important contributions to the study of minerals. The most prominent early Greek writers were Aristotle (384–322 B.C.) and Theophrastus (c. 370–287 B.C.). Aristotle in his work *Meteorologica* had a section entitled "stones" in which he discussed minerals, metals, and fossils. In a separate book entitled *Concerning Stones*, Theophrastus attempted to find the source of minerals. The famous Roman naturalist Pliny the Elder (A.D. 23–79) wrote a series of 37 books, 4 of which dealt with minerals such as gold, silver, copper, lead, and gemstones (Sinkankas, 1966:4). Pliny, in A.D. 79, while working on scientific observations and experiments, died when Mount Vesuvius erupted. Most mineralogists agree that the "Father of Modern Mineralogy" is Georgius Agricola. Among his many scientific works, *De Re Metallica* (1556) and *De Natura Fossilium* (1546) were his most significant. It was *De Natura Fossilium* that brought Agricola the most fame and is considered by many to be the first textbook of mineralogy. In it he described, in detail, all known minerals up to his time and became the first scientist to develop a mineral classification system. The search for the ideal mineral classification system occupied mineralogists for several hundred years after Agricola's pioneering work.

One hundred years after Agricola another important milestone was reached when Nicolaus Steno (1669) examined crystals in minerals. He analyzed quartz crystals and determined that despite their origin, size, and habitat, the angle between crystal faces remained constant. Steno's work with crystals led to the development of crystallography as the science of studying crystals, part of the science of mineralogy.

Slow but steady progress in mineralogy continued throughout the eighteenth century. Of particular significance was a technological advancement, the development of the goniometer by Carangeot in 1780 (Hurlburt and Klein, 1977: 3–4). The goniometer is a simple device that is used to measure crystal face angles. Perhaps the most important work of this period was done by the Frenchman R. J. Hauy (1822). Hauy showed how crystals change shape due to variations in chemical composition and that crystals were built by stacking components of the same size and shape. During this same period the German geologist Abraham G. Werner (1774) succeeded in standardizing methods and terms used in describing the outward appearance of minerals.

The nineteenth century saw many advances in the study of minerals. A very detailed classification system was developed by Friedrich Mohs (Burke, 1969), better known for his scale of hardness. The Swedish chemist Jakob Berzelius (1826) and several of his students were instrumental in developing a chemical classification of minerals. But perhaps the most important and long-lasting contribution of this time was made by James D. Dana (1837). Dana's *System of Mineralogy* developed a very detailed system of chemical classification in which minerals of closely similar structures are grouped together. His system has now been published in its seventh edition and according to Sinkankas (1966: 43), it is "the most authoritative compilation of mineral data in existence." Many new minerals were discovered during the nineteenth century, and the development of the polarizing microscope put a new and powerful tool in the hands of the mineralogist.

One of the greatest developments of the twentieth century was the demonstration by Max von Laue in 1912 that crystal diffraction patterns through X-ray analysis can show actual positions of atoms in a crystal (Mason and Berry, 1968: 6). This work was extended further by the Braggs, a father and son team of English physicists, who attempted to show exactly where atoms were located in any crystal framework (Sinkankas, 1966: 43).

The past few decades have seen many advances in the study of minerals. New techniques like the electron microprobe, sophistication of X-ray technology, and advances in on-line computers have brought new ways to understand crystal and atom formation in minerals (Hurlburt and Klein, 1977: 7). Much of the current research is also directed toward economic and strategic uses of minerals (Holden,

1981; Owen, 1983). The OCEAN floor is also a vast new frontier in research, and mineralogists are actively involved in assessing the nature and amount of mineral resources (Clark and Neutra, 1983).

References

Agricola, G. 1546. *De Natura Fossilium*. Translated from Latin by M. C. and J. A. Bandy. Special Paper 63. New York: Geological Society of America. A pioneering work by the "Father of Mineralogy."

———. 1556. *De Re Metallica*. English translation by H. C. and L. H. Hoover (1950). New York: Dover Publications.

Berry, L. G., and Mason, B. 1959. *Mineralogy: Conceptions, Descriptions, Determinations*. San Francisco: W. H. Freeman and Co.

Berzelius, Jakob. 1826. "Des Changemens Dans le Systeme de Mineralogie Chimique." *Annales de Chimie et de Physique* 31: 34–35. An early chemical analysis of minerals.

Burke, John G. 1969. "Mineral Classification in the Early Nineteenth Century." In Cecil J. Schneer, ed. *Toward a History of Geology*. Cambridge, Mass.: The M.I.T. Press, pp. 62–77. A good discussion of the historical development of mineralogy.

Clark, J. P., and Neutra, M. K. 1983. "Mining Manganese Nodules—Potential Economic and Environmental Effects." *Resources Policy* 9: 99–109.

Dana, J. D. 1837. *A System of Mineralogy*. Rewritten by E. S. Dana. New York: John Wiley and Sons. A classic book, revised many times and currently the "bible" of mineral classification.

Hauy, R. J. 1822. *Traite de Mineralogie*. 4 vols. Paris: Bachelier.

Holden, C. 1981. "Getting Serious about Strategic Minerals." *Science* 212: 305–307. A review of U.S. policies toward minerals and mining.

Hurlburt, Cornelius, and Klein, Cornelis. 1977. *Manual of Mineralogy*. New York: John Wiley and Sons.

Mason, Brian, and Berry, L. G. 1968. *Elements of Mineralogy*. San Francisco: W. H. Freeman and Co.

Owen, A. D. 1983. "The Economics of Uranium Demand." *Resources Policy* 9: 110–121. An analysis of the uranium market.

Sinkankas, John. 1966. *Mineralogy: A First Course*. Princeton, N.J.: D. Van Nostrand Co.

Steno, Nicolaus. 1669. *The Prodomus of Nicolaus Sterno's Dissertation* (trans. J. G. Winter). London: Macmillan, 1916. The first scientific study of crystals.

Werner, Abraham Gottlob. 1774. *Von Den Auszerlichen Kennzeichen der Fossilien*. Leipzig.

Sources of Additional Information

Short discussions and definition of mineral can be found in William Lee Stokes and David Varnes, *Glossary of Selected Geologic Terms* (Denver: Colorado Scientific Society, 1955), p. 165; Keith Frye, ed., *The Encyclopedia of Mineralogy*, vol. 4 (Stroudsburg, Pa.: Hutchinson Ross Publishing Co. 1981); John Challinor, *A Dictionary of Geology* (Oxford: Oxford University Press, 1978); and Arthur M. Strahler and Alan H. Strahler, *Modern Physical Geography*, 2d ed. (New York: John Wiley and Sons, 1983), pp. 192–195. Much of the literature in mineralogy is found in scholarly journals. The most important scientific journals in English are the *American Mineralogist*, published by the Mineralogical Society of America, and the *Mineralogical Magazine*, published by the Mineralogical Society of Great Britain. An interesting book for the amateur mineralogist is John Sinkankas, *Mineralogy for Amateurs* (Princeton, N.J.: D. Van Nostrand Co., 1964).

N

NATURAL HAZARD. 1. The interaction of natural events with human use systems that cause major negative disruptions to the affected areas and people. 2. An extreme geophysical event, in terms of magnitude and frequency, that greatly exceeds human expectations and causes major human hardship or loss of life, with significant material damage to human artifacts.

The study of natural hazards is associated primarily with the research of geographers. This geographic tradition began more than a half-century ago in the United States. As an area of geographic concern, the philosophical basis for natural hazards is attributed to Harlan Barrows (1923) and to his observations on the concept of human adjustment to the environment. Similar views on the concept and implicit assumptions are found in the writings of the French geographer Jean Brunhes (1920).

The primary thrust of the development of the natural hazards concept, however, is attributed to the work of Gilbert White and his colleagues at the University of Chicago. White's original work (1942) on natural hazards deals with the physical aspect of flood losses. During this period, most of the nongeographic research focused on technology, particularly the relationship of dam construction to flood loss (see FLOOD PLAIN). White questioned the validity of flood prevention as a means of reducing flood damage. In 1956 the University of Chicago, with the aid of federal funding, examined the question of the nature of human adjustments to floods. The results of that study (White et al., 1958) confirmed the suspicion that flood control had not significantly affected the amount of damage from floods.

Out of this same study emerged what is termed the Chicago school of natural hazard research. Research continued in the area of riverine flooding (Burton, 1962; Kates, 1962), but the scope of the research broadened into other areas of flooding, such as coastal floods (Burton and Kates, 1964) and other types of natural hazards.

One of the other types of natural hazard studies was undertaken by Thomas F. Saarinen (1966) when he studied drought-hazard perception in the Great Plains. Because his work attempts to develop a systematic method and technique, it has remained a feature of natural hazards research. In addition, Saarinen was instrumental in developing the concept of environmental perception. This behavioral science methodology is widely used to analyze natural hazard adjustment decisions made by resource managers (see NATURAL RESOURCE) in areas susceptible to hazards. According to an analysis of natural hazard research by James K. Mitchell (1974), "Complex theoretical, methodological, and policy questions raised by this research have stimulated a significant degree of interdisciplinary liaison between geographers, psychologists, engineers, economists and community planners."

Natural hazard research has adopted all or part of the research paradigm presented by Ian Burton et al. (1968), who outlined five principal areas of investigation designed (1) to assess the extent of human occupance in hazard zones, (2) to identify the full range of possible human adjustments to these hazards, (3) to study the perception and estimation of hazards, (4) to describe the process of adopting hazard adjustments, and (5) to estimate the optional set of adjustments in terms of anticipated social consequences. The goals of natural hazard research are to develop a global view of the range and extent of hazards and to develop a systematic method of their analysis.

Throughout the 1960s and 1970s the scope of natural hazards research expanded. Funding from the National Science Foundation and the designation of natural research as one of the principal concerns of the International Geographic Union added to the emphasis on this type of research.

Considerable research has been done on natural hazards in the past several decades. The types of research and their context are outlined in the work of John R. Gold (1980). In the 1970s a change in the dimension of natural hazard research occurred. The concept of urban environmental hazard was added and included man-made hazards such as water POLLUTION, fire, and road traffic. A study by Kenneth Hewitt and Ian Burton (1971) of London (Ontario) examines these hazards in conjunction with other types of hazards.

A group of radical researchers (Baird et al., 1975) has criticized the direction of natural hazards research. Their primary concern is that little research has been done in Third World countries, the areas that probably suffer the most from natural disasters.

References

Baird, A.; O'Keefe, P.; Westgate, K.; and Wisner, B. 1975. "Towards an Explanation and Reduction of Disaster Proneness." Occasional Paper, no. 2. Bradford, Eng.: Disaster Research Unit, University of Bradford. A criticism of some areas of natural hazard research.

Barrows, Harlan. 1923. "Geography as Human Ecology." *Annals of the Association of American Geographers* 13: 1–14. Traces the development of the philosophical basis for natural hazard research.
Brunhes, Jean. 1920. *Human Geography*. Chicago: Rand McNally and Co.
Burton, Ian. 1962. "Types of Agricultural Occupance of Flood Plains in the United States." Research Paper, no. 75. Chicago: Department of Geography, University of Chicago. A study of the natural hazards of flood plains.
Burton, Ian, and Kates, Robert W. 1964. "The Flood Plain and the Seashore: A Comparative Analysis of Hazard Zone Occupance." *Geographical Review* 54: 366–385. A comparative study of two types of natural hazards.
Burton, Ian; Kates, Robert; and White, Gilbert F. 1968. "The Human Ecology of Extreme Geographical Events." Working Paper, no. 1. Toronto: Natural Hazard Research Unit, Department of Geography, University of Toronto. An outline of the basic paradigm of natural hazards research.
Gold, John R. 1980. *An Introduction to Behavioral Geography*. New York: Oxford University Press.
Hewitt, K., and Burton, I. 1971. "The Hazardousness of a Place: A Regional Ecology of Damaging Events." Working Paper, no. 6. Toronto: Natural Hazard Research Unit, Department of Geography, University of Toronto. A case study of hazards in Ontario, Canada.
Kates, Robert W. 1962. "Hazard and Choice Perception in Flood Plain Management." Research Paper, no. 78. Chicago: Department of Geography, University of Chicago.
Mitchell, James K. 1974. "Natural Hazards Research." In Manners, Ian, and Mikesell, Marvin, eds., *Perspectives on Environment*. Washington, D.C.: Association of American Geographers, pp. 311–341. An excellent review article on the development of the natural hazards concept. Includes an extensive bibliography.
Saarinen, Thomas F. 1966. "Perception of Drought Hazard on the Great Plains." Research Paper, no. 106. Chicago: Department of Geography, University of Chicago. A study of environmental perceptions and hazards.
White, Gilbert F. 1942. "Human Adjustments to Floods: A Geographical Approach to the Flood Problem in the United States." Ph.D. dissertation, University of Chicago.
White, Gilbert F. et al. 1958. "Changes in Urban Occupance of Flood Plains in the United States." Working Paper. Chicago: University of Chicago, Department of Geography.

Sources of Additional Information

Good analyses of the different areas of natural hazards research are in Gilbert White, ed., *Natural Hazards: Local, National, Global* (New York: Oxford University Press, 1974), and Jan Burton, Robert W. Kates, and Gilbert F. White, *The Environment as Hazard* (New York: Oxford University Press, 1968). Discussions of the historical development of natural hazards research are in Gilbert White, "Natural Hazards Research," in R. J. Chorley, ed., *Directions in Geography* (New York: Methuen & Co., 1968), pp. 193–216, and John R. Gold, *An Introduction to Behavioral Geography* (New York: Oxford University Press, 1980). The literature of natural hazards research is presented in Anita Cochran, "A Selected Annotated Bibliography of Natural Hazards," *Natural Hazards Research Working Paper*, no. 22 (Toronto: Department of Geography, University of Toronto, 1972).

NATURAL RESOURCE. 1. Materials found in the physical environment that serve some function within a given cultural context; a response to human needs. 2. Elements that are derived from the physical and biotic conditions of the land and upon which people are dependent for aid and support.

Geographers have long been interested in natural resources and their relationship to cultural development. George P. March (1864) studied resource development more than 100 years ago, and at the turn of the century Nathaniel Shaler focused on natural resources in *Man and the Earth* (1905).

After World War I geographers involved with the interrelation of various items in resource complexes became aware of resource management as an important concern of geography (Parkins and Whitaker, 1936). Carl Sauer's (1947) work is of particular significance. Sauer emphasized the cultural component of natural resources and stated that the concept of a natural resource "implies the determination that the thing is useful and therefore a cultural achievement" (1947: 18). This cultural aspect of natural resources is important because it emphasizes their dynamic nature. Not static, they expand and contract in response to human needs and human actions. They are, therefore, "an expression or reflection of human appraisal, and without people, there would be no resources" (Larkin et al., 1981: 170).

Sauer's idea of cultural involvement was expanded in Erich W. Zimmerman's (1951) work on natural resources, *World Resources and Industries*. According to Zimmerman, resources are "living phenomena, expanding and contracting in response to human effort and behavior.... To a large extent, they are man's own creation" (1951: 7); therefore, natural resources exist only in conjunction with human culture.

Geographers classify natural resources in a variety of ways. The most common method is to classify them into two categories: renewable resources and nonrenewable resources. *Renewable resources* can be readily renewed or reproduced, such as a forest. Fossil fuels and minerals, such as gold, zinc, or copper, obviously cannot renew themselves and once extracted from the earth and consumed cannot be replaced. These are called *nonrenewable resources*.

In the past several decades geographers have again become interested in natural resources issues. *Man's Role in Changing the Face of the Earth* (1956), a collection of readings edited by William L. Thomas, has helped generate much of this interest.

Another area of concern to geographers is the relationship between population and natural resources. A definitive article on this topic by Edward A. Ackerman calls for "more carefully coordinated population-resource technology studies" (1959: 648).

Geographers have investigated the role of natural resources in the process of economic development. Norton Ginsberg (1957) was among the first to outline problems in this area. He said that the following questions should be asked:

"How important are natural resources in the course of economic development, and what relationships do they bear to other factors which enter into the developmental complex?" (1957: 197).

Ian Burton and Robert W. Kates (1965) have edited a useful book of readings on resource management and conservation practices. They stated that in the period since World War II "the United States has probably used more nonrenewable resources than all the rest of the world's population has used during the history of man on earth prior to that time." In their book, they sought "to explore some of the dimensions of this topic" (1965: 1).

More recently, Leonardo Scholars (1975) has analyzed policy decisions for natural resources that concern economic, social, and political issues.

References

Ackerman, Edward A. 1959. "Population and Natural Resources." In Philip M., Hauser and Otis Dudley, Duncan, eds., *The Study of Population*. Chicago: University of Chicago Press. Discusses the relationship between natural resources and population change.

Burton, Ian, and Kates, Robert W. 1965. *Readings in Resource Management and Conservation*. Chicago: University of Chicago Press. Collection of readings on natural resources and conservation management.

Ginsberg, Norton. 1957. "Natural Resources and Economic Development." *Annals of the Association of American Geographers* 47: 197–212. Analyzes the role of natural resources in economic development.

Larkin, Robert P.; Peters, Gary L.; and Exline, Christopher. 1981. *People, Environment, and Place: An Introduction to Human Geography*. Columbus, Ohio: Charles E. Merrill Publishing Co. Discussion of resources in Chapter 9.

Marsh, George P. 1864. *Man and Nature*. New York: Scribner's. A classic book on the relationship between people and nature.

Parkins, A. E., and Whitaker, J. R., eds. 1936. *Our Natural Resources and Their Conservation*. New York: Harper and Brothers.

Sauer, Carl O. 1947. "Early Relations of Man to Plants." *Geographical Review* 37: 1–25. Discusses the relationship of culture to natural resources.

Scholars, Leonardo. 1975. *Resources and Decisions*. North Scituate, Mass.: Duxbury Press/Wadsworth Publishing Co.

Shaler, N. S. 1905. *Man and the Earth*. New York: Duffield & Co.

Thomas, William L., ed. 1956. *Man's Role in Changing the Face of the Earth*. Chicago: University of Chicago Press. An excellent collection of readings on natural resources from a geographic perspective.

Zimmerman, Erich W. 1951. *World Resources and Industries*. 2d ed. New York: Harper and Brothers, Publishers.

Sources of Additional Information

Short definitions of the concept are in Douglas Greenwald, *Dictionary of Modern Economics*, 2d ed. (New York: McGraw-Hill Book Co., 1973), p. 395; Harold L. Sloan and Arnold J. Zurcher, *Dictionary of Economics*, 5th ed. (New York: Barnes and Noble

Books, 1970), p. 303; and Daniel N. Lapedes, ed., *Dictionary of the Life Sciences* (New York: McGraw-Hill Book Co., 1976), p. 568. A good discussion of the geographic aspects of natural resources appears in Peter Haggett, *Geography: A Modern Synthesis*, 3d ed. (New York: Harper and Row, Publishers, 1979), Chapter 9.

O

OCEAN. The interconnected body of salt water that covers nearly three-fourths of the earth's surface or one of the large water bodies into which this interconnected ocean is divided (Atlantic, Pacific, Indian, Arctic).

The PHYSICAL GEOGRAPHY of the earth cannot be understood without a knowledge of the oceans and their relationship to other physical and biological processes. Many aspects of ocean science or oceanography are of interest to geographers. The study of the earth in general must start with an understanding of the changing relationship between the land and water bodies. The geography of the ocean floor reflects processes associated with mountain building, sedimentation, PLATE TECTONICS, and a host of other important earth processes. There is a very significant relationship between the ocean and the atmosphere. To understand CLIMATOLOGY, a knowledge of the ocean is essential. Movement of water within oceans (currents) or at the shoreline (waves) are important processes that have a major impact on the earth's surface. The BIOGEOGRAPHY of oceans and the nature and distribution of ocean resources have important implications for mankind's future.

Although the science of oceanography is of relatively recent origin, the study of the sea can be traced to antiquity when early peoples took to the sea in boats to fish, fight, or trade (Deacon, 1971). Practical, rather than abstract, knowledge was first sought.

The scientists and scholars of ancient Greece and Rome tried to understand and explain various oceanic phenomena. In the fourth century B.C. Pytheas, a citizen of the Greek colony located at the present site of Marseilles in France, wrote an essay entitled "About the Ocean" in which he was able to relate tidal changes to the motion of the moon (Kish, 1978: 67–71). Another Greek geographer, Strabo (63 B.C. to A.D. 24) made important oceanographic observations and recorded data on tides and depth soundings (Jones, 1917). In the 1,500 years following these early Greek and Roman discoveries, little new knowledge has been added to the understanding of the oceans, except for the discoveries of the Vikings around the tenth century and the introduction of the magnetic compass by Arabic traders in the thirteenth century.

Toward the end of the fifteenth century ocean exploration began in earnest. Primarily responsible for this upsurge in interest about oceanographic principles was Prince Henry the Navigator of Portugal, who established a school and encouraged exploration. He established the first geographic research institute (James and Martin, 1981: 67). This exploratory period culminated with the global circumnavigation by Magellan in the early sixteenth century.

In the seventeenth and eighteenth centuries a need for more systematic oceanographic studies arose to improve the safety and speed of navigation along trade routes. The first compilation of information specifically written to aid sea travel was published in 1700 and entitled "A Discourse of the Winds," by W. Dampier. The postmaster general for the colonies before the American Revolution, Benjamin Franklin, was the first to chart and publish information on the Gulf Stream in 1786. The first voyages primarily done for the benefit of science, however, were conducted between 1768 and 1779 by Captain James Cook of the British Royal Navy. Captain Cook was commissioned to lead a scientific expedition to observe the transit of Venus in 1768, and subsequently, he was assigned to search for the undiscovered "great southern continent" (Stowe, 1979: 14).

The nineteenth century saw considerable interest in a more scientific understanding of the oceans. Systematic ocean voyages for scientific purposes were led by Charles Darwin, with his famous voyage of the *Beagle*, and by James Ross aboard the ships *Erebus* and *Terror*. The most thorough of these voyages was the voyage of the *Challenger* led by John Murray in 1872 to 1876. This voyage "represented the first systematic attempt to examine the depth and breadth of the world's ocean from the chemical, physical, and biological points of view" (Von Arx, 1962: 9).

Although these nineteenth-century voyagers made significant contributions to ocean science, one of the most important contributions was made by Matthew Fontaine Maury, a career officer in the U.S. Navy. Maury carefully analyzed data in log books and from that information produced a compilation of wind and current patterns. He was primarily responsible for the standardization of nautical and meteorological observations, which he summarized in his book *The Physical Geography of the Sea*, published in 1855. Because of his pioneering scientific work, many refer to Maury as the father of oceanography (Charlier and Charlier, 1970).

Many new discoveries and advances in oceanography have been made in the twentieth century. In the first few decades Fridtjof Nansen, a Norwegian, explored the Arctic and made important contributions about currents, salinity, temperature variations, and the origin of the Arctic Sea. From 1925 to 1927 the German ship *Meteor* crisscrossed the South Atlantic and, using the first echo sounder, developed an important methodology for determining sea floor topography.

The past few decades have seen much interest in ocean research. The United States, through a variety of government-supported programs, has supported much recent research. Among the many government agencies involved are the U.S. Navy Oceanographic Office, Office of Naval Research, National Ocean Survey, National Oceanic Atmospheric Administration, and the Office of Sea Grant. Aside from federal government programs three important marine research laboratories, Scripps Institution of Oceanography, Woods Hole Oceanographic Institution, and the Lamont-Doherty Geological Observatory, have been pioneers in twentieth-century oceanographic research. New technologies such as the floating instrument platform, submersibles, oceanographic research buoy, and the use of remote sensors on satellites and airplanes are responsible for many new breakthroughs in understanding the ocean (Thurman, 1985).

References

Charlier, R. H., and Charlier, P. A. 1970. "Matthew Fontaine Maury, Cyrus Field, and *The Physical Geography of the Sea*," *Sea Frontiers* 16: 5, 272–281. A discussion of one of the pioneering works in oceanography.

Deacon, Margaret. 1971. *Scientists and the Sea, 1650–1900: A Study of Marine Science*. New York: Academic Press. A good historical study of the development of marine science.

James, Preston E., and Martin, Geoffrey J. 1981. *All Possible Worlds: A History of Geographical Ideas*. New York: John Wiley and Sons.

Jones, H. L. 1917. *The Geography of Strabo*. London: W. Heinemann.

Kish, George, ed. 1978. *A Source Book in Geography*. Cambridge: Harvard University Press. A good discussion of Greek and Roman contributions to geography.

Stowe, Keith S. 1979. *Ocean Science*. New York: John Wiley and Sons.

Thurman, Harold V. 1985. *Introductory Oceanography*. 4th ed. Columbus, Ohio: Charles E. Merrill Publishing Co. An excellent book on the various aspects of oceanography.

Von Arx, William S. 1962. *An Introduction to Physical Oceanography*. Reading, Mass.: Addison-Wesley Publishing Co.

Sources of Additional Information

Many good books are available to introduce the science of oceanography, for example, P. K. Weyl, *Oceanography: An Introduction to the Marine Environment* (New York: John Wiley and Sons, 1970); A. C. Duxbury, *The Earth and Its Oceans* (Reading, Mass.: Addison-Wesley Publishing Co., 1971); and Gordon Pirie, *Oceanography: Contemporary Readings in Ocean Sciences* (New York: Oxford University Press, 1977). Much recent research has centered on marine pollution; see Jerome Williams, *Introduction to Marine Pollution Control* (New York: John Wiley and Sons, 1979), and W. Bascom, "Disposal of Waste in the Ocean," *Scientific American* 231 (1974): 16–25.

OROGRAPHIC PRECIPITATION. Precipitation that is created when a moist air mass is forced to rise over higher ground, such as a ridge, hill, or mountain.

Orographic refers specifically to mountains, but the orographic effect that causes precipitation to occur may be caused by lower barriers, including low hills or even ridges. As air is forced to rise over any barrier, it cools. If it cools to below the dew point (see LAPSE RATE), precipitation will occur.

A major geographic feature of orographic precipitation is that it is asymetric. If a hill or mountain barrier stands in the way of the flow of moist air, that hill or mountain will have a wet side on the downwind or windward slope and a much drier side on the upwind or leeward slope. That drier side, caused as the air moves downslope and warms as a result of compression, is often referred to as a rain shadow.

The rain-shadow effect can be seen almost anywhere that air moves in a constant direction. Throughout the world's trade-wind belts, for example, wet coastal climates exist where the moist trade winds flow onshore and upslope, whereas dry interior climates exist on the lee sides of these slopes. William Marsh (1987) provided an excellent example by looking at the distribution of precipitation in Hawaii. He pointed out that ''on the windward slope of Waialaele, annual average rainfall changes at a rate of 118 inches per mile . . . over a distance of 2.5 miles'' (1987: 100). Thus orographic barriers can indeed have a considerable impact on the distribution of rainfall, even over short distances. On a larger scale, many of the earth's deserts are at least partially a result of their locations in rain shadows. On the other hand, places like Cherrapunji, India, are extremely wet because an orographic effect increases the rainfall that is being generated by another precipitation mechanism. Cherrapunji averages more than 400 inches of rainfall each year, because the warm moist air flowing inland with the summer monsoon in India is forced to rise up the slopes of the Himalayan Mountains.

References

Fairbridge, Rhodes W. 1967. "Orographic Precipitation, Rain Shadow." In Fairbridge, Rhodes W., ed., *The Encyclopedia of Atmospheric Sciences and Astrogeology.* Encyclopedia of Earth Sciences Series. Vol. 2. New York: Reinhold Publishing Co., pp. 717–718.

Marsh, William M. 1987. *Earthscape: A Physical Geography.* New York: John Wiley and Sons. An excellent physical geography text.

Sources of Additional Information

Orographic precipitation is discussed in most elementary texts in physical geography and meteorology, including the following: Arthur N. Strahler and Alan H. Strahler, *Elements of Physical Geography*, 3d ed. (New York: John Wiley and Sons, 1984); Tom L. McKnight, *Physical Geography: A Landscape Appreciation* (Englewood Cliffs, N.J.: Prentice-Hall, 1984); Herbert J. Spiegel and Arnold Gruber, *From Weather Vanes to Satellites: An Introduction to Meteorology* (New York: John Wiley and Sons, 1983); and Joe R. Eagleman, *Meteorology: The Atmosphere in Action* (New York: D. Van Nostrand Co., 1980). Orographic precipitation and its climatic consequences are discussed in most climatology texts, including the following: John E. Oliver and John J. Hidore, *Clima-*

tology: An Introduction (Columbus, Ohio: Charles E. Merrill Publishing Co., 1984), and Glenn T. Trewartha and Lyle H. Horn, *An Introduction to Climate* (New York: McGraw-Hill Book Co., 1980). The role of orographic precipitation is also discussed in Glenn T. Trewartha, *The Earth's Problem Climates* (Madison: University of Wisconsin Press, 1961).

OZONE. 1. The triatomic form of oxygen that is formed when oxygen is irradiated with high energy ultraviolet radiation. 2. The gas that forms the protective layer in the stratosphere against ultraviolet radiation 3. A component of photochemical smog.

During the evolution of the atmosphere, the level of ozone in the air has been directly related to the levels of other forms of oxygen. Ozone is created from oxygen molecules through photochemical action of solar energy on those molecules (Levine, 1985: 34). In the current atmosphere most of the ozone exists in the STRATOSPHERE between 15 and 30 kilometers above the earth's surface. Less than 100 parts per billion of the TROPOSPHERE is ozone, whereas 90 percent of all ozone exists in the stratosphere (Levine, 1985: 6–9). The two areas of concern in studying ozone seem paradoxical. One area is that ozone is a photochemical pollutant at the earth's surface; the other is that ozone in the stratosphere is essential for life as we know it.

A. J. Haagen-Smit (1952) first described the photochemical links needed to produce ozone as a pollutant near the surface of the earth. ELECTROMAGNETIC RADIATION, especially the ULTRAVIOLET (UV), causes a reaction between nitrogen dioxide and a variety of organic compounds in the air. Ozone is one of the end products of this reaction. Haagen-Smit thus showed why the smog of a city like London and the smog of a city like Los Angeles are very different. Los Angeles has a much more intensive radiation environment, thus creating these photochemical smogs that are almost nonexistent in London (Graedel, 1985: 40). H. Levy (1971) provided further evidence about reactions involving the hydroxyl ion in the air that essentially is the state of the art today in urban atmospheric photochemistry. Work continues in laboratories to try to determine exact chemical reactions that are possible among the myriad compounds in urban air. Ozone, for its part, is now used as the indicator molecule for other types of photochemical oxidants in smog (Graedel, 1985: 70).

Conversely, stratospheric ozone is a requisite for life on earth as it exists today. As oxygen was being produced in early earth history, ozone was also produced. "Increasing levels of atmospheric ozone began to shield the Earth's surface from lethal solar UV radiation. This shielding by ozone eventually permitted life to leave the safety of the oceans and go ashore for the first time" (Levine, 1985: 5). Ozone shielding of UV radiation continues today, albeit with accelerating destruction of this ozone shield.

The attenuation of UV radiation by the atmosphere was described by A. Cornu (1879). The presence of large quantities of ozone in the stratosphere was demonstrated two years later by W. N. Hartley (1881a, 1881b). The presence of ozone in significant quantities (4.0×10^{12} molecules/cm^3) in the stratosphere could not be explained at that time. It took nearly a half-century before the photochemical processes of ozone production in the stratosphere were explained. In 1930 S. Chapman devised the photochemical scheme that explained ozone production in the atmosphere at altitudes between 10 and 50 kilometers (Chapman, 1930). His creation of an outline for ozone production soon became known as the Chapman Ozone Cycle. This cycle was one of the few significant advances in stratospheric photochemistry until the 1970s and 1980s.

In the mid-1970s it was discovered that the chlorine cycle in the stratosphere was coupled to all other photochemical cycles in that part of the atmosphere. The connections between chlorine and ozone depletion were first put forward by R. S. Stolarski and R. J. Cicerone (1974) and P. J. Crutzen (1974). Initial concern for the ozone layer concentrated on things such as the emissions of the space shuttle rocket motors. It was soon shown that industrial chlorofluorocarbons were the primary cause of ozone depletion (Molina and Rowland, 1974). The removal of chlorine from the atmosphere is a slow process, and it appears that ozone decomposition will continue until all chlorine is removed from the chemical cycle of the stratosphere (Demore et al., 1983). There is presently a huge amount of chlorine in the atmosphere. "The total chlorine budget of the present-day stratosphere is roughly 5×10^5 metric tons of chlorine per year. Chlorine is, for the most part, transported into the stratosphere in the form of the stable source compounds and is removed as HCL, eventually to be scavenged by precipitation below [about] 5Km" (Turco, 1985: 102).

The chemistry of the stratosphere is very complex with more than 130 major photochemical processes identified and several hundred more possible. There will be considerable work in the future to try to explain more thoroughly the total chemical nature of the atmosphere. However, the main focus of research for geographers, climatologists, and meteorologists will be on atmospheric measurements and collecting meteorological and climatic data relevant to ozone creation, movement, and depletion.

References

Chapman, S. 1930. "A Theory of Upper Atmospheric Ozone." *Mem. Royal Meteorological Society* 3: 103. This is the seminal work on ozone photochemistry in the stratosphere.

Cornu, A. 1879. "Sur la Limite Ultra-Violette du Spectre Solaire." *Comptes Rendus des Seances Academie Sciences* 88: 1101. This work first described the reduced ultraviolet radiation caused by atmospheric attenuation.

Crutzen, P. J. 1974. "Review of Upper Atmospheric Photochemistry." *Canadian Journal of Chemistry* 52: 1569–1581. Along with the work of R. S. Stolarski and R. J. Cicerone, this paper was one of the first to show the deterioration of ozone by chlorine reactions.

Demore, W. B.; Watson, R. T.; Golden, D. M.; Hampson, R. F.; Kurylo, M.; Howard, C. J.; Molina, M. J.; and Ravishankara, A. R. 1983. *Chemical Kinetics and*

Photochemical Data for Use in Stratospheric Modeling. Jet Propulsion Laboratory Publication, no. 83-62. Pasadena, Calif.: Jet Propulsion Laboratory.

Graedel, T. E. 1985. "The Photochemistry of the Troposphere." In Levine, J. S., ed., *The Photochemistry of Atmospheres.* New York: Academic Press, pp. 39-786.

Haagen-Smit, A. J. 1952. "Chemistry and Physiology of Los Angeles Smog." *Industrial and Engineering Chemistry.* 44: 1342-1351. First described the process of photochemical smog.

Hartley, W. N. 1881a. "On the Absorption of Solar Rays by Atmospheric Ozone." *Journal of the Chemical Society* 39: 111.

———. 1881b. "On the Absorption Spectrum of Ozone." *Journal of the Chemical Society* 39: 57. Along with the above reference, established the spectral absorption bands of ozone.

Levine, J. S. 1985. "The Photochemistry of Early Atmosphere." In J. S. Levine, ed., *The Photochemistry of Atmospheres.* New York: Academic Press, pp. 3-38.

Levy, H., II. 1971. "Normal Atmosphere: Large Radical and Formaldehyde Concentrations Predicted." *Science* 173: 141-143.

Molina, M. J., and Rowland, F. S. 1974. "Stratospheric Sink for Chlorofluoromethanes: Chlorine Atom-Catalyzed Destruction of Ozone." *Nature* 249: 810-812. Showed that chlorofluoromethanes were the primary cause of ozone depletion.

Stolarski, R. S., and Cicerone, R. J. 1974. "Stratospheric Chlorine: A Possible Sink for Ozone." *Canadian Journal of Chemistry* 52: 1610-1615.

Turco, R. P. 1985. "The Photochemistry of the Stratosphere." In J. S. Levine, ed., *The Photochemistry of Atmospheres.* New York: Academic Press, pp. 77-128.

Sources of Additional Information

General references to the atmospheric sciences and ozone include R. W. Fairbridge, ed., *The Encyclopedia of Atmospheric Sciences and Astrogeology*, Encyclopedia of Earth Sciences Series, vol. 2 (New York: Van Nostrand-Reinhold Co., 1967); D. V. Hunt, ed., *Energy Dictionary* (New York: Van Nostrand-Reinhold Co., 1979); and J. S. Levine, ed., *The Photochemistry of Atmospheres* (New York: Academic Press, 1985). For additional work on ozone and atmospheric evolution see L. V. Berkner and L. C. Marshall, "On the Origin and Rises of Oxygen Concentration in the Earth's Atmosphere," *Journal of Atmospheric Science* 22 (1965): 225-261; V. M. Canuto, J. S. Levine, T. R. Augustsson, and C. L. Imhoff, "UV Radiation from the Young Sun and Oxygen and Ozone Levels in the Prebiological Paleoatmosphere," *Nature* 305 (1982): 281-286; and J.C.G. Walker, *Evolution of the Atmosphere* (New York: Macmillan, 1977). Additional work on ozone and atmospheric pollution can be seen in O. Hutzinger, ed., *Handbook of Environmental Chemistry* vol. 2A. (New York: Springer-Verlag, 1980), and A. C. Lloyd, R. Atkinson, F. W. Lurmann, and B. Nitta, "Modeling Potential Ozone Impacts from Natural Hydrocarbons," *Atmospheric Environment* 17 (1983): 1931-1950. For additional references for ozone in the atmosphere see D. A. Brewer, T. R. Augustsson, and J. S. Levine, "The Photochemistry of Anthropozenic Nonmethane Hydocarbons in the Troposphere," *Journal of Geophysical Research* 88 (1983): 6683-6695; R. J. Cicerone, R. S. Stolarski, and S. Walters, "Stratospheric Ozone Destruction by Manmade Chlorofluoromethanes," *Science* 185 (1974): 1165-1167; R. J. Cicerone, S. Walters, and R. S. Stolarski, "Chlorine Compounds and Stratospheric Ozone," *Science* 188 (1975): 378-379; P. J. Crutzen, "Estimates of Possible Future Ozone Reductions from

Continued Use of Fluorochloromethanes (CF_2Cl_2, $CFCl_3$)," *Geophysical Research Letter* 1 (1974): 205–208; and National Academy of Sciences, series of reports on ozone and stratosphere (Washington, D.C.: National Academy of Sciences, 1975, 1976, 1982, 1983).

P

PARALLELS. See LATITUDE/LONGITUDE.

PEDOLOGY. See SOIL.

PERIGLACIAL. 1. Terrestrial and nonglacial processes and landforms of colder climates that entail intensive frost actions. 2. The climate and climatically controlled features near the pleistocene ice sheets. 3. Cold-climate processes and landscapes regardless of age or proximity to glacier.

There is no single, fully accepted idea of just what the term *periglacial* means; the only aspect in the definition of periglacial that is universally accepted is the belief that intensive frost action must be involved. A. L. Washburn stated it well: "By far the most widespread and important periglacial process is frost action. Actually, frost action is a 'catch-all' term for a complex of processes involving freezing and thawing including, especially, frost cracking, frost wedging, frost heaving, and frost sorting" (1980: 6).

It has been long held that there are specific geomorphic phenomena related to cold-weather environments. J. Geikie's volume *The Great Ice Age* (1874) noted nonglacial features readily apparent that were presumed to have been deposited under former cold-climate conditions. The process of "nivation," which demands freeze-thaw in a heavy snow environment without glaciation, was proposed by F. E. Matthes (1900). J. G. Andersson (1906) used the term *solifluction* for the form of MASS WASTING occurring under subglacial conditions. All of these and more are part of the encompassing term *periglacial*. It was not until 1909, however, that W. Lozinski (1909: 10–18) proposed to replace the term *subglacial* with a new term, *periglacial*. Lozinski failed explicitly to include nivation and solifluction in his term *periglacial* and used it only to refer to areas near the margins of the PLEISTOCENE ice that were intensively frost weathered.

Lozinski's narrow definition notwithstanding, other periglacial phenomena were quickly encompassed in the definition including equiplanation altiplanation, and frost heaving (Embleton and King, 1975: 1).

Despite the fact that the concept has been in use for almost eight decades, there has never been an accepted set of quantitative parameters of temperature and PRECIPITATION that describes the limits of periglacial activity. There are several proposed schemes to solve this problem. L. C. Peltier (1950: 215), for example, used a mean annual temperature range from $-1°C$ to $-15°C$ and a mean annual precipitation range (excluding snow) from 127 millimeters to 1397 millimeters. He also devised a periglacial erosion cycle similar to the one outlined by W. M. Davis (1909). Peltier's cycle, deemed too idealized because it ignores running water, is given little credence in research today (French, 1976: 164). Lee Wilson (1968: 723, 1969: 308) used the mean annual temperature range from $+2°C$ to $-12°C$ and the mean annual precipitation range from 50 millimeters to 1,250 millimeters. H. M. French defined the periglacial domain as "all areas where the mean annual air temperature is less than $+3°C$" (1976: 5). Jean Tricart (1967: 29–30) emphasized permafrost as an important component of periglacial regions, but he did allow for periglacial features outside of PERMAFROST areas, and he did not require that all periglacial domains have permafrost.

Considerable debate has focused on the question of whether or not permafrost is a necessary condition for periglacial activity. T. L. Péwé (1969: 2–4) came close to regarding permafrost as necessary for an area to be classed as periglacial. French (1976: 2) used permafrost as an indicator for a periglacial environment. Alfred Jahn (1975: 11), however, specifically excluded permafrost as a valid criterion. He did accept the mean annual temperature of $-1°C$ as diagnostic (this is also the temperature often used to define permafrost regimes) for periglacial environments. A. L. Washburn stated that:

> Clearly, the periglacial and permafrost environments have much in common but to make them synonymous is overly restrictive, since many features such as gelifluction, frost creep, and several forms of patterned ground that are related to frost action, are commonly regarded as periglacial but are not necessarily associated with permafrost. Furthermore, it is common to speak of former periglacial environments; yet in the present state of our knowledge there are very few criteria by which a former permafrost condition can be proved. (1980: 4)

One significant problem with the study of periglacial processes and landforms is that much of the best work has been done in Russian and has not been translated. This causes considerable slowing of the research process. Jahn (1975) has been especially helpful since he has an extensive discussion of the Russian literature; however, more detailed translations are needed.

A great impetus to the study of periglacial processes and environments in recent years has been the development of petroleum resources in high latitude (see LATITUDE/LONGITUDE) areas such as the north shore of Alaska and the Alberta tar sands. Most of this literature deals with engineering and construction problems in permafrost zones. Examples of this research are the work of the Research Group on Experimental Roadbed Research (1979), the Research Group on Pile Foundation in Permafrost (1979), and G. H. Johnston (1981). Specific work on permafrost in periglacial regions includes examples such as that of R.J.E. Brown (1978), K. M. Scott (1978), and J. R. Mackay et al. (1979).

References

Andersson, J. G. 1906. "Solifluction, a Component of Subaerial Denudation." *Journal of Geology* 14: 91–112. This is the article that originated the concept of solifluction.
Brown, R.J.E. 1978. "Map 32. Permafrost." In *Hydological Atlas of Canada*. Ottawa: Environment Canada.
Davis, W. M. 1909. "The Geographical Cycle." In W. M. Davis, *Geographical Essays*. Boston: Ginn and Co., pp. 249–278. The most detailed essay on the cycle of erosion theory of William Morris Davis.
Embleton, C., and King, C.A.M. 1975. *Periglacial Geomorphology*. New York: John Wiley and Sons. An excellent reference on the work up to 1975 in periglacial environments.
French, H. M. 1976. *The Periglacial Environment*. New York: Longman. Another very good general reference for use in periglacial research.
Geikie, J. 1874. *The Great Ice Age*. London: Dalby and Ibister. The first implication of unique geomorphic activity in periglacial environments.
Jahn, Alfred. 1975. *Problems of the Periglacial Zone*. Warsaw: Panstwowe Wydawnictwo Naukowe. An excellent work summarizing much of the work done in the Russian and Polish languages.
Johnson, G. H., ed. 1981. *Permafrost Engineering Design and Construction*. Toronto: John Wiley and Associate Committee for Geotechnical Research, National Research Council of Canada.
Lozinski, W. 1909. "Uber Die Mechanische Verwitterung Der Sandsteine im Gemassigten Klima." *Academie des Sciences de Cracovie Bulletin International, Classe des Sciences et Mathematic et Naturelles* 1: 1–25. Place where the term *periglacial* was first used.
Mackay, J. R.; Konishev, V. N.; and Popov, A. I. 1979. "Geologic Controls of the Origin, Characteristics, and Distribution of Ground Ice." In *Proceedings of the Third International Conference on Permafrost* 2: 1–18.
Matthes, F. E. 1900. *Glacial Sculpture of the Big Horn Mountains, Wyoming*. U.S. Geological Survey 21st Annual Report. Washington, D.C.: U.S. Geological Survey, pp. 167–190. First described freeze-thaw cycles as a specific erosional phenomenon.
Peltier, L. C. 1950. "The Geographic Cycle in Periglacial Regions as It Is Related to Climatic Geomorphology." *Association of American Geographers Annals* 40: 214–236. First to propose the now discredited idea of cyclic behavior in periglacial processes and landscapes.
Péwé, T. L. 1969. "The Periglacial Environment." In T. L. Pewe, ed. *The Periglacial Environment*. Montreal: McGill-Queen's University Press, pp. 1–9.
Research Group on Experimental Roadbed Research. 1979. "Experimental Roadbed in an Area with a Thick Layer of Ground Ice." In *Proceedings of the Third International Conference on Permafrost* 2: 187–198.

Research Group on Pile Foundation in Permafrost. 1979. "Testing of Pile Foundations in Permafrost." In *Proceedings of the Third International Conference on Permafrost* 2: 179–186.

Scott, K. M. 1978. "Effects of Permafrost on Stream Channel Behavior in Arctic Alaska." U.S. Geological Survey Professional Paper, no. 1068. Washington, D.C.: U.S. Geological Survey.

Tricart, Jean. 1967. "Le Modele des Regions Periglaciaires." In J. Tricart and A. Cailleux, eds. *Traite de Geomorphologie*. Vol. 2. Paris: Sedes.

Washburn, A. L. 1980. *Geocryology: A Survey of Periglacial Processes and Environments*. New York: John Wiley and Sons. The best synthesis on periglacial phenomena available.

Wilson, Lee. 1968. "Morphogenetic Classification." In Fairbridge, R. W., ed., *The Encyclopedia of Geomorphology*. New York: Reinhold Book Corp., pp. 717–729.

———. 1969. "Les Relations entre les Processus Geomorphologiques et le Climat Moderne Comme Methode de Paleoclimatologie." *Revue de Geographie Physique et de Geologie Dynamique* 11: 303–314.

Sources of Additional Information

The journal that specifically deals with periglacial processes and landforms is *Biuletyn Peryglacjalny* (1954-present). The terminology in periglacial studies is large and complex. For definitions see R.J.E. Brown and W. O. Kupsch, "Permafrost Terminology," Association Committee on Geotechnical Research Technical Memorandum, no. 111, (Ottawa: Canada National Research Council, 1974); J. S. Fyodorov (compiler) and N. S. Ivanov, ed., *English-Russian Geocryological Dictionary* (Yakutsk: Yakutsk State University, 1974); and V. Poppe and R.J.E. Brown, "Russian-English Glossary of Permafrost Terms," Association Committee on Geotechnical Research Technical Memorandum, no. 117 (Ottawa: Canada National Research Council, 1976). For bibliographic information of recent work see C. Harris, *Periglacial Mass-Wasting: A Review of Research*, GeoAbstracts, British Geomorphological Research Group Research Monograph, no. 4 (Norwich, Eng.: British Geomorphological Research Group, 1981), and A. M. Brennan, "Permafrost: A Bibliography 1978–82," *Glaciological Data* 14 (1983): 162, 1972. To understand past research and future research directions see The International Conferences on Permafrost, 1966, 1973, 1978, and 1983. For a history of geomorphology in general and periglacial processes in particular see Keith J. Tinkler, *A Short History of Geomorphology* (Totowa, N.J.: Barnes and Noble Books, 1985). A very good collection of the seminal papers in periglacial research is Cuchlaine A. M. King, ed., *Periglacial Processes* (Stroudsburg, Pa.: Dowden, Hutchinson and Ross, 1976).

PERMAFROST. 1. Any surficial deposit, including soil and subsoil, found in an alpine, arctic, or subarctic region, in which a temperature below freezing has existed continuously for at least two years. **2.** Permanently frozen ground found in an arctic, subarctic, or alpine area including the active layer above the frost zone.

Between 20 and 25 percent of the earth's land surface, according to Cuchlain King (1976: 3), is underlain by permafrost. The largest areas are found in the USSR and Canada, where nearly half of the total land area of these two countries is underlain by permafrost. This perennially frozen ground covers approximately 8.6 million square miles in the northern hemisphere and 5.1 million square miles in the southern hemisphere (Price, 1972: 10). Much of the circumpolar area of the northern hemisphere is underlain by permafrost, but Antarctica is the only major area of occurrence in the southern hemisphere. Consequently, most of the knowledge of permafrost conditions at this time has been gained from the north.

Since much of the USSR is underlain with permafrost, it is not surprising that the USSR is the forerunner in the compilation of permafrost data and in research and development on perennially frozen ground. Although general descriptions of ground ice and other PERIGLACIAL processes were reported as early as 1806 in the Soviet Union, the first detailed study of frozen ground was undertaken by A. F. Von Middendorf from 1844 to 1846 (Britton, 1973: 31). He studied the temperature regime of a well shaft and analyzed the temperature changes with depth.

The early and steady northerly settlement of the USSR Arctic coast and the development of OCEAN shelf mineral resources created a growing interest in the properties of subsea permafrost and its adverse effect on construction. Expeditions to study subsea permafrost occurrences were made as early as the beginning of this century. In the decades of 1926–1946 the offshore areas surrounding the Soviet coastline were again the subject of much research. These studies, as outlined by Michael Vigdorchik (1980), have provided a continuous and centralized accumulation of experience and data for Soviet scientists.

In 1930 the USSR, pushing for further settlement of the sparsely populated North and East, established the first formal agency to assemble data and to investigate permafrost further. This was the Commission for the Study of Permafrost, part of the Soviet Academy of Sciences. Since 1930 all governmental entities must make a thorough and systematic survey of permafrost conditions before any permanent structures can be erected in a permafrost region. By 1940 there were special permafrost laboratories at Moscow and Leningrad, as well as four field stations in the north (Vigdorchik, 1980: 3).

According to King (1976: 153), irregular surface features indicating the presence of permafrost conditions had been observed before the 1940s in North America, but the true significance of permafrost as a major phenomena was not realized until World War II. Attempts by the American armed forces to obtain permanent water supplies and construct roads, buildings, and runways in permafrost regions focused attention on permafrost problems and the lack of research on permafrost development. Realizing the strategic importance of arctic areas and the need for military and civilian construction projects, the need to solve permafrost problems was addressed. Also, construction of the Alaska Highway and other civilian construction projects further highlighted the importance of understanding permafrost. Thus, according to Max Britton (1973: 31), the late 1940s saw serious efforts get underway to investigate the frozen ground conditions in the Alaskan arctic.

A landmark study at this time was undertaken by Siemon W. Muller (1947), who is credited with first using the term *permafrost*. In his 1947 study Muller found that the largest temperature changes in permafrost areas took place directly above the permafrost zone in the active layer. Further thermal studies of permafrost were initiated at the Naval Arctic Research Laboratory at Point Barrow, Alaska, in 1949. Temperature observations continued throughout the 1950s resulting in numerous papers that document results, explore heat-flow energy, and interpret geomorphic Events (Britton, 1973: 31). Also, a study by Robert Black (1954) looks at the geographic distribution of permafrost and outlines the major role played by climate in determining that distribution.

If it was the influx of civilian and military personnel into Alaska in the mid-1940s that brought about the first major American research thrust, it was the 1969 discovery of oil at Prudhoe Bay in Alaska that was the catalyst for much of the current research on permafrost. To transport the oil from Prudhoe Bay it was necessary to build the Alaska pipeline. Oil companies, therefore, became interested in the research and preplanning associated with the building of the pipeline (Kane, 1975: 18–19). According to A. L. Washburn (1973: 43), the mechanical properties of permafrost associated with freezing and thawing became of paramount importance.

Within the past twenty years there have also been three international conferences dealing exclusively with permafrost. The first conference was held in the United States at Purdue University in 1963. Attended by research scientists and technologists, the conference was devoted to scientific and developmental problems associated with perennially frozen ground. The Second Internal Conference was held at Yakutssk, USSR, in July 1973, with the objective of furthering interdisciplinary exchange of information in the field. Accelerated developmental activities in these areas greatly increased the need for knowledge of both sound construction practices and the associated environmental implications. For this reason, the scope of the second conference was broadened to include the consideration of environmental, ecological, and resource development issues (National Academy of Sciences, 1973). The Third International Conference was held in Edmonton, Canada, in July 1978 with engineers and scientists from more than thirteen countries in attendance.

Current research in permafrost covers a variety of topics in many places around the world. From the analysis of core drillings in the Swiss Alps by Dielrich Barsch (1979) to the research on periglacial wedges in Wyoming by Brainerd Means (1981), geographers and other earth scientists are trying to understand the complex processes associated with permafrost development.

References

Barsch, Dielrich, 1979. "Shallow Core Drilling and Bore-Hole Measurements in the Permafrost of Active Rock Glaciers Near the Grubenglelscher, Wallis, Swiss Alps." *Arctic and Alpine Research* 11: 215–228. A recent analysis of permafrost and rock glaciers.

Black, Robert F. 1954. "Permafrost: A Review." *Geological Society of America Bulletin*

65: 839–855. A somewhat outdated, although useful, article reviewing current research up to 1954.
Britton, Max E., ed. 1973. *"Alaskan Arctic Tundra." Arctic Institute of North America, Technical Paper*, no. 25. Washington, D.C.: D. Saulis Lithograph.
Kane, J. T., 1975. "Alaskan Pipeline Begins Major Thrust toward Petroleum Basins in Arctic Region." *Professional Engineer* 45: 18–19. A discussion of the Alaskan pipeline from the engineering viewpoint.
King, Cuchlaine, ed. 1976. *Periglacial Processes*. Stroudsburg, Pa.: Dowden, Hutchinson and Ross. An excellent textbook on periglacial processes with a good discussion on permafrost.
Mears, Brainerd, Jr. 1981. "Periglacial Wedges and the Late Pleistocene Environment of Wyoming's Intermontane Basins." *Quarterly Research* 15: 171–198.
Muller, Siemon W. 1947. *Permafrost of Permanently Frozen Ground annd Related Engineering Problems*. Ann Arbor, Mich.: Edwards Bros. One of the early studies in English by the person who coined the term *permafrost*.
National Academy of Sciences. 1973. *North American Contribution to Permafrost: Second International Conference*. Washington, D.C.: National Academy of Sciences.
Price, Larry W., 1972. "The Periglacial Environment, Permafrost and Man." Resource Paper, no. 14. Washington, D.C.: Association of American Geographers. A good synopsis of research from a geographic point of view.
Vigdorchik, Michael E., 1980. *Arctic Pleistocene History and the Development of Submarine Permafrost*. Boulder, Colo.: Westview Press. An excellent Analysis of submarine permafrost by a former member of the USSR Academy of Sciences and one of the world's foremost experts on permafrost.
Washburn, A. L. 1973. *Periglacial Processes and Environments*. New York: St. Martin's Press.

Sources of Additional Information

For a short definition of the concept see Robert L. Bates and Julia A. Jackson, *Glossary of Geology*, 2d ed. (Falls Church, Va.: Geological Institute, 1980); W. C. Moore, *A Dictionary of Geography* (New York: Harper and Row, Publishers, 1978); and R. W. Fairbridge, *The Encyclopedia of Geomorphology* (New York: Reinhold Book Corp., 1968). A discussion of the concept can be found in J. Tricart, *Geomorphology of Cold Environments* (London: Macmillan, 1970); A. L. Washburn, *Geocryology: A Survey of Periglacial Processes and Environments* (New York: John Wiley and Sons, 1980); Clifford Embleton, *Glacial and Periglacial Geomorphology* (New York: John Wiley and Sons, 1975); and Hugh M. French, *The Periglacial Environment* (London: Longmans, Green and Co., 1976. Detailed studies of particular aspects of permafrost are Kevin M. Scott, *Effects of Permafrost on Stream Channel Behavior in Arctic Alaska* (Washington, D.C.: U.S. Government Printing Office, 1978); R. B. Williams, "Frost and the Works of Man," *Antiquity* 45 (1973): 19–31; Glenn Brown, "Palsas and Other Permafrost Features in the Lower Rock Creek Valley, West Central Alberta," *Arctic and Alpine Research* 12 (1980): 31–40; and Louis Ray, *Permafrost* (Washington, D.C.: U.S. Government Printing Office, 1973). A discussion of the importance of permafrost can be found in these popular press articles: "North Slope, Will Oil and Tundra Mix?" *National Geographic* 140 (1971): 485–515, and "Alaska Embarks on Its Biggest Boom as Oil Pipeline Gets Under Way," *Smithsonian* 5 (1974): 38–48.

PETROLOGY. See ROCK.

PHYSICAL GEOGRAPHY. 1. The geographical study of the earth's natural features including land, water, and air. 2. The science that deals with the natural or physical features of the earth's surface and the processes that determine or explain their aerial distribution.

The history of geographic study can be traced to prehistoric times. People in early hunting and gathering societies had to develop a sense of distance, direction, and location and were intimately associated with the natural features of the landscape. Much of their understanding of the physical world, however, was based on myths and legendary tales. The first significant contributions to geographic knowledge were made by people from the ancient Mediterranean world (Pearson, 1968).

Both the Egyptians and Babylonians developed exact surveying methods by the year 2000 B.C. Early large-scale maps (see MAP PROJECTION) were developed to locate boundaries and locations of quarries and mines, but small-scale maps of large areas were not developed until the sixth century B.C., when the Ionian Greeks initiated the scientific study of geography.

During the six centuries that preceded the Christian Era many Greek scholars were concerned with the description and explanation of the earth's surface features. Thales of Miletus (624–546 B.C.) was foremost among these early scholars. He recognized the earth's curvature, and according to tradition, he accurately predicted the occurrence of the solar eclipse of May 27, 585 B.C. (Kish, 1978: 10). Thales's pupil and friend Anaximandes (c. 611–547 B.C.) further developed these geographic concepts and is credited with being the true founder of geography and the person who compiled the first known map of the inhabited Western world (Kahn, 1960). Anaximandes is also credited with inventing the sun dial and gnomon for determining latitude.

Two other early Greek scholars, Hecataeus (c. 500 B.C.) and Herodotus (c. 485–428 B.C.), were noted for their wide travels and descriptions of Mediterranean lands. The most noted of Greek scholars, Aristotle (c. 384–322 B.C.), also made important contributions to physical geography. He defined the climatic zones—torrid, temperate, and frigid—and developed basic concepts about weather, EARTHQUAKES, and EROSION by STREAMS and waves (Warmington, 1934). It was during Aristotle's time that the word *geography*, meaning "ge" (the earth) and "graphein" (to write), became an accepted part of the Greek language.

The Greeks were primarily concerned with theoretical advances in geography, whereas the Roman geographers who followed them were primarily interested in encyclopedic descriptions. Foremost among the descriptive works by Roman geographers was the seventeen-volume *Geographica* in which Strabo (c. 63 B.C.– c. A.D. 25) presented a detailed survey of the habitable world (Jones, 1924) and the eight-volume work by Ptolemy (c. A.D. 100–c. A.D. 170), which included the first atlas of maps and a summary of Greek knowledge about geography (Stevenson, 1932).

Most of the geographical advances made by the Greeks and Romans practically disappeared during the medieval period. Most scholars devoted their energies to theological topics, and the study of "pagan science" was forbidden by the Christian church. Interest in geography, however, was maintained in the Moslem world. In a study of Moslem geography, George Kish noted:

> The Moslem contribution to geography took many forms: the translation and absorption of learning from both West and East—from Greek, Persian, and Sanskrit sources, and possibly from Chinese ones as well; the preservation of writings that were essential, later on, to the revival of geography in the Christian world; the compilation of geographical compendia on the world of Islam, which extended from the Atlantic to the Pacific; and original contributions in mathematical geography and surveying. (1978: 199)

The revival of learning in Europe that took place after 1250 brought about an increased interest in understanding the nature and processes associated with the earth's physical environment. Much of this interest was revived by the Crusades and the travels of Marco Polo and others (Parks, 1929). A growing interest in lucrative trade routes to the Orient led men such as Prince Henry the Navigator of Portugal to make many voyages of discovery during the fifteenth and sixteenth centuries. The general outlines of the continents were determined, and for several centuries the primary task of geographers was to make accurate maps of discoveries as well as descriptions of these new lands (Pearson, 1968: 3).

Until the latter part of the eighteenth century most geographers were concerned with description and mapping of newly explored lands. They were primarily engaged in the collection and compiling of data and geographic descriptions as well as filling in the details of the world map. A change in emphasis to the formulation of general principles in physical geography is usually credited to the German geographer and philosopher Immanuel Kant. Kant offered a course of lectures on physical geography at the University of Konegsberg, in Prussia, for forty years from 1757 to 1797. He claimed physical geography to be "a summary of nature." Physical geography thus embraced the outer physical world, the earth's surface, and its cover of plant and animal life as well as man and his works (Dickinson, 1969: 11).

Most modern physical geographers look upon another German geographer, Alexander von Humboldt, as one of the first founders of modern physical geography (Troll, 1960). In his great work *Kosmos*, von Humboldt advanced the principle of the unity of nature and attempted to document the harmony of nature. As George Kish said regarding Humboldt's works, "These are pioneer works in geography and in natural history; monumental in their scope, innovative in their concepts, original in their organization, impressive in their magnificent illustrations" (1978: 403).

The latter part of the nineteenth century and early twentieth century saw many major advances in physical geography. Foremost among these advances was the development of landscape studies in GEOMORPHOLOGY, with pioneering work by William Morris Davis (1899), and advancements in CLIMATOLOGY resulting from the work of Wladimir Köppen (1931).

The science of physical geography has, like many other sciences, made enormous advances in the twentieth century. Three major trends can be differentiated within the several branches of physical geography in the past several decades. First, new technologies relative to data-collection and technical processing systems have been applied to physical distributions. Sophisticated techniques of aerial photography (Parry, 1967) and REMOTE SENSING (McKee, 1979) have been instrumental in adding new dimensions to geographical studies. Second, instead of just emphasizing the distribution of physical features, more emphasis is now being placed on process-oriented approaches to understanding the physical environment (Gregory, 1985). Third, the interrelationship between humans and the physical environment has been reactivated as a main area of study. Of particular importance has been the development of the application of physical geography principles to human problems (Tufnell, 1984; Craig and Craft, 1982).

References

Craig, R. C., and Craft, J. L. 1982. *Applied Geomorphology*. Binghamton: State University of New York. A good presentation of recent work in applications of geography.
Dickinson, Robert E. 1969. *The Makers of Modern Geography*. London: Routledge and Kegan Paul. A good book for the background to modern geography.
Gregory, K. J. 1985. *The Nature of Physical Geography*. London: E. Arnold. An excellent book outlining the development of modern physical geography.
Jones, H. L. 1924. *The Geography of Strabo*. London: W. Heinemann.
Kahn, Charles H. 1960. *Anaximander and the Origins of Greek Cosmology*. New York: Columbia University Press.
Kish, George. 1978. *A Source Book in Geography*. Cambridge: Harvard University Press. An excellent book for understanding the developments in geography before the twentieth century.
Köppen, W. 1931. *Grundriss der Klimakunde*. Berlin: Walter DeGruyter Co. A classic study and development of a climate classification system.
McKee, E. D., ed. 1979. "Global Sand Seas," U.S. Geological Survey Professional Paper, no. 1052. Washington, D.C.: U.S. Government Printing Office. A good discussion of remote-sensing techniques.
Parks, G. B. 1929. *The Book of Sir Marco Polo*. New York: Macmillan.
Parry, J. T. 1967. "Geomorphology in Canada." *Canadian Geographer* 9: 280–311. A discussion of the importance of aerial photography.
Pearson, Ross N. 1968. *Physical Geography*. New York: Barnes and Noble Books. An outline of the major concepts of physical geography is presented.
Stevenson, Edward L. 1932. *The Geography of Ptolemy*. New York: New York Public Library.
Troll, Carl. 1960. "The Work of Alexander von Humboldt and Carl Ritter: A Centenary Address." *The Advancement of Science* 64: 441.

Tufnell, L. 1984. *Glacier Hazards*. London: Longmans, Green and Co. A recent book on hazards associated with glaciers.

Warmington, E. H. 1934. *Greek Geography*. New York: E. P. Dutton.

Sources of Additional Information

For more detailed discussions on the nature of geography and its historical development see William Warntz, *Geography Now and Then* (New York: American Geographical Society, 1964); Richard Hartshorne, *Perspective on the Nature of Geography* (Chicago: Rand McNally and Co., 1959); R. J. Chorley, *Directions in Geography* (London: Methuen & Co., 1973); R. J. Chorley, A. J. Dunn, and R. P. Beckinsale, *The History of the Study of Landforms. Vol. I, Geomorphology before Davis* (London: Methuen & Co., 1964); R. J. Chorley, R. P. Beckinsale, and A. J. Dunn, *The History of the Study of Landforms. Vol. II, The Life and Work of William Morris Davis* (London: Methuen & Co., 1973), and E. H. Brown, ed., *Geography Yesterday and Tomorrow* (Oxford: Oxford University Press, 1980). For a historical account of the development of the subfields of physical geography in the United States see Preston E. James and Clarence F. Jones, ed., *American Geography: Inventory and Prospect* (Syracuse, N.Y.: Syracuse University Press, 1954).

PHYSICAL WEATHERING. See WEATHERING.

PHYTOGEOGRAPHY. See BIOGEOGRAPHY.

PLANT SUCCESSION/CLIMAX. 1. The progressive, unidirectional natural development of vegetation (plant succession) toward a final form (climax), during which one community is gradually replaced by others. 2. The progressive change in a biologic community as a result of the response of the member species to the environment (succession). 3. The terminal community in ecological succession capable of self-replacement under prevailing environmental conditions (climax).

Although *plant succession* and *climax* are two separate terms, they will be treated together here because in general use they are a part of the same process. Climax vegetation is the end product of successional development. Therefore, plant succession leads to climax vegetation in an area, and the two terms essentially form a continuum.

Changes in plant communities in a given area over time have been observed for centuries. King, for example, observed the vegetational changes in Irish bogs in 1685 (Whittaker, 1953: 268) and Georges Buffon noted the progression of shrub species in France in 1742 (Daubenmire, 1968: 99). Excellent descriptions of plant succession in sand dunes, burned forest land, and other habitats were published by Kerner in 1863 (Westhoff and van der Maarel, 1978: 290).

The modern concept of succession was largely developed by two botanists, J.E.B. Warming (Whittaker, 1953: 276) and H. C. Cowles (1901). They both observed the successional stages of vegetation development in sand dunes. The classical ideas of succession were elaborately discussed by F. E. Clements (1916,

1936). There are two general hypotheses of plant succession and community development in the classical model. The first is the *monoclimax hypothesis*, which describes a highly integrated succession to a single end point called the climatic climax. This succession is gradual and progressive from simple pioneer communities to the ultimate climax stage. Retrogression is not possible unless some outside disturbance occurs such as fire or erosion. Joseph Connell and Ralph Slayter (1977: 1120) coined the term *facilitation* for this model of succession.

The second hypothesis is called the *initial floristic composition model* (Egler, 1954). This hypothesis shows succession as being very heterogenous and dependent upon which species is the pioneer in an area. This is a more individualistic, less predictable succession and has been called the inhibition model (Connell and Slayter, 1977: 1123).

A third major hypothesis intermediate between the above classical hypotheses was proposed by Connell and Slayter (1977: 1122–1123). This *tolerance model* states that no matter which species is the pioneer, there are others that may become dominant. Some species may be replaced by others that are more tolerant of the environmental conditions. Charles Krebs put these three models into perspective:

> The critical distinction among the three hypotheses is in the mechanisms that determine subsequent establishment. In the classical facilitation model, species replacement is facilitated by the previous stages. In the inhibition model, special replacement is inhibited by the present residents until they are damaged or killed. In the third model, species replacement is not affected by the present residents. (1985: 488)

The final stage of succession is the climax stage. There are three basic schools of thought about the climax stage. First is the *monoclimax* (same as above) *theory* proposed by Clements (1916, 1936). According to this theory there is only one possible climax community for any given region. Given enough time and freedom from interference, the climax vegetation will be produced and stabilized despite earlier conditions. This theory assumes that climate is the one true determinant of climax vegetation. Clements allowed for vegetation equilibrium that is nonclimax and depends upon other factors such as SOIL, topography, and biotic activity.

A *polyclimax theory* developed in response to the monoclimax theory. A. G. Tansley (1939) was one of the original proponents of this newer theory. He believed that the nonclimax equilibrium communities were truly climax communities in their own right and were based on soil moisture, soil nutrients, animal disturbance, and other factors. R. F. Daubenmire (1966) is also a strong proponent of this view. Probably, the real difference between these two theories is the time factor allowable for climax generation (Krebs, 1985: 503). The monoclimax enthusiasts would state that "given enough time," a single climax community would develop. The polyclimax enthusiasts believe that there is seldom enough time since environmental conditions change over long periods.

Robert Whittaker (1953) proposed a variation in the polyclimax theory often called the *climax-pattern theory*. This theory allows for several climaxes varying gradually according to environmental conditions or gradients. This approach has also been termed *gradational analysis*. For example, various climaxes in a region are said to be dependent upon "gradation" of climate, soil, topography, fire, chance, and other environmental conditions.

The more recent research stresses several lines of inquiry but usually deals with succession mechanisms rather than the general theory of succession and climax. Examples of this kind of research are in the work of Connell and Slayter (1977), Jeremy Jackson (1981), V. K. Brown (1984), and S. E. Hodgkin (1984). Modelling of ecological succession has also become a major research topic as seen in the work of M. B. Usher (1981) and C.W.D. Gibson et al. (1983). With the proliferation of computer technology, modeling of succession should become an even more common line of inquiry.

References

Brown, V. K. 1984. "The Ecology of Secondary Succession: Insect-Plant Relationships." *Bioscience* 34: 710–716.
Clements, F. E. 1916. *Plant Succession: An Analysis of the Development of Vegetation.* Publication no. 242. Washington, D.C.: Carnegie Institute. This is the seminal work for modern successional theory.
———. 1936. "Nature and Structure of the Climax." *Journal of Ecology* 24: 252–284.
Connell, Joseph H., and Slayter, Ralph O. 1977. "Mechanisms of Succession in Natural Communities and Their Role in Community Stability and Organization." *The American Naturalist* 111: 1119–1144.
Cowles, H. C. 1901. "The Physiographic Ecology of Chicago and Vicinity." *Botanical Gazette* 31: 73–108, 145–181. This work helped develop the modern idea of succession.
Daubenmire, R. F. 1966. "Vegetation: Identification of Typal Communities," *Science* 151: 291–298.
———. 1968. *Plant Communities.* New York: Harper and Row, Publishers.
Egler, F. E. 1954. "Vegetation Science Concepts. I. Initial Fluristic Composition, a Factor in Old-Field Vegetation Development." *Vegetatio* 14: 412–417.
Gibson, C.W.D.; Guilford, T. C.; Hambler, C.; and Sterling, P. H. 1983. "Transition Matrix Models and Succession after Release from Grazing on Aldabra Atoll." *Vegetatio* 52: 151–159.
Hodgkin, S. E. 1984. "Shrub Encroachment and Its Effect on Soil Fertility on Newborough Warren, Anglesey, Wales." *Biological Conservation* 29: 99–119.
Jackson, Jeremy B. C. 1981. "Interspecific Competition and Species Distributions: The Ghosts of Theories and Data Past." *American Zoologist* 21: 889–901.
Krebs, Charles J. 1985. *Ecology.* 3d ed., New York: Harper and Row, Publishers. This is an excellent text with sections dealing with the history of succession and climax.
Tansley, A. G. 1939. *The British Islands and Their Vegetation.* Cambridge: Cambridge

University Press. This work proposed the polyclimax theory of vegetation communities.

Usher, M. B. 1981. "Modelling Ecological Succession, with Particular Reference to Markovian Models," *Vegetatio* 46: 11–18.

Westhoff, Victor, and Eddy van der Maarel, 1978. "The Braun-Blanquet Approach." In Whittaker, Robert H., ed., *Classification of Plant Communities*. Boston: Dr. W. Junk bv. Publishers, pp. 287–399.

Whittaker, Robert H. 1953. "A Consideration of Climax Theory: The Climax as a Population and Pattern." *Ecological Monographs* 23: 41–78. Proposed the climax-pattern history theory of vegetation climax and introduced gradient analysis as a viable model.

Sources of Additional Information

For the general history and progress of the study of succession and climax see P. A. Werner, "Ecology of Plant Populations in Successional Environments," *Systematic Botany* 1 (1976): 246–268; F. B. Golley, *Ecological Succession* (Stroudsburg, Pa.: Dowden, Hutchinson and Ross, 1977); R. H. Whittaker, ed., *Classification of Plant Communities* (Boston: Dr. W. Junk bv. Publishers, 1978); and C.W.D. Gibson and V. K. Brown, "Plant Succession: Theory and Applications," *Progress in Physical Geography* 9 (1985): 473–493. For an excellent review of the contributions of Frederic Clements see D. Worster, *Nature's Economy: the Roots of Ecology* (San Francisco: Sierra Club Books, 1977. For an interesting history of the study of patterns of succession see V. E. Shelford, "Ecological Succession. II. Pond Fishes," *Biological Bulletin* 21 (1911): 127–151; W. S. Cooper, "The Recent Ecological History of Glacier Bay, Alaska: II. The Present Vegetation Cycle," *Ecology* 4 (1923): 223–246; W. M. Crofton and B. W. Wells, "The Old Field Prisere: An Ecological Study," *Journal of the Elisha Mitchell Science Society* 50 (1934): 225–246; and E. P. Odum, "Organic Production and Turnover in Old Field Succession," *Ecology* 41 (1960): 34–49. For an idea of the research done on mechanisms of succession see H. J. Oosting and M. E. Humphreys, "Buried Viable Seeds in a Successional Series of Old Field and Forest Soils," *Torre Botanical Club Bulletin* 67 (1940): 253–273; C. Keever, "Causes of Succession in Old Fields of the Piedmont, North Carolina," *Ecological Monographs* 20 (1950): 231–250; and P. L. Marks and F. H. Bormann, "Revegetation Following Forest Cutting: Mechanisms for Return to Steady State Nutrient Cycling," *Science* 176 (1972): 914–915. For insight into the interpretation of successional stages see E. P. Odum, "The Strategy of Ecosystem Development," *Science* 164 (1969): 262–270, and W. H. Drury and I.C.T. Nisbet, "Succession," *Journal of the Arnold Arboretum* 54 (1973): 331–368. Examples of work on the stability of vegetation communities include R. C. Lewontin, "The Meaning of Stability," *Diversity and Stability—Ecological Systems*, Brookhaven Symposium of Biology, 22 (1969): 13–24; O. L. Loucks, "Evolution of Diversity, Efficiency, and Community Stability," *American Zoologist* 10 (1970): 17–25; L. E. Hurd, M. V. Mellinger, L. L. Wolf, and S. J. McNaughton, "Stability and Diversity at Three Trophic Levels in Terrestrial Successional Ecosystems," *Science* 173 (1971): 1134–1135; and R. M. May, "Will a Large Complex System be Stable?" *Nature* 283 (1972): 413–414.

PLATE TECTONICS. A theory of global dynamics involving the movement of a dozen or so thin, rigid plates. These plates float on top of hot, semiplastic rocks beneath them and result in continental drift and changes in the shape and size of oceanic basins and continents.

Why the earth is shaped and formed as it is has been a subject of conjecture and speculation for thousands of years. Although many at the time probably wondered what they were talking about, it appears that the works of Aristotle, Strabo, and Pliny the Elder indicate they were beginning to wonder how it all came about.

By the 1600s countless courageous explorers had proven the world was not flat but round, and the major land masses and oceans had begun to take shape in the form of crude maps. Although these early maps were not very accurate by today's standards, there was enough detail to outline the general form of the continents we are so familiar with today.

These early maps were to lead Francis Bacon (1620) in his *Novum Organum* to comment on the general conformity of the outline of Africa and South America. He mentioned how both continents tapered southward and that there was a similarity between the eastern coast of South America and the western coast of Africa. Shortly after Bacon's speculations, the French moralist Francois Placet published a booklet in 1666 entitled *La Corruption du grand et petit Monde* in which he proposed that before the Noachian flood there was one earth (Carozzi, 1970). At the beginning of the nineteenth century the German explorer and geographer Alexander von Humboldt commented on the congruence of the African and South American coastlines and speculated that the Atlantic Ocean was a huge valley scooped out by the sea (Sullivan, 1974: 2).

The first real clear-cut indication of the breakup and movement apart of the continents came in 1858 with the publication of *La Creation et ses mysteres devoiles* by Antonio Snider-Pellegrini. Snider-Pellegrini theorized that the earth cooled from a molten mass and crystallized, thus causing catastrophic fracturing and pulling apart of the land masses. To support this catastrophic fracturing thesis, he cited the coastal fit of Africa and South America as supporting evidence.

The latter part of the nineteenth century brought about the rise of uniformitarian concepts and the decline of catastrophic ideas like those of Snider-Pellegrini. Near the end of the nineteenth century Eduard Suess (1900) noted that much of the geologic record in the southern continents of South America, Africa, Australia, and parts of India was nearly identical. He thought these continents had all been united at one time in a supercontinent that he called Gondwana-Land.

Others explained the breakup of the continents and the opening of the Atlantic OCEAN in terms of George Darwin's (1879) idea that the moon was thrown off from a fast spinning earth and the hole that was left formed the Pacific Ocean. Darwin speculated that to fill this hole, the flow of material dragged the Americas away from Europe, thus forming the Atlantic Ocean. This association of continental drift with the origin of the moon dominated much of the later ideas into the twentieth century (Baker, 1932; Pickering, 1907).

Although many European scientists were interested in what later came to be called continental drift, the first published paper presenting a logically worked out and coherent hypothesis involving what we now recognize as continental drift was that of the American scientist Frank B. Taylor (1910). Taylor introduced the idea of slow movement or crustal creep by the continents. Although he laid out a coherent theory, he paid little attention to the mechanism of continental movement and presented only scant evidence to support his ideas.

Even though many scientists noted the congruence of the African and South American coastlines and speculated on continental drift, the one version of the theory that stands out is that of Alfred Wegener (1912). Wegener is noted as the father of continental drift, and his hypothesis "stands not merely as a forerunner of the concept that now prevails, but as its true runner" (Hallam, 1975: 88). The first suspicion Wegener had that the continents might have moved laterally came to him in 1910 when he was amazed by the remarkable coincidence of the fitting of the African and South American coastlines. At the same time, however, he came across paleontological evidence that pointed to a former land connection between Brazil and Africa (Hallam, 1972).

Wegener first presented his ideas in 1912 and later fully developed the concept in *The Origins of Continents and Oceans*. During the last eighteen years of his life, until his death in 1930, he continued to refine and improve his original ideas.

According to Irene Keifer, in a study of the development of plate-tectonics theory, Wegener proposed

> that at one time all the continents were one huge land mass, which he named "Pangea." About 40 million years ago Pangea began breaking into fragments that, like giant ships, plowed through the ocean floor. North America moved west and the southern continents moved toward the equator, reaching their present positions about a million years ago. The continents moved... because of the same forces that produce great mountain ranges. These forces are related to the earth's daily rotation about its axis. (1978: 14–15)

During the 1920s, Wegener's theory drew great controversy. Partially because he was by trade a meteorologist and partially because he apparently mixed in good arguments with poor ones, his hypothesis was severely criticized by many geologists and earth scientists. Geologists showed that some of Wegener's suggestions for reassembling the continents into a single continent were certainly wrong and that drift was unnecessary to explain the coincidences of geology in many areas (Wilson, 1963). Another problem with Wegener's theory was that there was no force strong enough to move the continents through the ocean floors. So in 1930 the theory of continental drift died and remained dormant until it was reborn following World War II.

As a result of the war effort, new scientific techniques and principles were developed, techniques that were not available to Wegener. According to Keifer, however, "the most important parts of Wegener's work have been proved to be correct, although he was wrong on some of the details. Wegener had far fewer clues than modern scientists, but he looked at all the clues, not just those in his own field. With imagination and insight, he conceived a great scientific idea" (1978: 18).

Interest in continental drift began to revive in the 1950s and 1960s. This was largely due to the development of the concept of sea-floor spreading and rock magnetism. Sea-floor spreading involves the concept that the floor of the ocean is being continuously pulled apart or spreading along a narrow crack that is centered on a ridge that can be traced through the major ocean basins. In the early 1960s Harry Hess, a geologist at Princeton University, elaborated on a theory of sea-floor spreading. Hess (1962) believed that new ocean floor was continually manufactured at the mid-oceanic ridges, where it traveled slowly outward, eventually to be destroyed in ocean trenches where the old ocean floor sinks into the mantle.

During this time ocean scientists had been puzzled over their recent find that sensitive magnetic detectors onboard research ships had revealed magnetic irregularities across the ocean floor. When one field was weaker and another stronger, it showed up as magnetic stripes, or bands. Then Fred Vine, a 23-year-old research student at Cambridge University who was familiar with Hess's theory of ocean-floor spreading, reasoned that the ocean floor was partially molten at the time of its formation at the mid-ocean ridge. As it cooled, the rocks became permanently magnetized by the earth's prevailing field, which was sometimes reversed or flipped over from its present direction. This reasoning, backed up by more precise calculations, led to an extraordinary fact of nature. The ocean floor is like a huge tape recorder, telling the story of its own formation and growth. The magnetic field of the earth, therefore, imposes a series of time markers at the formation of each new ocean-floor segment (Vine and Matthews, 1963).

The discovery of symmetrical patterns of magnetized rock strips on either side of these mid-oceanic ridges was a critical factor in confirming the theory that new crustal material forms as the ridges spread apart (Simon, 1982). This confirmation of sea-floor spreading presents the mechanism or force that is strong enough to move the continents, an idea Wegener's theory lacked.

The concept of sea-floor spreading has now been linked to the earlier idea of continental drift in a single unifying theme now called plate tectonics. In 1965 J. Tuzo Wilson proposed that a network of several large rigid plates make up the crust of the earth. The size and disposition of the plates were thought to generate volcanic (see VOLCANO) and EARTHQUAKE activity as they moved. This activity, when plotted worldwide using recorded volcanic and earthquake-activity data, conformed to the edge outlines of the plates as they were thought to exist. If the boundaries of these plates could be seen, they would make the earth's surface resemble a cracked eggshell. The plates are in constant motion and can slide past or collide with one another. Thus we have arrived at the present theory of plate tectonics. The concept of plate tectonics, therefore, is the present theory used to explain continental drift and the terms are often used interchangeably.

The concept of plate tectonics is now widely accepted by earth scientists. The past several decades have seen a great deal of research in an effort to understand various aspects of plate dynamics. Included in these research activities have been studies on the reconstruction of past continental landmasses (McElhinny and Valencio, 1981), the implications of plate movement for volcanic activity, the nature of subduction zones where one plate moves under another (Wyatt, 1984), and the relationship between plate tectonics and geomagnetic changes (Cox, 1973).

Alfred Wegener and scientists like him knew nothing of the continents resting on movable plates, but they did see the similarities in continental coastal areas, and they came to the conclusion that the continents were formerly a single landmass. Although the cause of plate movement is still speculative, the theory of plate tectonics gives the evidence of how the supercontinent split apart and how the earth's crust behaves today.

References

Bacon, Francis. 1620. *Novum Organum* (trans. G. W. Kitchin). London: Oxford University Press, 1855.

Baker, H. B. 1932. *The Atlantic Rift and Its Meaning*. Detroit.

Carozzi, A. V. 1970. "A Propos de L'origine de la Theorie des Derives Continentales." *Compte Renda des Seances Societe Physique Histoire Naturella Geneva* 4: 171–179.

Cox, Allan, ed. 1973. *Plate Tectonics and Geomagnetic Reversals*. San Francisco: W. H. Freeman and Co.

Darwin, G. H. 1879. "The Precession of a Viscous Spheroid and the Remote History of the Earth." *Philosophical Transactions of the Royal Society of London* 170: 447–538. A discussion of the idea that the moon spun off from the earth and left the Pacific Ocean void.

Hallam, A. 1972. "Continental Drift and the Fossil Record." *Scientific American* 228: 87–95.

———. 1975. "Alfred Wegener and the Hypothesis of Continental Drift." *Scientific American* 323, no. 4: 88–97. A good discussion of the role of Alfred Wegener in the development of plate-tectonic theory.

Hess, H. H. 1962. "History of Ocean Basins." In A.E.J. Engel, H. L. James, and D. F. Leonard, eds., *Petrologic Studies*. New York: Geological Society of America. A. classic paper that first outlined the concept of sea-floor spreading.

Kiefer, Irene. 1978. *Global Jigsaw Puzzle*. New York: Atheneum. An excellent book on the development of plate-tectonics theory. Easy to read and not too technical.

McElhinny, M. W., and Valencio, D. A., eds. 1981. *Paleoreconstruction of the Continents*. Washington, D.C.: American Geophysical Union. A collection of recent papers dealing with many aspects of continental evolution.

Pickering, W. H. 1907. "The Place of Origin of the Moon—The Volcanic Problem." *Journal of Geology* 15: 23–38. A discussion of the origin of the moon.

Simon, Cheryl. 1982. "The Great Earth Debate." *Science News* 121: 178–179.

Snider-Pellegrini, A. 1858. *La Creation et ses Mysteres Devoiles*. Paris: Franck and Dentu. The first clear-cut statement of the splitting apart of the continents.

Suess, Eduard. 1900. *Das Antlitz der Erde*. Translated into English Hertha B. C. Sollas. Oxford: Clarendon Press. The first discussion of a supercontinent by one of the pioneers in the development of plate tectonics.

Sullivan, Walter. 1974. *Continents in Motion*. New York: McGraw-Hill Book Co.

Taylor, Frank B. 1910. "Bearing of the Tertiary Mountain Belt on the Origin of the Earth's Plan." *Geological Society of America Bulletin* 21: 179–226. The first complete statement of the continental drift idea by an American geologist.

Vine, F. J., and Matthews, D. H. 1963. "Magnetic Anomalies over Oceanic Ridges." *Nature* 199: 947–949. The first research results on paleomagnetism and oceanic ridges is presented.

Wegener, A. 1912. "Die Entstehung der Kontinente." *Petermanns Mitteilungen*, pp. 185–195, 253–256, 305–309. The classic work by the father of continental drift.

———. 1966. *The Origin of Continents and Oceans*. Translated from the 4th rev. German ed. J. Biram (1924). London: Methuen & Co. The fully developed concept of continental drift is presented.

Wilson, J. Tuzo. 1963. "Continental Drift." *Scientific American* 208: 86–100.

———. 1965. "A New Class of Faults and Their Bearing on Continental Drift." *Nature* 211: 676–681. The first discussion of rigid plates and their relationship to fault activity.

Wyatt, A. R. 1984. "Relationship between Continental Area and Elevation." *Nature*, pp. 370–372.

Sources of Additional Information

Short definitions of the concept can be found in Sir Dudley Stamp and Arthur M. Clark, *A Glossary of Geographical Terms* (New York: Longmans, 1979); Stella E. Steigeler, ed., *Dictionary of Earth Sciences* (New York: Pica Press, 1976); and Daniel Lapedes, ed., *McGraw-Hill Encyclopedia of the Geological Sciences* (New York: McGraw-Hill Book Co., 1978). Good discussions of the evolution of continental drift and plate-tectonics theory can be found in Ursula B. Marvin, *Continental Drift: The Evolution of a Concept* (Washington, D.C.: Smithsonian Institute Press, 1973); A. Hallam, *A Revolution in the Earth Sciences* (Oxford: Clarendon Press, 1973); John F. Dewey, "Plate Tectonics," *Scientific American* 226 (1972): 56–68; and Charles F. Kahle, *Plate Tectonics: Assessments and Reassessments* (Menasha: George Banta Co., 1974). More recent books on plate-tectonics theory are Lester C. King, *Wandering Continents and Spreading Sea Floors on an Expanding Earth* (Durban, S. Afr.: John Wiley and Sons, 1983); J. Bird, *Plate Tectonics* (Washington, D.C.: American Geophysical Union, 1982); and B. F. Windley, *The Evolving Continents* (New York: John Wiley and Sons, 1977).

PLEISTOCENE. 1. The first epoch of the Quaternary period in the Cenozoic era characterized by the rise and recession of continental glaciers. 2. The period from 1.8 million years ago to approximately 10,000 years ago, generally referred to as the Ice Age.

Although the derivation of the term *pleistocene* can be traced to the Greek words meaning "most recent," there has been disagreement over the precise boundaries and length of the time involved. Some scholars have included the postglacial period in the pleistocene, whereas others have deleted the postglacial period (Monkhouse, 1970: 272). The length of the Pleistocene is also a subject of dispute. Most scientists believed the Pleistocene was about 1 million years long, but recent radiometric and magnetostratigraphic methods tend to put the onset of the Pleistocene considerably further in time to more than 3 million years in some cases (Nilsson, 1983: 24).

Early observations at the close of the eighteenth century were instrumental in forming our present conception of the Pleistocene and Ice Age; however, it was not until the middle of the nineteenth century that the modern concept was formed (Rudwick, 1969). The surficial deposits now associated with glacial deposition were first attributed to the action of flood waters and became known as "drift" deposits. This Diluvial Theory associated with the Noachian flood of biblical origin was proposed by geologists such as William Buckland and Adam Sedgwick during the 1820s (North, 1943).

About the same time, geologists were trying to explain the origin of large erratic boulders found in Switzerland. Attributing the movement of these large rocks to glaciers was proposed by the British geologist John Playfair as well as the Swiss J. Venetz and J. de Charpentier (West, 1968: 3). Similar conclusions were reached by those who studied the mountainous regions of northern Europe (Playfair, 1802; Nilsson, 1983: 13). The first scientist explicitly to set forth the idea that a large glacier spread outward from Scandinavia over the North German area was the German geologist A. Bernhardi (1832). This work was further substantiated by the French geologist G. Martins from his experience with modern glaciers on Spitsbergen (Nilsson, 1983: 13).

The term *Ice Age* was first proposed by the South German botanist Karl Schimper (1937), but it was the British geologist Charles Lyell (1839) who first coined the term *pleistocene*. Lyell first used the term for those beds that contained more than 70 percent living species, but E. Forbes (1846) proposed that *pleistocene* be used only for the glacial epoch and the term *recent* for the postglacial. Lyell (1873) later concurred with Forbes, and *Ice Age* became synonymous with *pleistocene*.

A further refinement of the Ice Age concept and the evolution of glacier-formed landscapes was developed by the Swiss zoologist Louis Agassiz. Agassiz (1840), who recognized the distinction between Alpine glaciers (see VALLEY GLACIER) and continental ice sheets (see CONTINENTAL GLACIER), eventually moved to the United States to become a professor at Harvard. While at Harvard he studied glacial landscapes in the White Mountains of New Hampshire and the Great Lakes. He also had a major influence on the work of John Muir (1894), who studied glacial processes in the Sierra Nevada Mountains of California.

The modern Ice Age concept was clearly established after 1850 through a variety of research projects by British and Scandinavian geologists. Correct interpretation of glacial landscapes was undertaken in Scotland and England by Andrew Ramsay, in Sweden by Otto Torell, and in Norway by T. Kjerulf (Nilsson, 1983: 14).

Although the glacial theory is universally accepted today by all geologists and physical geographers, controversy in the twentieth century has focused on the causes of the drastic climatic changes that led to glacial advances and retreats (Hallam, 1983). One early theory attributed the cause of an ice age to a decrease in the amount of energy radiated from the sun. This sun-spot theory, however, has been supported by little evidence. "Most investigators believe that the only way to test the solar theory would be to develop a way of calculating how the intensity of solar radiation varies through time (Imbrie and Imbrie, 1979: 63). Other theories for the origin of ice ages have focused on phenomena such as the abrupt sliding of large portions of the Antarctic ice sheet into the ocean (Wilson, 1964), a worldwide uplifting of land areas (Dana, 1894), and an ice-free period in the Arctic Ocean (Ewing and Donn, 1956).

The most promising of ice age theories, however, had its origins in the latter part of the nineteenth century when James Croll (1875) hypothesized that ice ages were the result of a combination of astronomical variations associated with changes in the tilt of the earth's axis and variations in the earth's orbit. Croll's pioneering work was further extended in the twentieth century by the Serbian mathematical physicist Milutin Milankovitch (1920). Milankovitch developed a model of cyclic changes in the earth's orbital eccentricity, tilt, and precession, and his ideas have received support from recent research on deep-sea cores (Hayes, Imbrie, and Shackleton, 1976). Continued research on the pleistocene and its role on landscape evolutions not only provides insights into the historical evolution of landscapes but provides important information needed to solve problems associated with the management of NATURAL RESOURCES, urban development, and NATURAL HAZARDS.

References

Agassiz, L. 1840. "Etudes sur les Glaciers." Neuchatel: Privately published. A classic study of glacial theory.
Bernhardi, A. 1832. "Wiekamen die aus dem Norden Stammenden Felsenbruchsteine." *Min. Geognosie* 3: 257–267.
Croll, J. 1875. *Climate and Time.* New York: Appleton and Co. Croll was the first scientist to look at astronomical changes as an explanation for the advent of the pleistocene.
Dana, J. D. 1894. *Manual of Geology.* New York: American Book Co.
Ewing, M., and Donn, W. L. 1956. "A Theory of the Ice Ages." *Science* 123: 1061–1066.
Forbes, E. 1846. "On the Connection..." *Geologic Survey of Great Britain*, 1: 336–342.
Hallam, A. 1983. *Great Geological Controversies.* Oxford: Oxford University Press. This book contains an excellent chapter on the evolution of research on the ice age.
Hayes, J. D.; Imbrie, J.; and Shackleton, N. J. 1976. "Variations in the Earth's Orbit: Pacemaker of the Ice Ages." *Science* 194: 1121–1132.
Imbrie, J., and Imbrie, K. 1979. *Ice Ages: Solving the Mystery.* Short Hills, N.J.: Enslow Publishers. An excellent book; easy to read and has a good bibliography.
Lyell, C. 1839. *Nouveaux Elements de Geologie.* Paris: Pitois-Levrault.
———. 1873. *The Geological Evidences of the Antiquity of Man.* 4th ed. London: J. Murray.

Milankovitch, M. 1920. *Theorie Mathematique des Phenomenes Thermiques Produits per la Radiation Solaire.* Paris: Gauthier-Villars.

Monkhouse, F. J. 1970. *A Dictionary of Geography.* 2d ed. Chicago: Aldine Publishing Co. A short, concise definition of the term *geography* is presented.

Muir, John. 1894. *The Mountains of California.* Berkeley, Calif.: Ten Speed Press. An excellent book by one of America's foremost naturalists and glacial pioneers.

Nilsson, Tage. 1983. *The Pleistocene: Geology and Life in the Quaternary Ice Age.* Boston: D. Reidel Publishing Co. A thorough in-depth analysis of recent research pertaining to the pleistocene. An excellent bibliography is included.

North, F. J. 1943. "Centenary of the Glacial Theory." In *Proceedings of the Geological Association of London* 54: 1–28. A good discussion of the historical evolution of glacial theory.

Playfair, J. 1802. *Illustrations of the Huttonian Theory of the Earth.* Edinburgh: W. Creech.

Rudwick, M.J.S. 1969. "The Glacial Theory." *History of Science* 8: 136–157.

Schimper, K. 1837. "Uber die Eiszeit." *Soc. Helv. Sci. Nat.,* Actes 22: 38–51.

West, R. G. 1968. *Pleistocene Geology and Biology.* New York: John Wiley and Sons. An easy-to-read synopsis of Pleistocene research.

Wilson, A. T. 1964. "Origin of Ice Ages: An Ice Shelf Theory for Pleistocene Glaciation." *Nature* 201: 147–149.

Sources of Additional Information

For recent developments in our understanding of the Pleistocene see Stephen C. Porter, ed., *Late-Quaternary Environments in the United States. Vol. 1, The Late Pleistocene* (Minneapolis: University of Minnesota Press, 1984); M. S. Stoker, "Paleomagnetic Evidence for the Early Pleistocene in the Central and Northern North Sea," *Nature* 304 (1983): 332–334; Stanley E. Chericoff, "Glacial Characteristics of a Pleistocene Ice Lobe in East-Central Minnesota," *Geological Society of America Bulletin* 94 (1984): 1401–1414; and Sidney White and Salvatore Valastro, Jr., "Pleistocene Glaciation of Volcano Ajusco, Central Mexico, and Comparison with the Standard Mexican Glacial Sequence," *Quaternary Research* 21 (1984): 21–35. For a more thorough discussion of climatic changes and their role in understanding the Pleistocene see Roger G. Barry, "Late-Pleistocene Climatology," in Stephen C. Porter, ed., *Late-Quaternary Environments in the United States* (Minneapolis: University of Minnesota Press, 1984), Chapter 20; G. Denton and T. Hughes, eds., *The Last Great Ice Sheets* (New York: John Wiley and Sons, 1981); and A. B. Pittock, L. A. Frakes, D. Jenssen, J. Peterson, and J. Zillman, *Climatic Changes and Variability* (Cambridge: Cambridge University Press, 1978). For recent information on glacial processes and a good review of research see Martin Sharp, "Glacial Geomorphology," *Progress in Physical Geography* 9 (1985): 291–301.

POLLUTION. An unfavorable change or alteration in the chemical, biological, or physical properties of the environment that can have harmful effects on humans and other living organisms.

Pollution is nothing new; it has plagued humankind for thousands of years. Undesired changes to the environment due to human actions, primarily resulting in air and water pollution, have taken place for a long time. According to Marshall

Goldman (1972: 4): "Pollution has plagued mankind for centuries. There is virtually no naturally pure water and air. The mere presence of humans and animals is enough to alter whatever conditions existed before."

Unfavorable changes in the water supply were noted by the Romans before the first century B.C., when they realized that the city's sewage was affecting the taste of the drinking water. In response, the Romans built one of the first municipal sewers in history. Venice disposed of its sewage twice a day by using the natural flow of the tides. This led Goldman to call it "a sewer in search of a city" (1972: 4).

Pollution became such a problem in England during the Middle Ages that King Richard II, disturbed by the increasing pollution of the air and water in London, banned the dumping of garbage in ditches, rivers (see STREAM), and other water systems. London had such stringent laws about air pollution that at least one man was hanged for burning coal (Emmel, 1977: 6). Several centuries later another Englishman, John Evelyn, wrote a pamphlet about the evils of air pollution. He emphasized the unhealthful aspects, particularly the damage to lungs, as well as the general unpleasantness of polluted air (Parkinson, 1985: 98). Despite Evelyn's protests, "In seventeenth-century London rivers of filth coursed down the streets and a pall of smoke obscured buildings and fouled clothing" (Emmel, 1977: 6).

The pollution of air and water was not restricted to the European continent. When Juan Cabrillo visited present-day California in 1542 he noted that while anchored in San Pedro Bay he could see the mountain peaks in the distance but not their bases. A thermal inversion in the area trapped the smoke from Amerindian fires, bringing about air pollution as well as human health hazards (Peters and Larkin, 1983: 258).

Geographers have long been concerned with the interaction of organisms and their environment. George Perkins Marsh's *Man and Nature* (1864) is a classic treatise on the relationship between humans and the environment. Marsh analyzed the causes for the decline of ancient civilizations and forecast a similar doom for modern societies unless people took what we would today call an ecological approach (see ECOSYSTEM) to humans and nature. Other early writers interested in pollution problems were the German Karl Mobius (1877) and the Americans S. A. Forbes (1887) and John Muir (1901).

Although geographers were interested in pollution problems in the early part of this century, particularly with the role of pollution in human-ecology concepts (Barrows, 1923), it was not until 1956 and the publication of the symposium proceedings "Man's Role in Changing the Face of the Earth" (Thomas, 1956) that there was renewed interest in pollution issues.

The decades of the 1960s and 1970s saw a marked increase in the number of scientists concerned with pollution issues. Following the publication of *Silent Spring* by Rachel Carson (1962) there was an increased awareness of pollution problems by both the general public and scientists interested in the problem. The 1960s was an important decade in pollution awareness because "two forces converged to confront mankind with a literally threatening dilemma: unrestricted modern technology accelerated the affronts to the natural world by enhancing man's ability to alter biochemical cycles and deplete natural resources [see NATURAL RESOURCES]; at the same time the global population began increasing at a phenomenal rate, and more was asked of the earth than it was able to give'' (Emmel, 1977: 7–8).

We are currently polluting our planet at an unprecedented rate. The reasons for this are many and complex, but as Kenneth Boulding summarized it, "Our desire to conquer nature often means simply that we diminish the probability of small inconveniences at the cost of increasing the probability of very large disasters" (1966: 14).

Pollution can be either biological or chemical. Human population density often leads to increased biological pollution, as evidence from early civilizations reveals. However, even today the crowding of large numbers of people into small places has resulted in increased pollution. As a population increases, so does the accumulation of its human organic waste. City water supplies may be contaminated as it becomes increasingly difficult to dispose of large volumes of waste (Brown, et al., 1976; Burmaster, 1982; National Academy of Sciences, 1983).

Chemical pollution is another by-product of a rapidly growing population and increased industrialization. Many STREAMS and lakes have been polluted by the addition of toxic chemicals (Stoker and Seager, 1976). Pollution problems that were once primarily local are now becoming global concerns. Chemical wastes that break down very slowly can ultimately reach the oceans, which unfortunately are becoming major dumping grounds. Compounds such as DDT can be found in virtually all parts of the oceans as they increasingly become a common sink for industrial wastes and garbage (Marx, 1981).

Air pollution is probably the most obvious form of pollution. It can be seen in almost every major metropolitan area as well as felt in the eyes or lungs of their inhabitants. Although it is more obvious in urban areas, large portions of relatively uninhabited regions are now afflicted to some degree. Perhaps the best view of air pollution was that of the Apollo astronauts who were unable to see southern California (Fagan, 1974: 10). As Reid Bryson and John Ross commented: "Both gases and suspended solids are dramatically present in the urban atmosphere. We have already pointed out that there is from 4 to 1,000 times more dust in the city air than in the country air. What is in the city dust? One thing is lead. Inner areas of cities typically have 3 times as much lead in the air as the suburbs. These figures are matched by the figures for lead concentration in the blood of the inhabitants" (1972: 66).

An air pollution problem that has received much recent attention has been that of acid rain pollution. Acid deposition is a serious and growing problem in many parts of the world (Luoma, 1984).

Another serious problem that has recently received much attention is the transportation, disposal, deactivation, and storage of hazardous industrial wastes.

Hazardous waste expert Samuel S. Epstein said that hazardous wastes have created "the environmental problem of the century, second only to nuclear war. . . . I've never really been scared until I started working on this issue" (1982: 27).

With the increases in the human population and in industrial output over the past century, persistent contamination or pollution of the BIOSPHERE has become a serious national and global problem. Continued population growth and the accompanying new demand for food, shelter, and other goods and services will make it difficult to bring pollution under control.

References

Barrows, H.H. 1923. "Geography as Human Ecology." *Annals of the Association of American Geographers* 13: 1–14.
Boulding, Kenneth E. 1966. *Human Values on the Spaceship Earth*. New York: National Council of Churches.
Brown, Lester R.; McGrath, Patricia L.; and Stokes, Bruce. 1976. "Twenty-Two Dimensions of the Population Problem," Worldwatch Paper, no. 5. New York: Worldwatch Institute.
Bryson, Reid A., and Ross, John E. 1972. "The Climate of the City." In Detwyler, Thomas R., and Marcus, Melvin G., eds. *Urbanization and Environment*. Belmont, Calif.: Duxbury Press, pp. 51–68.
Burmaster, David E. 1982. "The New Pollution: Groundwater Contamination," *Environment* 24: 2, 4–12, 33–36.
Carson, Rachel. 1962. *Silent Spring*. Boston: Houghton Mifflin.
Emmel, Thomas C. 1977. *Global Perspectives on Ecology*. Palo Alto, Calif.: Mayfield Publishing Co.
Epstein, Samuel S. 1982. *Hazardous Waste in America*. San Francisco: Sierra Club Books. An excellent analysis of the problem with suggested solutions.
Fagan, J. J. 1974. *The Earth Environment*. Englewood Cliffs, N.J.: Prentice-Hall.
Forbes, S. A. 1877. "The Lake as a Microcosm." Reprinted in *Illinois Natural History Survey Bulletin* 15 (1925): 537–550.
Goldman, Marshall I. 1972. "Pollution: The Mess Around Us." In Marshall Goldman, ed. *Ecology and Economics: Controlling Pollution in the 70's*. Englewood Cliffs, N.J.: Prentice-Hall, pp. 17–36. An excellent essay on pollution problems.
Luoma, Jon R. 1984. *Troubled Skies, Troubled Water: The Story of Acid Rain*. New York: Viking Press. An excellent overview of the acid rain problem.
Marx, Wesley. 1981. *The Oceans: Our Last Resource*. San Francisco: Sierra Club Books. Excellent discussion of ocean resources and how to preserve them.
Marsh, George P. 1864. *Man and Nature*. Reprint (1965). Cambridge: Harvard University Press. A classic study on the interaction of humans and the environment.
Mobius, Karl. 1877. "Die Auster und Die Austernwirtschaft." Translated in *Report, U.S. Fish Commission*, 1880, pp. 683–751.
Muir, John. 1901. *Our National Parks*. Boston: Houghton Mifflin. An early voice in the U.S. conservation movement discusses preservation and parks.
National Academy of Sciences, 1983. *Drinking Water and Health*. Vol. 5. Washington, D.C.: National Academy Press. Analysis and discussion of 21 drinking water contaminants.

Parkinson, Claire L. 1985. *Breakthrough: A Chronology of Great Achievements in Science and Mathematics*. Boston: G. K. Hall and Co.

Peters, Gary L., and Larkin, Robert P. 1983. *Population Geography: Problems, Concepts, and Prospects*. 2d ed. Dubuque, Ia.: Kendall/Hunt Publishing Co.

Stoker, H. S., and Seager, Spencer L. 1976. *Environmental Chemistry: Air and Water Pollution*. 2d ed. Glenview, Ill.: Scott, Foresman and Co. Good summary of air- and water-pollution problems.

Thomas, William L., ed. 1956. *Man's Role in Changing the Face of the Earth*. Chicago: University of Chicago Press.

Sources of Additional Information

For good lengthy discussions on pollution problems and the historical evolution of ideas about ecology see John Passmore, *Man's Responsibility for Nature* (New York: Scribner's, 1974), and Lynn White, "The Historical Roots of our Ecologic Crisis," *Science* 155 (1967): 1204. For a nontechnical discussion of pollution problems see G. Tyler Miller, Jr., *Living in the Environment: An Introduction to Environmental Science*, 4th ed. (Belmont, Calif.: Wadsworth Publishing Co., 1985); Barry Commoner, *The Closing Circle: Nature, Man and Technology* (New York: Alfred A. Knopf, 1971); Erik Eckholm, *Down to Earth: Environment and Human Needs* (New York: W. W. Norton and Co., 1982); and Council on Environmental Quality and U.S. Department of State, *The Global 2000 Report to the President*, vols. 1–3 (Washington, D.C.: U.S. Government Printing Office, 1980).

POTENTIAL EVAPOTRANSPIRATION. See EVAPOTRANSPIRATION.

PRECIPITATION. 1. A term that encompasses all forms of moisture deposited on the earth's surface by atmospheric sources. 2. Includes rain and drizzle (liquid precipitation) and ice crystals, snow, ice pellets, and hail (frozen precipitation).

The amounts and forms of water vapor in the atmosphere are inextricably linked to the types and intensities of precipitation during any given meteorological event. J. S. Gardner stated, "The formation of clouds and fog through condensation of water vapor and the precipitation of that moisture back to the earth's surface are important meteorological and hydrological processes" (1977: 125). To discuss precipitation adequately, we must first discuss the level of water vapor in the atmosphere; this level is termed *humidity*.

There are at least three accepted ways of expressing humidity: absolute humidity, specific humidity, and relative humidity. R. W. Fairbridge (1967: 444) gave a very good explanation of them. *Absolute humidity* is the ratio of the mass of water vapor to the volume of the parcel of air being measured. This term is little used in meteorology because of the lack of reference to ATMOSPHERIC PRESSURE. *Specific humidity* is the ratio of the mass of water vapor to the mass of air being measured. This figure gives the actual amount of water vapor in a mass of air and is useful in meteorology. The most useful term is *relative humidity*: the ratio of the water vapor in the air to the maximum water vapor possible at a given temperature. Since cold air can hold less water vapor than warm air, the relative humidity of a parcel of air increases with cooling even without increasing the specific humidity. When the term *humidity* is used in meteorology or CLIMATOLOGY, it almost always refers to relative humidity.

The condensation of water vapor can occur only when relative humidity reaches 100 percent. The growth of water droplets into larger drops that produce precipitation occurs in clouds where relative humidity is at 100 percent and the temperature is at or below dew point. An area of considerable research in the past concerns just how this process works. J. Dalton (1793) first tried to theorize on raindrop formation. Only a few of his observations have proven to be valid. His theories and the myriad other theories of raindrop formation are described in detail by W.E.K. Middleton (1965). None of the early efforts was able to provide an adequate explanation of raindrop or ice-crystal formation in clouds.

Two tentatively accepted theories are currently proposed. The first, devised by T. Bergeron (1933), is the *cold-cloud theory* suggesting that ice crystals and water droplets occur simultaneously in a cloud. They have sublimated and condensed because the temperature is at or below dew point. Because vapor pressure around the droplets is greater than around the ice crystals, water vapor moves from the droplet to the crystal. When the crystal grows sufficiently large, it will overcome updrafts and fall to earth in solid form or melt and fall as rain or drizzle.

The second theory currently used is the *warm-cloud theory* (Crowe, 1971: 114–116). It is also called the *coalescence theory* (Atkinson, 1979: 24) because larger droplets will move through the cloud mass and coalesce with smaller droplets until they are large enough to overcome updrafts and fall. F. H. Ludlam (1951) discussed the *double-sweep process* whereby droplets moved both up and down through the cloud, gaining mass in both directions until they fall.

Most recent research views precipitation patterns as an integral part of other climatological and meteorological studies. J. E. Oliver (1973: 30–44) reviewed research on the spatial and temporal distributions of precipitation around the world. An example of current phenomenon specific research can be seen in the review article by B. Yarnal (1985) on the El Niño phenomenon. Effects of precipitation patterns on agriculture can be seen in the work of L. W. Hanna (1983). For continuing information on these and other related topics refer to the journal *Progress in Physical Geography*.

References

Atkinson, B. W. 1979. "Precipitation." In K. J. Gregory and D. E. Walling, eds., *Man and Environmental Processes*. Boulder, Colo.: Westview Press, pp. 23–37.

Bergeron, T. 1933. "On the Physics of Cloud and Precipitation." *Memo Meteorologie U.G.G.I.* 2: 156–178. First proposed the cold-cloud theory of precipitation formation.

Crowe, P. R. 1971. *Concepts in Climatology*. New York: St. Martin's Press. Has a large section on precipitation patterns and generation.

Dalton, J. 1793. *Meteorological Observations and Essays*. London: Richardson. First to theorize on the formation of rain.

Fairbridge, R. W. 1967. "Humidity." In R. W. Fairbridge, ed., *The Encyclopedia of Atmospheric Sciences and Astrogeology*. Encyclopedia of Earth Science Series. Vol. 2. New York: Reinhold Publishing Co., pp. 444–447.

Gardner, J. S. 1977. *Physical Geography*. New York: Harper's College Press. A good general reference for precipitation and humidity.

Hanna, L. W. 1983. "Agricultural Meteorology." *Progress in Physical Geography* 7: 329–344.

Ludlam, F. H. 1951. "The Production of Showers by the Coalescence of Cloud Droplets." *Quarterly Journal of the Royal Meteorological Society* 77: 402–417. First proposed the double sweep process of raindrop formation.

Middleton, W.E.K. 1965. *A History of the Theories of Rain*. London: Oldbourne. An excellent reference for the early theories of rain.

Oliver, J. E. 1973. *Climate and Man's Environment—An Introduction to Applied Climatology*. New York: John Wiley and Sons. Has a very good section on spatial and temporal precipitation distributions.

Yarnal, B. 1985. "Extratropical Teleconnections with El Niño/Southern Oscillation (ENSO) Events." *Progress in Physical Geography* 9: 315–352.

Sources of Additional Information

For good general discussions and definitions of precipitation and humidity see J. Whittow, *Dictionary of Physical Geography* (New York: Penguin Books, 1984); R. W. Fairbridge, ed., *The Encyclopedia of Atmospheric Sciences and Astrogeology*, Encyclopedia of Earth Science Series, vol. 2 (New York: Reinhold Publishing Co., 1967); and F. A. Berry, Jr., E. Bollay, and N. R. Beers, eds., *Handbook of Meteorology* (New York: McGraw-Hill Book Co., 1973). Detailed and extended discussions of precipitation and humidity can be seen in texts such as A. N. Strahler and A. H. Strahler, *Modern Physical Geography*, 3d ed. (New York: John Wiley and Sons, 1987); J. F. Griffiths and D. M. Driscoll, *Survey of Climatology* (Columbus, Ohio: Charles E. Merrill Publishing Co., 1982); and F. K. Lutgens and E. J. Tarbuck, *The Atmosphere—An Introduction to Meteorology*, 3d ed. (Englewood Cliffs, N.J.: Prentice-Hall, 1986).

R

RAIN SHADOW. See OROGRAPHIC PRECIPITATION.

REMOTE SENSING. 1. The acquisition of information about objects without physical contact with those objects. 2. Information acquisition through the use of photographic and nonphotographic instruments detecting some portions of the electromagnetic spectrum.

The quest for permanently recorded images has been progressing for more than 2,000 years. Aristotle's "camera obscura" was the first recorded experiment with images. The chemistry of photosensitive materials provided the knowledge necessary for progress toward true photography, which was finally achieved in 1839 by Louis Jacques Monde Daguerre and Joseph Nicephoce Niepce (Reeves et al., 1983: 2). J. M. Sturge (1977) provided an excellent summary of the history of photography.

Some early attempts to use cameras from elevated vantage points included the first balloon photograph taken in 1859 of the French village Petit Becetre and balloon photographs of Boston in 1860. The use of balloons as platforms for photography advanced considerably during the Civil War. The advent of fixed-wing aircraft as platforms gave a large boost to aerial photography during World War I. Although there was somewhat of a hiatus in the advancement of remote sensing between World War I and World War II, some work dealing with nonmilitary uses of the technique (e.g., GEOLOGY, agriculture, cartography) was accomplished. World War II saw the evolution of remote sensing from merely photography of the visible range of the electromagnetic spectrum (EMS) to remote sensing using many of the other areas of the EMS such as thermal INFRARED, near infrared, and radar (active microwave). Many of the pioneers in the science of remote sensing during the past four decades have come from the specialists trained during this war (Reeves et al., 1983: 2–3).

During the late 1940s and the 1950s the development of remote sensing in the nonvisible portions of the EMS flourished (e.g., Colwell, 1956; Krinov, 1947; Sherwin et al., 1962). These other techniques were in the areas of color infrared photography, thermal infrared, and radar, both side-looking air-borne radar (SLAR) and synthetic aperature radar (SAR).

Since 1960 space has evolved as the dominant theater for remote-sensing research. The impetus of Sputnik I created a space-minded United States and destined the greatest advances in remote sensing to be in this realm. The pioneering efforts in systematic earth observation from space were concentrated in weather-observation satellites. The Television Infrared Observation Satellite (TIROS-I) was launched in the early 1960s. Since this first effort, dozens of meterological satellites have been launched by the Environmental Sciences Services Administration, which became part of the National Oceanic and Atmospheric Administration (NOAA) in 1970. The latest NOAA satellites include systems such as the Advanced Very High Resolution Radiometer, the TIROS Operational Vertical Sounder, the Data Collection and Platform Location System, and the Space Environment Monitor. Many of the manned Gemini and Apollo missions also contributed to the increasing knowledge of remote sensing from space (Lawman et al., 1967).

The Earth Resources Technology Satellite (ERTS–1) was the first satellite specifically designed to collect data on the resources and the surface of the earth. ERTS–1 was put into a nearly polar orbit in July 1972. The satellite carried two sensors: a three-channel Return Beam Vidicon (RBV) sensor and a four-channel multispectral sensor (MSS). All bands on both sensors are in the visible or near infrared range of the EMS. The RBV failed soon after launch, but the MSS worked far longer than its designed one-year lifespan. An almost identical satellite was launched in January 1975, and the program was renamed LANDSAT. LANDSAT 3 was launched in March 1978 and included a fifth MSS band in the thermal infrared range ($10.4–12.6 \mu m$).

From LANDSAT 1 through LANDSAT 3, there were several limitations in the system. They included spectral and radiometric resolution, aiming accuracy, mechanical limitations, and limited spectral channel choice. LANDSAT 4 was designed to eliminate many of these problems and was launched in July 1982. Additionally, a new generation sensor was included in LANDSAT 4, the thematic mapper (TM). There are seven channels on this sensor with greatly improved spectral and spatial resolution. This satellite was designed specifically by specialists in many disciplines to alleviate problems and enhance performance for given areas of interest (e.g., SOIL, geology, agriculture). The new system includes a system of real-time data transmission, including the DOMSAT and Tracking and Data Relay Systems, thus eliminating the need for on-board mechanical or electronic recording of data.

Several other space systems including Skylab, the Heat Capacity Mapping Mission, and SEASAT have been developed and operated in the late 1970s and early 1980s. The future of remote sensing, however, seems to include more and more sophisticated systems from countries other than the United States. They include the European Space Agency's (ESA) SPOT satellite program, Japan's Marine Observation Satellite series, ESA and NASA combining for the SPACE-LAB project, Canada's RADARSAT, and continuing efforts in the U.S. Space Shuttle program (Reeves et al., 1983: 7–13).

References

Colwell, R. N. 1956. "Determining the Prevalence of Certain Crop Diseases by Means of Aerial Photography." *Hilgardia* 26: 223–286.

Krinov, E. L. 1947. *Spectral Reflectance of Natural Formations.* Moscow: Akad. Nank. USSR, Laboratorica Aerometodov. Translated from NEC of Canada.

Lawman, P. D., Jr.; McDivitt, J. A.; and White, E. H., II. 1967. *Terrain Photography on the Gemini IV Mission.* Technical Report, no. D-3982. Washington, D.C.: NASA.

Reeves, Robert G.; Estes, John E.; Bertke, Susan E.; and Sailer, Charlene T. 1983. "The Development and Principles of Remote Sensing." In Colwell, R. N., ed., *Manual of Remote Sensing.* Falls Church, Va.: American Society of Photogrammetry, pp. 1–35. This is an excellent chapter on the history and future prospects of remote sensing.

Sherwin, C. W.; Ruina, J. P.; and Rawclisse, R. D. 1962. "Some Early Developments in Synthetic Aperture Radar Systems." *IRE Transactions on Military Electronics, MIL-6* 2: 111–115.

Sturge, J. M., ed. 1977. *Neblette's Handbook on Photography and Reprography.* 7th ed. New York: Van Nostrand-Reinhold Co.

Sources of Additional Information

Two papers relevant to the future prospects and present conditions in remote sensing are R. N. Colwell, "Remote Sensing of Natural Resources—Retrospect and Prospect," in *Proceedings of Remote Sensing for Natural Resources* (Moscow: University of Idaho, 1979), pp. 48–68, and R. N. Colwell, "The Remote Sensing Picture in 1984," *Technical Papers*, American Congress of Surveying and Mapping-American Society of Photogrammetry Fall Convention (Washington, D.C.: American Society of Photogrammetry, 1984), pp. 1–34. General philosophies and policies concerning the use of remote sensing include American Society of Photogrammetry, "Photo Interpretations in the Space Sciences," *Photogrammetic Engineering* 31 (1965): 1060–1075; R. N. Colwell, "The 'Multi' Concept as Applied to the Acquisition and Analysis of Remote Sensing Data," in R. G. Reeves, ed., *Manual of Remote Sensing* (Falls Church, Va.: American Society of Photogrammetry, 1975), pp. 5–11; J. E. Estes, "United States Remote Sensing Policy," in *Proceedings of the International Geoscience and Remote Sensing Symposium* (Munich, 1982); J. E. Estes, J. R. Jensen, and D. S. Simonett, "Impacts of Remote Sensing on U.S Geography," *Remote Sensing of the Environment* 10 (1980): 1–72; J. Everett and D. S. Simonett, "Principles, Concepts, and Philosophies in Remote Sensing," in J. Lintz and D. S. Simonett, eds., *Remote Sensing of Environment* (Reading, Mass.: Addison-Wesley Publishing Co., 1976), pp. 85–127; and W. A. Fischer and C. J. Robinove, "A Rationale for a General Purpose Earth Resources Observation Satellite," in *Proceedings of the Remote Sensing Symposium* (Seattle: University of Washington, 1968). The legal implications of remote sensing are discussed in S. D. Estep, "Legal and Social Policy Ramifications of Remote Sensing Techniques," in *Proceedings of the Fifth Symposium on Remote Sensing of Environment*, 1968), pp. 197–217, and R. F. Stowe, "Legal Implications of Remote Sensing," *Photogrammetric Engineering and Remote Sensing* 44 (1978): 183–188. The applications of remote sensing to the problems in developing countries are discussed in K. Paul and C. Mascarenhas, "Remote Sensing in Development," *Science* 214 (1981): 139–145, and D. S. Simonett, "Remote Sensing and the Developing World: Examples of Major Benefits," in J. C. Ma and A. G. Noble, eds., *The Environment: Chinese and American Views* (New York: Methuen & Co., 1981). Technical aspects of remote sensing are published widely; the following is the best single source for information in all aspects of remote sensing: R. N. Colwell, *Manual of Remote Sensing*, vols. *1 and 2*, 2d ed. (Washington, D.C.: American Society of Photogrammetry, 1983). The classic work on remote sensing and land-use classification is J. R. Anderson, E. E. Hardy, and J. T. Roach, "A Land Use Classification System for Use with Remote-Sensor Data," *U.S. Geological Survey Circular*, no. 671 (Washington, D.C.: U.S. Geological Survey, 1972). The newest work in remote sensing deals with integrating data into a system of geographic information; the following are examples of this work: W. G. Brooner, "An Overview of Remote Sensing Input into Geographic Information Systems," in *Proceedings* (Sioux Falls, S.Dak.: Pecora VII Symposium, 1981), pp. 318–329, and J. E. Estes, "Remote Sensing and Geographic Information Systems: Coming of Age in the Eighties," in *Proceedings* (Sioux Falls, S.Dak.: Pecora VII Symposium, 1981), pp. 23–40.

RIVER. See STREAM; FLOODPLAIN.

ROCK. A natural assemblage or aggregate of minerals that is usually hard and in the solid state.

Rocks can be organized or classified in a variety of schemes, but the most common classification method is one based on the manner and environment of rock formation. Under this system rocks can be divided into three major categories: igneous, sedimentary, and metamorphic.

An *igneous rock* is composed of MINERAL crystals that form as hot, molten liquid rock material, called MAGMA, cools. The depth at which the cooling takes place can be used to subdivide igneous rocks further into extrusive rocks that cool at or very near the surface and intrusive rocks that cool below the surface. *Sedimentary rocks* are primarily composed of small residues such as sand or clay-sized particles that accumulate in low areas or basins and over time get compressed and cemented together to form rock. Sedimentary rocks are usually divided into two types: detrital and chemical. *Detrital rocks* are formed from the consolidation of small particles. *Chemical sedimentary rocks* like limestone form from the precipitation of minerals out of water. *Metamorphic rocks* form when a rock undergoes changes due to heat, pressure, or other related factors. For example, the igneous rock granite, through heat and pressure changes, is metamorphosed into the rock gneiss. The sedimentary rock shale, through heat and pressure changes becomes the metamorphic rock shale (Strahler and Strahler 1983: 192).

The study of rocks, called the science of petrology, has a long history starting with the writings of the ancient Greeks. Although many of the early Greek writings were a combination of legends, superstitions, guesses and facts, some Greek scholars tried to analyze the nature of rocks. In the 400s B.C. the Greek historian Herodotus observed fossils in Lower Egypt and believed that the sea had once covered the land (Schneer, 1969: 81). The most prominent Greek writers, however, were Aristotle (384–322 B.C.) and his pupil Theophrastus (370–287 B.C.). Aristotle, in his work *Meteorologica*, had a section entitled "Stones" in which he discussed the nature of rocks. In a separate book entitled *Concerning Stones*, Theophrastus gathered together for the first time a great deal of information about rocks, minerals, and fossils (McKeon, 1941).

The Romans had an interest in rocks and minerals primarily as they related to their mining enterprises in the far-reaching Roman Empire. Detailed information on rocks and minerals is found in the 37-volume *Historia Naturalis* by Pliny the Elder (A.D. 23–79). Pliny died in A.D. 79 in the eruption of Mount Vesuvius while he was working on scientific observations of the rocks.

The Roman Empire ended in the A.D. 400s, and there was little advancement in the scientific study of rocks until the eleventh-century work of the Arab physician Avicenna. Avicenna (Ibn Sina) wrote an important book on erosion and on the origins of meteorites, rocks, and mountains (Avicenna, 1927).

The Renaissance brought about an increasing interest in the study of rocks. Important contributions were made by Georgius Agricola (1546) and Nicolaus Steno (1669). Steno was the first to demonstrate that layers (strata) of rocks are always deposited with the oldest layer at the bottom.

The late 1700s and early 1800s were times of much disagreement among geologists about the origin of rocks. German mineralogist Abraham Werner believed that an ocean once covered the entire surface of the earth, and chemicals in the water slowly settled on the ocean floor where they formed granite and other types of rocks (Ospovat, 1971). Although Werner and his followers believed rocks formed in layers, they also believed that the earth had been formed completely at that time and no other changes would take place. The scientists who believed these ideas were called Neptunists, after the Roman god of the sea Neptune.

A much different idea on the origin of rocks was held by the Scottish physician James Hutton (1795) and French geologist Nicolas Desmarest (Taylor, 1969: 339–356). These men believed that rocks were formed by the cooling of hot rock from volcanos. These scientists were called Plutonists, after Pluto, the Greek god of the Lower World. The argument over rock origins was settled in the early 1800s, when two of Werner's most famous students, Leopold von Buch and Alexander von Humboldt, switched sides after looking at regions with volcanic rocks. Also, according to a study by A. Hallam, "the principal reason why the Neptunist-Plutonist controversy faded so quickly in the early nineteenth century is because of the development of a marvelous new research technique, the correlation of strata by fossils" (Hallam, 1983: 25).

The nineteenth century saw many advances in the study of rocks. In 1791–1793 the Englishman William Smith was involved with surveying for the building of canals in England. He observed the constant relationship between certain strata and fossil succession. In his *Geologic Map of England and Wales with Part of Scotland* (1815), Smith's idea of fossil succession was demonstrated as a valid scientific principle, and since that time it has become a widely accepted procedure for studying sedimentary rocks (Eyles, 1969: 142–158). Several years after Smith's publication the French naturalist Baron Georges Cuvier and the French geologist Alexandre Bronigniart published a book that described the rocks and fossils in the Paris region. They realized that certain fossils were found in specific strata and they could trace these strata throughout the Paris basin (Cuvier, 1825). Before the end of the nineteenth century, most studies of sedimentary rocks were concerned with fossil identification and strata correlation. Work was beginning, however, on an analysis of the actual physical properties of the rocks. Pioneering work in 1879 and 1880 by H. C. Sorby focused on the structure and origin of limestone and noncalcareous stratified rock. Following Sorby, T. G. Bonney (1900) and H. H. Thomas (1902, 1909) looked in more detail at detrital rocks.

Prompted by the Neptunist-Plutonist controversy, Sir John Hall undertook a series of experiments with igneous and metamorphic rocks (1805). His experiments laid the foundations for the experimental study of rocks. Following these initial experiments of Hall, experimental studies in petrology developed rapidly, particularly in Europe. Much work was done by French, German, and Russian scientists (Loewinson-Lessing, 1954). Important advances in equipment and the rapid development of physical chemistry in the latter half of the nineteenth century led to many important discoveries.

Another important event was the founding of the Geophysical Laboratory of the Carnegie Institution of Washington in 1907. According to Edgar, this "marked the most important milestone in the history of experimental petrology, and shifted the emphasis from Europe to North America" (1973: 5).

The twentieth century has seen many advances in the science of petrology. With the foundation of the Geophysical Laboratory, many advances were made in igneous petrology. Experimental studies and meticulous field observations by N. L. Bowen (1913, 1915) resulted in his reaction principles. With the publication of *The Evolution of Igneous Rocks* (1928) Bowen firmly established the importance of experiments in rock analyses as well as emphasizing the importance of combining laboratory and field methods. Although the majority of experimental studies were concerned with igneous rocks, studies were also being carried out on systems important in metamorphic rock formation (Edgar, 1973: 7).

The twentieth century has also seen major advancements in the study of sedimentary rocks. Resource shortages that occurred during World War I led to important studies of sands and rocks associated with petroleum resources. The first textbook devoted to the use of sedimentary rock analysis in stratigraphic correlation was that of H. B. Milner (1922). The past several decades have seen a consolidation of these early ideas as well as revolutionary developments in the study of clays (Fairbridge, 1978: 669).

The past ten years have seen revolutionary advances in the science of petrology. These advances have been the result of in-depth analysis of the role of PLATE TECTONICS in rock formation. Plate tectonics has been found to play an important role in the origin and distribution of igneous and metamorphic rocks. The occurrence of sedimentary rocks has also been related to plate tectonics theory (Ehlers and Blatt, 1982). Isotopic studies can now more accurately determine the date of formation of igneous and metamorphic rocks as well as determine the temperatures at which organisms can live in sedimentary environments (Ehlers and Blatt, 1982).

As the future unfolds and research in petrology continues, it is not unreasonable to anticipate even further advances in understanding the nature of rocks.

References

Agricola, G. 1546. *De Natura Fossilium*. Translated from Latin by M. C. and J. A. Bandy. Special Paper, no. 63. New York: Geological Society of America. A very early study of rocks.

Avicenna (Sina, Ibn). 1927. *Avicennae de Congelatione et Conglutinatione Lapidum* (trans. E. J. Holmyard and D. C. Mandeville). Paris: Librarie Orientaliste.

Bonney, T. G. 1900. "The Bunter Pebble Beds of the Midlands and the Source of their Materials." *Quarterly Journal, Geological Society of London* 71: 536–591. A pioneering study of detrital sedimentary rocks.

Bowen, N. L. 1913. "The Melting Phenomena of the Plagioclase Feldspar." *American Journal of Science* 35: 577–599. An important study used to help define reaction principles.

———. 1915. "The Crystallization of Haplobasaltic Haplodioritic, and Related Magmas." *American Journal of Science* 40: 161–185.

———. 1928. *The Evolution of Igneous Rocks*. Princeton, N.J.: Princeton University Press. A classic work combining laboratory experimentation and field methods.

Cuvier, G. 1825. *Essays on the Theory of the Earth*. Edinburgh: Blackwood.

Edgar, Alan D. 1973. *Experimental Petrology*. Oxford: Clarendon Press. A good book that looks at experimental rock analyses.

Ehlers, Ernest G., and Blatt, Harvey. 1982. *Petrology: Igneous, Sedimentary, and Metamorphic*. San Francisco: W. H. Freeman and Co. A recent textbook on the study of rocks.

Eyles, Joan M. 1969. "William Smith: Some Aspects of His Life and Work." In Cecil J. Schneer, ed., *Toward a History of Geology*. Cambridge, Mass.: The M.I.T. Press, pp. 142–158. A short study dealing with the scientific contributions of William Smith.

Fairbridge, Rhodes W., ed. 1978. *The Encyclopedia of Sedimentology*. Encyclopedia of Earth Sciences Series. Vol. 6. Stroudsburg, Pa.: Dowden, Hutchinson and Ross.

Hall, Sir J. 1805. "Experiments on Whinstone and Lava." *Transactions, Royal Society of Edinburgh* 5: 43–75. A pioneering work in experimental petrology.

Hallam, A. 1983. "Neptunists, Vulcanists, and Plutonists." In A. Hallam, *Great Geological Controversies*. New York: Oxford University Press, Chapter 1. An easy-to-read discussion of the Neptunist-Plutonist controversy.

Hutton, James. 1795. *Theory of the Earth*. 2 vols. Edinburgh. A monumental work by the "father of modern geology."
Loewinson-Lessing, F. Y. 1954. *A Historical Survey of Petrology*. Edinburgh: Oliver and Boyd. A good discussion of the historical roots of rock studies.
McKeon, Richard, ed. 1941. *The Basic Works of Aristotle*. New York: Random House.
Milner, H. B. 1922. *An Introduction to Sedimentary Petrography*. London: Murby.
Ospovat, A. M. 1971. *Abraham Gottlob Werner, Short Classification and Description of the Various Rocks*. New York: Hafner Publishing Co.
Schneer, Cecil J., ed. 1969. *Toward a History of Geology*. Cambridge, Mass.: The M.I.T. Press. A good analysis of important milestones in the history of geology.
Sorby, H. C. 1879. "The Structure and Origin of Limestones." *Quarterly Journal, Geological Society of London* 35: 56–95. An early study of the development of sedimentary rocks.
——. 1880. "The Structure and Origin of Non-Calcareous Stratified Rocks." *Quarterly Journal, Geological Society of London* 36: 46–92.
Steno, Nicolaus. 1669. *The Prodromus of Nicolaus Sterno's Dissertation* (trans. J. G. Winter). London: Macmillan, 1916.
Strahler, Arthur N., and Strahler, Alan H. 1983. *Modern Physical Geography*. 2d ed. New York: John Wiley and Sons. A good short discussion of the three rock types.
Taylor, Kenneth. 1969. "Nicolas Desmarest and Geology in the Eighteenth Century." In Cecil J. Schneer, ed., *Toward a History of Geology*. Cambridge, Mass.: The M.I.T. Press, pp. 339–356.
Thomas, H. H. 1902. "The Mineralogical Constitution of the Finer Material of the Bunter Pebble Beds in the West of England." *Quarterly Journal, Geological Society of London* 58: 620–632.
——. 1909. "A Contribution to the Petrography of the New Red Sandstone in the West of England." *Quarterly Journal, Geological Society of London* 65: 229–245.

Sources of Additional Information

Brief discussions of the concept can be found in Daniel Lapedes, ed., *McGraw-Hill Encyclopedia of the Geological Sciences* (New York: McGraw-Hill Book Co., 1978), pp. 696–698; Robert Bates and Julia Jackson, eds., *Glossary of Geology*, 2d ed. (Falls Church, Va.: American Geological Institute, 1980), p. 542; William I. Rose, Jr., *McGraw-Hill Encyclopedia of Science and Technology* (New York: McGraw-Hill Book Co., 1982), pp. 102–108; and Anthony Wyatt, *Challinor's Dictionary of Geology* (New York: Oxford University Press, 1986), p. 270. Historical evolution of the concept can be found in W. F. Bynum, E. J. Browne, and Ray Porter, ed., *Dictionary of the History of Science* (Princeton, N.J.: Princeton University Press, 1981); R. Porter, *The Making of Geology* (Cambridge: Cambridge University Press, 1977); and H. B. Milner, *Sedimentary Petrography*, 2 vols. (London: Allen and Unwin, 1962). Recent detailed analyses of igneous and metamorphic rocks are in Sven Maaloe, *Principles of Igneous Petrology* (New York: Springer-Verlag, 1985); A. R. McBirney and R. M. Noyes, "Crystallization and Layering of the Skaergaard Intrusion," *Journal of Petrology* 20 (1979): 487–554; B. D. Marsh, "Mechanics and Energetics of Magma Formation and Ascention," in F. R. Boyd, ed., *Studies in Geophysics* (Washington, D.C.: National Academy Press, 1984), pp. 67–83; and H. S. Yoder, ed., *The Evolution of Igneous Rocks* (Princeton, N.J.: Princeton University Press, 1979).

S

SEA BREEZE. See LAND AND SEA BREEZES.

SEISMOLOGY. See EARTHQUAKE.

SOIL. 1. The unconsolidated surface material of the earth in which plants grow. 2. All loose, unconsolidated material at the earth's surface. 3. A natural body of loose material at the earth's surface that interacts with the environment and usually supports some biologic activity.

The concept of the soil is especially elusive because the term *soil* has meant different things to different people throughout time. Even now the term *soil* carries different meanings depending on which science or field of study is using the soil. Throughout prehistory and history, the concept of a soil has evolved through at least four levels of understanding. These four levels include soil as a medium for plant growth, soil as part of the basic matter of the universe, soil as the mantle of loose and weathered rocks (see WEATHERING), and soil as an independent, organized, natural body (Simonson, 1968). Today there are four equally accepted concepts for soil that are used by different fields of study: the edaphic, the pedologic, the geographic (Butler, 1985), and the engineering.

The earliest view of soil by humans must have been that of soil as a medium for plant growth. From the time of transition from a hunting and gathering lifestyle to that of farming, the soil has been considered an important, if misunderstood, component of plant growth. Distinctions between good and bad soil (for plant growth) were made even in Neolithic times (Evans, 1956: 223–226). This attitude was nurtured throughout the Roman period, the Middle Ages, and the late nineteenth and early twentieth centuries (Whitney, 1904; Leibig, 1843).

The concept of soil as part of the basic matter of the universe is also a long-held view. Empedocles (c. 400 B.C.) held that fire, water, earth, and air made up all matter. At that time *earth* and *soil* were synonymous terms (Simonson, 1968:10). The view that soil is part of the basic matter of the universe is not necessarily exclusive of other concepts of soil. Many have believed that both of the above views, and possibly others, can be and are held simultaneously.

The concept of soil as a product of ROCK weathering was predominant during the nineteenth century when GEOLOGY was an infant but rapidly growing discipline. Most soil surveys of that time were really analyses of the *regolith* (unconsolidated surface material) at the earth's surface (e.g., Ruffin 1832; Eaton and Beck, 1820; Coffey 1912; Hilgard 1860; Fallou 1862). To a certain extent this view is held to this day by civil engineers and some geologists (Hough, 1957; Hunt 1967; Jumikis, 1962; Leggett, 1967; Peck, Hanson, and Thornburn, 1974.)

The most comprehensive, complex, and scientific concept of soil is that of soil as an organized natural body that is studied and analyzed in and of itself. Much of the development of this concept parallels that done in SOIL CLASSIFICATION. The classic initial development of this concept began with V. V. Dokuchaiev's (1883) work on Russian chernozems. Many of his students continued the progressive development of this concept. After much hesitance on the part of Western soil scientists to accept these newly evolving ideas, K. D. Glinka (1914; 1927) finally convinced the Western world of the value of this way of looking at soil. In the United States C. F. Marbut (1922, 1927), building his work upon the foundation left by others, provided the impetus for the diffusion of this concept. The amalgamation of new knowledge and older soil ideas was stated succinctly by Marbut in 1927. That year can be used as the beginning of modern soil science in the United States. Since then the main thrust of soil science has been to fill in the voids of knowledge about how the soil system functions and the interrelationship between a soil and its environment (e.g., Jenny, 1941; Soil Survey Staff, 1975).

Today there is a set of four concepts acceptable to varying degrees by various disciplines and scientists. The *edaphic concept* is used primarily by agronomists and studies methods to increase or maintain soil fertility and structure for crop production (e.g., Donahue, Miller and Shickluna, 1983; Foth, 1984). The *pedologic concept* is the center of the study of the soil as an independent natural body (Soil Survey Staff, 1951; Buol, Hole, and McCracken, 1980). Pedologists concern themselves with the upper surface of the LITHOSPHERE that has (usually) been altered into horizons that differ from one another and from the material that lies below. The *geographic concept* relates to the areal extent and distribution of soils and is related inextricably to the edaphic and pedologic concepts. The *engineering concept* correlates somewhat with the pedologic view above. In addition to these concepts, soil mechanics is fundamental to the engineering view (e.g., Lee , 1968).

References

Buol, Stanley W.; Hole, Francis D.; and McCracken, Ralph J., 1980. *Soil Genesis and Classification*. Ames: Iowa State University Press. Excellent text for looking at historical development of soil science.

Butler, B. E. 1985. "The Diversity of Concepts about Soils." *Journal of the Australian Institute Agricultural Science*, 24: 14–20. First to elucidate the presently accepted views on the concepts of soil as endaphic, pedologic, and geographic.

Coffey, C. N. 1912. "The Development of Soil Survey Work in the United States with a Brief Reference to Foreign Countries." In *Journal of the American Society of Agronomy* 3: 115–129. One of the first to propose the complexity and interactions of soils and environments.

Dokuchaiev, V. V. 1883. "Russian Chernozem." In *Collected Works* (1967). Vol. 3. Jerusalem: Israel: Program for Scientific Translations. The classic work on soil as a natural, independent body.

Donahue, Roy L.; Miller, Raymond W.; and Shickluna, John C. S. 1983. *Soils: An Introduction to Soils and Plant Growth*. Englewood Cliffs, N.J.: Prentice-Hall.

Eaton, A., and Beck, T. R. 1820. *A Geological Survey of the County of Albany*. Albany, N.Y.: Agronomy Society of Albany County, New York.

Evans, E. E. 1956. "The Ecology of Peasant Life in Western Europe." In William A. Thomas, Jr., ed., *Man's Role in Changing the Face of the Earth*. Chicago: University of Chicago Press, pp. 217–239.

Fallou, F. A. 1862. *Pedologie Oder Allgemeine und Besondere Bodenkunde*. Dresden.

Foth, Henry D. 1984. *Fundamentals of Soil Science*. 7th ed. New York: John Wiley and Sons.

Glinka, K. D. 1914. *Die Typen der Baderbildung, Ihre Klassifikation und Geographische Verbertung*. Berlin: Gebruder Bountraeger. Introduced to the Western world the working concepts of soil from the Russian work.

———. 1927. *The Great Soil Groups of the World*. Translated from German by C. F. Marbut. Ann Arbor, Mich.: Edward Brothers.

Hilgard, E. W. 1860. *Report on the Geology and Agriculture of the State of Mississippi*. Jackson, Miss.: E. Barksdale, State Printer.

Hough, B. K. 1957. *Basic Soils Engineering*. New York: Ronald Press.

Hunt, C. B. 1967. *Physiography of the United States*. San Francisco: W. H. Freeman and Co.

Jenny, Hans. 1941. *Factors of Soil Formation*. New York: McGraw-Hill Book Co.

Jumikis, A. R. 1962. *Soil Mechanics*. New York: D. Van Nostrand Co.

Lee, I. K. 1968. *Soil Mechanics*. Sydney: Butterworths.

Leggett, R. F. 1967. "Soil—Its Geology and Use." *Geological Society of America Bulletin* 78: 1433–1460.

Liebig, J. 1843. *Chemistry in its Application to Agriculture and Physiology*. Philadelphia: Campbell. Showed the need for at least minimal nutrient levels in plant growth.

Marbut, C. F. 1922. "Soil Classification." *American Association of Soil Workers, Second Annual Report, Bulletin* 3: 24–32.

———. 1927. "A Scheme of Soil Classification." *First International Congress of Soil Science, Transactions* 4: 1–31.

Peck, R. B.; Hanson, W. E.; and Thornburn, T. H. 1974. *Foundation Engineering*. New York: John Wiley and Sons.

Ruffin, E. 1832. *An Essay on Calcareous Manures*. Petersburg, Va.: Campbell.

Simonson, Roy W. 1968. "Concept of Soil." *Advances in Agronomy* 20: 1–47. An excellent extensive review of the historical development of the concept of soil.

Soil Survey Staff. 1951. *Soil Survey Manual*. Handbook, no. 18. Washington, D.C.: U.S. Department of Agriculture.

———. 1975. *Soil Taxonomy*. Handbook, no. 436. Washington, D.C.: U.S. Department of Agriculture.

Whitney, M. 1904. "Soil Surveys, Extension and Practical Application of." Office Experimental Station Bulletin, no. 142. Washington, D.C.: U.S. Department of Agriculture, pp.111–117.

Sources of Additional Information

For more background on the models used in soil analysis see M. G. Cline, "The Changing Model of Soil," In *Soil Science Society of America, Proceedings* 25 (1949): 442–446; G. C. Nikiforoff, "Reappraisal of the Soil," *Science* 129 (1959): 186–196; and R. W. Simonson, "Outline of a Generalized Theory of Soil Genesis," *Soil Sci. Soc. America Proc.* 23 (1959): 161–164. For more information on the concept of pedologic analysis of soil see P. W. Birkeland, *Soils and Geomorphology* (New York: Oxford University Press, 1984); E. M. Bridges and D. A. Davidson, *Principles and Applications of Soil Geography* (London: Longmans, Green and Co., 1982); E. A. Fitzpatrick, *Pedology* (Edinburgh: Oliver and Boyd, 1971); A. J. Gerrard, *Soils and Landforms* (London: Allen and Unwin, 1981); R. D. Hole and J. B. Campbell, *Soil Landscape Analysis* (Totowa, N.J.: Rowman and Allanheld, Publishers, 1985); and J. S. Jaffe, *Pedology* (New Brunswick, N.J.: Pedology Publications, 1949). For an example of the geologic concept see R. Brewer, *Fabric and Mineral Analysis of Soils* (New York: John Wiley and Sons, 1964). A selection of works dealing with soil properties and the environment include C. A. Black, *Soil-Plant Relationships* (New York: John Wiley and Sons, 1957); H. O. Buckman and N. C. Brady, *The Nature and Properties of Soils* (Toronto: Macmillan, 1969); C. E. Kellog, "Climate and Soils," in *Climate and Men*, Yearbook (Washington, D.C.: U. S. Department of Agriculture, 1941); pp. 276–277; and D. H. Yaalon, "Soil-Forming Processes in Time and Space," in D. H. Yaalon, *Paleopedology* (Jerusalem: Israel University Press, 1971), pp. 29–39.

SOIL CLASSIFICATION. 1. Method by which soils are placed into categories based upon some specific characteristics for a given purpose. 2. Taxonomy of soils according to natural relationships.

Many bodies of knowledge (e.g., botany, zoology) have as their basis an established and comprehensive taxonomic (classification) system. John S. Mill (1925) provided a logical explanation of why and how classification processes work. In general, he stated that natural phenomena are classed so that we can organize knowledge, understand relationships, discover new relationships, and predict behavior. These same underlying principles are relevant to the past and present classification of SOILS and to the ever-evolving systems of classification in the future.

The earliest period of soil classification began in the mid-nineteenth century. Thaer, in 1853, published a sixfold classification system that, by present standards, was really only a soil-classification scheme. This system included six types of soil: clay, loam, sandy loam, loamy sand, sand, and HUMUS. He also broke these types into sublevels that corresponded with soil suitability for certain crops. Fallou in 1862 and Richtofer in 1886 each developed a system of soil classification dependent only upon geologic origin of the parent material. These systems were probably less useful than the Thaer system but were logical considering the rapid growth of the science of GEOLOGY during that period (Buol et al., 1980: 184–185).

In Russia toward the end of nineteenth century an environment was created that proved very fertile for the growth of soil science. It was during this period that the study of soil became an independent body of knowledge. Before this time most of the emphasis on soil development was concentrated on the single soil-formation factor of parent material. Other inputs (e.g. climate, biota, TIME, topography) were ignored until V. V. Dokuchaev (1846–1903) introduced a more systematic and comprehensive classification of soils (Bridges and Davidson, 1982: 63). In 1883 Dokuchaev published the first classification of soils based on environmental factors (Dokuchaev, 1883). This publication began the study of soil as an independent natural body. It also provided a new descriptive language including the terms *soil horizon, soil profiles*, and *soil morphology*. Dokuchaev (1886) also used his classification system for more practical purposes such as tax assessment.

K. D. Glinka became Dokuchaev's most articulate and influential protégé. In a series of books, Glinka (1914, 1927, 1931) established soil classification and soil science in Russia. Many of the currently recognizable names of soil types, such as chernozem and podzol, come from his writings.

The early American efforts at soil classification were plagued by the same geologic outlook as the earlier Russian systems (Buol et al., 1980: 186). At the turn of the century, several scientists began looking at soils as natural bodies worthy of independent study (Whitney, 1909; Coffey, 1912). Soil classification in the United States really started to become a serious undertaking with the work of C. F. Marbut in the 1920s (1922, 1927). This work culminated with a comprehensive system of soil classification published in the *Atlas of American Agriculture* (Marbut, 1935). Although this was a milestone in American soil science, it was essentially replaced only three years later by the system of M. Baldwin, C. E. Kellogg, and J. Thorp (1938) published in the U.S. Department of Agriculture Yearbook of Agriculture. Revisions of this 1938 system have been ongoing for decades and have culminated in an almost entirely new, comprehensive, and rigorous soil taxonomy published in 1960 (Soil Survey Staff, 1960). This *Seventh Approximation* has been circulated and critiqued numerous times before and since its publication. The most recent system was eventually published in 1975 as *Soil Taxonomy* (Soil Survey Staff, 1975). This system replaced the qualitative descriptions of former systems and established a quantitative, mutually exclusive system of soil classes. There is an ongoing review of this current U.S. Department of Agriculture classification scheme that will undoubtedly bring about revisions as new techniques and information become available.

The term *soil classification* is broader than a soil taxonomic system. Classification systems can be produced for specific practical purposes, whereas a taxonomic system is a description of soil-character relationships. Other classification systems include schemes such as the Unified Soil Classification System (Craig, 1983: 20–21) and the California Storie Index (Donahue et al., 1983: 561–562) to name only two of the perhaps dozens of systems used for specific purposes.

There are also individual soil classification or taxonomic systems for many other countries and even one for worldwide use being produced by the Food and Agriculture Organization of the United Nations (Bridges and Davidson, 1982: 75–79; Birkeland, 1984: 42). An excellent short review of these other national systems can be found in the work of E. M. Bridges and D. A. Davidson (1982: 74–90).

References

Baldwin, M.; Kellogg, C. E.; and Thorp, J. 1938. "Soil Classification." In *Soils and Men, Yearbook of Agriculture*. Washington, D.C.: U.S. Department of Agriculture. The standard U.S. soil-classification system from 1938 to 1960.

Birkeland, Peter W. 1984. *Soils and Geomorphology*. New York: Oxford University Press. An excellent text on the geomorphic applications of soils with a section of soil-classification terminology.

Bridges, E. M., and Davidson, D. A. 1982. *Principles and Applications of Soil Geography*. New York: Longmans. An excellent comprehensive chapter on soil classifications of the past and present.

Buol, Stanley W.; Hole, Francis D.; and McCracken, Ralph J. 1980. *Soil Genesis and Classification*. Ames: Iowa State University Press. The standard text used for studying the historical trends and current states of soil classification.

Coffey, G. N. 1912. "A Study of the Soils of the United States." Soil Bulletin, no. 85. Washington, D.C.: U.S. Department of Agriculture.

Craig, R. F. 1983. *Soil Mechanics*. Wokingham, Eng.: Van Nostrand-Reinhold Co.

Dokuchaev, V. V. 1883. "Russian Chernozem." In *Collected Works* (1967). Vol. 3. Jerusalem: Israeli Program for Scientific Translations. Classic monograph on classification according to soil-forming factors.

———. 1886. "Report to the Provincial Zenstvo of Nizhnii-Norgo-Rod, No. 1." In *Collected Writings* (1950). Vol. 4. Moscow: Academy of Science, USSR.

Donahue, Roy L.; Miller, Raymond W.; and Shickluna, John C. 1983. *Soils: An Introduction to Soils and Plant Growth*. Englewood Cliffs, N.J.: Prentice-Hall.

Glinka, K. D. 1914. *Die Typen der Bodenbildung, Ihre Klassifikation und Geographische Verbreitung*. Berlin: Gebruder Borntraeger. Introduced to the world the new Russian concepts of soils, soil classification, and terminology.

———. 1927. *The Great Soil Groups of the World*. Translated from German by C. F. Marbut. Ann Arbor, Mich.: Edward Brothers.

———. 1931. *Treatise on Soil Science*. 4th ed. Translated from Russian by A. Gourevich (1963). Jerusalem: Israeli Program for Scientific Translations.

Marbut, C. F. 1922. "Soil Classification." *Soil Survey Workers, Second Annual Report, Bulletin* 3: 24–32.

———. 1927. "A Scheme for Soil Classification." In *Proceedings and Papers, First International Congress of Soil Science* 4: 1–31.

———. 1935. "Soils of the United States." In *Atlas of American Agriculture*. Part 3. Washington, D.C.: U.S. Department of Agriculture. The classic American soil-classification work that brought together all of Marbut's previous work.

Mill, John S. 1925. *A System of Logic*. 8th ed. London: Longmans, Green and Co.
Soil Survey Staff. 1960. *Soil Classification, a Comprehensive System—Seventh Approximation*. Washington, D.C.: U.S. Department of Agriculture. Established the current classification system for the United States based on observed and usually quantifiable data.
———. 1975. *Soil Taxonomy*. Handbook, no. 436. Washington, D.C.: U.S. Department of Agriculture. The most recent draft of the classification system initiated in 1960 (see above). A must document for workers in the area of soils.
Whitney, M. 1909. "Soils of the United States." *Bureau of Soils Bulletin*, no. 55. Washington, D.C.: U.S. Department of Agriculture.

Sources of Additional Information

For the modern soil-classification systems for foreign countries see (Canada) Canada Soil Survey Committee, *The Canadian System of Soil Classification*, Report, no. 2. (Ottawa: Department of Environment, 1978; (USSR) N. N. Rozov and E. N. Ivanova, "Classification of the Soils of the USSR," *Soviet Soil Science* 2 (1967): 147–156; (France) Commission de Pedologie et de Cartographie des Sols, *Classification des Sols* (Grignon: Ecole Nationale Superieure Agronomique, 1967), Mimeographed; (Great Britain) B. W. Avery, "Soil Classification in the Soil Survey of England and Wales," *Journal of Soil Science* 24 (1973): 324–338; (Australia) H.C.T. Stace, G. D. Hubble, R. Brewer, K. H. Northcote, J. R. Sleeman, M. J. Mulcohy, and E. G. Hallsworth, *A Handbook on Australian Soils* (S. Australia: Rellim Technical Publications, 1968) and (New Zealand) The U.S. system was adopted on a trial basis by New Zealand in 1977. For the viewpoint contrary to the U.S. classification system see R. Webster, "Fundamental Objections to the Seventh Approximation," *Journal of Soil Science* 19 (1968): 354–366. This same author produced a volume concerned with quantitative methods of soil classification: R. Webster, *Quantitative and Numerical Methods in Soil Classification and Survey* (New York: Clarendon Press, 1977). An interesting ecological and world perspective can be seen in P. Duchaufour, *Ecological Atlas of Soils of the World*, translated from French by G. R. Mehuys, C. R. DeKimpe, and Y. A. Martel (New York: Masson Publications, 1978). Two edited volumes of papers on soil classification include J. V. Drew, ed., *Selected Papers in Soil Formation and Classification* (Madison, Wis.: Soil Science Society of America, 1967); and C. W. Finkl, Jr., ed., *Soil Classification* (Stroudsburg, Pa.: Hutchinson Ross Publishing Co., 1982). A selection of regional soil-classification publications include F.S.C.P. Kalpage, *Tropical Soils: Classification, Fertility, and Management* (New York: St. Martin's Press, 1976); E. I. Ivanova, *Genesis and Classification of Semidesert Soils*, translated from Russian by A. Gourevitch (Jerusalem: Israeli Program for Scientific Translations, 1970); J.C.F. Tedrow, *Soils of the Polar Landscapes* (New Brunswick, N.J.: Rutgers University Press, 1977); and A. Young, *Tropical Soils and Soil Survey* (New York: Cambridge University Press, 1976).

SOIL HORIZON. See SOIL CLASSIFICATION.

SOIL PROFILE. See SOIL CLASSIFICATION.

SOLIFLUCTION. See PERIGLACIAL.

STRATOPAUSE. See STRATOSPHERE.

STRATOSPHERE. A layer in the atmosphere that lies immediately above the tropopause (see TROPOSPHERE) and extends from about 10 kilometers (6 miles) outward to about 50 kilometers (30 miles).

The stratosphere extends outward from the tropopause to its outer limits at the stratopause, which marks the end of the stratosphere. R. M. Goody (1954) presented the first major discussion of the stratosphere in a book. The stratosphere is virtually without water vapor and clouds, but it does contain ozone.

The OZONE layer in the stratosphere reaches its maximum concentration around 25 kilometers (15 miles) above the earth's surface. W. L. Godson (1960) was among many who studied the ozone layer. More recently J. K. Angell and J. Korshover (1976) studied fluctuations in the ozone layer. As Arthur Strahler and Alan Strahler noted, "The ozone layer serves as a shield, protecting the troposphere and earth's surface by absorbing most of the ultraviolet radiation found in the sun's rays" (1987: 43). In the 1970s it became apparent that chlorofluorocarbons were impacting the ozone layer negatively, and these materials were banned in the United States.

Outward from the tropopause temperatures remain steady at first and then begin to increase until they reach the stratopause. According to Glenn Trewartha and Lyle Horn, "The relatively warm temperatures found near the top of the stratosphere are a consequence of the photochemical absorption of sunlight, especially its ultraviolet wavelengths, in the processes which produce and maintain the gas ozone . . . in the middle and upper stratosphere" (1980: 4).

References

Angell, J. K., and Korshover, J. 1976. "Global Analysis of Recent Total Ozone Fluctuations." *Monthly Weather Review* 104: 63–75. An interesting look at how ozone levels fluctuate around the world.

Godson, W. L. 1960. "Total Ozone and the Middle Stratosphere over Arctic and Subarctic Areas in Winter and Spring." *Quarterly Journal of the Royal Meteorological Society* 86: 301–317. An original and useful look at the ozone layer as it appears in the high latitudes.

Goody, R. M. 1954. *The Physics of the Stratosphere*. London: Cambridge University Press. The first major book-length study of the stratosphere.

Strahler, Arthur N., and Strahler, Alan H. 1987. *Modern Physical Geography*. 3d ed. New York: John Wiley and Sons. Perhaps the best of the physical geography texts that are currently available and an excellent reference book for anyone interested in physical geography.

Trewartha, Glenn T., and Horn, Lyle H. 1980. *An Introduction to Climate*. 5th ed. New York: McGraw-Hill Book Co. This has long been the standard book in climatology, although in recent years several competitors have entered the market.

Sources of Additional Information

Among the various introductory physical geography texts, the following present useful discussions of the stratosphere and the ozone layer: William M. Marsh, *Earthscape: A Physical Geography* (New York: John Wiley and Sons, 1987); E. Willard Miller, *Physical Geography: Earth Systems and Human Interactions* (Columbus, Ohio: Charles E. Merrill Publishing Co., 1985); and Tom L. McKnight, *Physical Geography: A Land-*

scape Appreciation (Englewood Cliffs, N.J.: Prentice-Hall, 1984. Similarly, most climatology texts contain discussions of the stratosphere, stratopause, and ozone layer, and the following are worth considering: Paul E. Lydolph, *The Climate of the Earth* (Totowa, N.J.: Rowman and Allanheld, Publishers, 1985); John E. Oliver and John J. Hidore, *Climatology: An Introduction* (Columbus, Ohio: Charles E. Merrill Publishing Co., 1984); and A. S. Monin, *An Introduction to the Theory of Climate* (Boston: D. Reidel Publishing Co., 1986).

STREAM. A body of flowing water that is confined within a channel.

The concept of a stream is all-encompassing and takes into account all forms of running water that occur in channelized flows, from the smallest rills to the world's giants. For the latter, although the dividing line is not always perfectly defined or understood, the term *river* is often used. As Marie Morisawa noted, "Most geologists use the terms interchangeably to denote running water of any size, but the word *river* is often reserved to denote the main stream or larger branches of a drainage system" (1968: 11).

For streams to flow they must be fed from some water source, and this source, simple as it now appears to us, eluded the study of early stream scholars. Aristotle, the well-known Greek philosopher, conceived the idea of the hydrologic cycle, in which water is evaporated from the ocean and other exposed water surfaces by heat, carried into the atmosphere, and then precipitated back to the earth's surface, although he believed that coolness within the earth was a major factor in the condensation of water vapor. He believed that streams depended not just on rain water for their flows but, more importantly, also on underground sources from within the earth itself. He argued that there was not enough rain to sustain the world's streams. Some 400 years later this same view was echoed by Seneca, a Roman, who argued that precipitation only added to "normal" stream flows.

Throughout the Middle Ages, despite the occasional efforts of Leonardo da Vinci (see Alexander, 1982) and a few others, natural scientists were satisfied to hold the view that water from within the earth was the major contributor to stream flow. However, in the sixteenth century Palissy wrote, with respect to streams, "I finally understood that they could not come from or be produced by anything but rains" (Morisawa, 1968: 3). Although more convincing arguments were needed before natural scientists of the day were to embrace this view, we recognize today that it is a correct perspective on the subject. Those interested in more details about early contributions to the study of streams should turn to Morisawa (1968, 1981), who has succinctly summarized these contributions.

The geomorphic work of streams was slow to gain recognition among natural scientists also, partially because of the predominant religious views that were held with respect to creation of the earth and all that was upon it. According to Morisawa (1968), as late as the first part of the nineteenth century many geologists still subscribed to the notion that stream canyons were structural features and not features carved over the centuries by the work of running water. James

Hutton presented a paper in 1785 called "Concerning the Systems of the Earth, its Duration, and Stability." That paper was the nucleus of his ideas. He followed them up himself, and then they were subsequently popularized by John Playfair (1802). Gradually, with the help of Playfair and others, Hutton's view that rivers erode the valleys within which they flow became accepted.

As nineteenth-century natural scientists accepted the view that streams carved their valleys, they began to look more closely at streams and their dynamics. Among the pioneers in these studies were a number of Americans who were beginning to unravel the geology of the western United States, an undertaking that made many lasting contributions to the geological literature.

G. K. Gilbert (1875) and John Wesley Powell (1876) were major contributors. Gilbert studied stream erosion and transportation, whereas Powell introduced the idea of BASE LEVEL and wrote extensively on drainage systems, introducing the concepts of antecedent and consequent streams into the geological literature.

Soon afterward, William Morris Davis (1899) introduced the idea of a cycle of erosion into the literature. The erosion cycle built upon the notion of base level, and Davis expanded it (1902) and extended it to arid regions (1905). Early in the twentieth century the progress that had been made in the study and interpretation of landforms was chronicled by H. E. Gregory (1918).

The cycle of erosion formed a basis for the study and understanding of fluvial landscapes well into the twentieth century, and it still finds its adherents. However, by the middle of the twentieth century stream studies began to take on new aspects, primarily geared toward the study of fluvial processes or fluvial geomorphology. This work incorporated studies by hydrologists and hydraulic engineers, and it was characterized by an increase in the use of quantitative methods.

Perhaps the first major contribution to the study of fluvial processes was the work of R. E. Horton (1945). Horton began to use quantitative methods to describe and analyze stream drainage patterns. His study was followed by a number of significant studies of stream patterns and processes. J. H. Mackin (1948) wrote at length on the concept of grade and the graded river. Arthur Strahler (1950) presented a seminal study of erosional slopes and equilibrium theory. Stream-channel morphology was explored by L. B. Leopold and Thomas Maddock (1953). River-channel patterns were also studied in detail by Leopold and M. G. Wolman (1957). Finally, J. T. Hack (1960) restated and expanded on Gilbert's earlier contributions and emphasized the concept of dynamic equilibrium in the landscape as an alternative to the cycle of erosion that had been introduced earlier by Davis (1899, 1902, 1905).

The quantitative study of streams reached a new respectability with the publication of the work of Luna B. Leopold, M. Gordon Wolman, and John P. Miller (1964), the first text on fluvial processes and the best summary at that time of the work that had been going on during the previous two decades before its publication.

Subsequent to that time considerable work has been done, and new books have been published to help keep pace with it. Among these new books are those of Morisawa (1968, 1981) and Stanley Schumm (1972, 1977).

References

Alexander, David. 1982. "Leonardo da Vinci and Fluvial Geomorphology." *American Journal of Science* 282: 735–755. A detailed account of the contributions that da Vinci made to our understanding of streams, their dynamics and the features associated with them.

Davis, William Morris. 1899. "The Geographical Cycle." *Geographical Journal* 14: 481–504. The classic statement on the cycle of erosion.

———. 1902. "Base Level, Grade, and Peneplain." *Journal of Geology* 10: 77–111. Extends and clarifies Davis's earlier work on the cycle of erosion.

———. 1905. "The Geographical Cycle in an Arid Climate." *Journal of Geology* 13: 381–407. Carries the idea of an erosion cycle into arid environments.

Gilbert, G. K. 1875. "Report on the Geology of Portions of Nevada, Utah, California and Arizona." *Report upon Geographical and Geological Explorations and Surveys of the 100th Meridian* 3: 21–187. A premier contribution to American geological studies in the nineteenth century.

Gregory, H. E. 1918. "Steps of Progress in the Interpretation of Landforms." *American Journal of Science* 46: 104–117, 127–131. A useful review of the work on streams that had been done up until the article was written.

Hack, J. T. 1960. "Interpretation of Erosional Topography in Humid-Temperate Regions." *American Journal of Science* 258A: 80–97. A major statement with respect to an alternate view to Davis and the cycle of erosion.

Horton, R. E. 1945. "Erosional Development of Streams and Their Drainage Basins: Hydrophysical Approach to Quantitative Morphology." *Geological Society of American Bulletin* 56: 275–370. The first work to introduce quantitative analysis into the study of streams and stream patterns.

Leopold, L. B., and Maddock, Thomas, Jr. 1953. "The Hydraulic Geometry of Stream Channels and Some Physiographic Implications." U.S. Geological Survey Professional Paper, no. 252: Washington, D.C.: U.S. Geological Survey, pp. 1–4, 9–16. One of the first early works on quantitative aspects of stream patterns and morphology.

Leopold, L. B., and Wolman, M. G. 1957. "River Channel Patterns: Braided, Meandering and Straight." U.S. Geological Survey Professional Paper, no. 282-B: Washington, D.C.: U.S. Geological Survey, pp. 39–40, 43–50, 52–63. A major study of different types of river-channel patterns.

Leopold, Luna B., Wolman, M. Gordon, and Miller, John P. 1964. *Fluvial Processes in Geomorphology.* San Francisco: W. H. Freeman. A classic and still highly respected work on fluvial dynamics.

Mackin, J. H. 1948. "Concept of the Graded River." *Geological Society of America Bulletin* 59: 463–511.

Morisawa, Marie. 1968. *Streams: Their Dynamics and Morphology.* New York: McGraw-Hill Book Co. A minor classic on the dynamics of streams, including hydrology, fluvial processes, channel patterns, and river basins.

———. 1981. *Fluvial Geomorphology.* London: Allen and Unwin. Updates and expands many of the materials covered in her earlier work. Required reading for those who really want to understand streams and the landforms associated with them.

Playfair, John. 1802. *Illustrations of the Huttonian Theory of the Earth*. London: Cadell and Davies. One of the great early classics in the literature of geology.

Powell, John Wesley. 1876. "Report on the Geology of the Eastern Portion of the Uinta Mountains." *Geographical and Geological Survey of the Rocky Mountain Region*. Washington, D.C.: U.S. Government Printing Office. A classic in nineteenth-century geology in the United States.

Schumm, Stanley A. ed. 1972. *River Morphology*. Stroudsburg, Pa.: Dowden, Hutchinson, and Ross. An extraordinary collection of papers on quantitative studies of fluvial processes, including many of the classic papers.

———. 1977. *The Fluvial System*. New York: John Wiley and Sons. An excellent sourcebook for anyone interested in fluvial processes.

Strahler, Arthur N. 1950. "Equilibrium Theory of Erosional Slopes Approached by Frequency Distribution Analysis." *American Journal of Science* 248: 687–690. A precise and influential look at erosional slopes and equilibrium theory.

Sources of Additional Information

Most introductory physical geography and geology textbooks dedicate at least one chapter to streams and the work that they do in the landscape. Excellent examples include the following: Arthur N. Strahler and Alan H. Strahler, *Modern Physical Geography*, 3d ed. (New York: John Wiley and Sons, 1987); Tom L. McKnight, *Physical Geography: A Landscape Appreciation* (Englewood Cliffs, N.J.: Prentice-Hall, 1984); William M. Marsh, *Earthscape: A Physical Geography* (New York: John Wiley and Sons, 1987); and Edgar W. Spencer, *Physical Geology* (Reading, Mass.: Addison-Wesley Publishing Co., 1983). Historical perspectives on the study of streams and stream-related landforms and patterns can be found in F. D. Adams, *The Birth and Development of Geological Sciences* (New York: Dover Publications, 1938). An excellent study of the ecology of streams can be found in H.B.N. Hynes, *The Ecology of Running Waters* (Toronto: University of Toronto Press, 1970). For a look at different views that people have toward the uses and abuses of streams see Tim Palmer, *Stanislaus: The Struggle for a River* (Berkeley: University of California Press, 1982). Fluvial processes and fluvial landscapes are studied in reasonable detail in most geomorphology texts, and the following are particularly useful: R. J. Chorley, S. A. Schumm, and D. E. Sugden, *Geomorphology* (New York: Methuen & Co., 1985), and H. F. Garner, *The Origin of Landscapes: A Synthesis of Geomorphology* (New York: Oxford University Press, 1974).

T

TECTONICS. 1. A branch of earth science that deals with changes in rock structures resulting from deformation of the earth's crust through faulting and/or folding. 2. A study of the earth's structural and deformational features and the processes that produced them.

Why the earth is shaped and formed as it is has been a subject of conjecture and investigation for thousands of years. Speculations about changes in rock structures through faulting (breaking) or folding (bending) can be traced to the Greeks and Romans. A number of Greek philosophers offered explanations for earth movements based on their experience with EARTHQUAKES and VOLCANOES along the Aegean coast (Bolt, 1978).

Scientific study of these earth movements and resultant formations primarily took place during the nineteenth and twentieth centuries. During the 1830s the idea of crustal movements and their relationship to folding and mountain building was very popular among Russian geologists (Tikhomirov, 1969: 379). They studied tectonic processes and, noting that positive movements were followed by negative movements in the earth's crust, coined the term *oscillating movements* to describe this phenomenon. By the middle of the nineteenth century Russian scientists had defined some of the characteristics of tectonic processes.

At the end of the nineteenth century and into the early twentieth century an American scientist, G. K. Gilbert, classified tectonic movements based on the speed of movement and the size or area of the movement region. A more precise definition was further developed by G. Stille, who defined slow uplift and subsidence as *epeirogenesis* and the crumpling of strata into folds and their displacement or movement along faults as *orogenic movement* (Beloussov, 1980: 1).

The early twentieth century brought about some interest in global or large-scale earth movements. Although the apparent fit or coincidence associated with continental shapes had been noted earlier, it was Alfred Wegener who proposed large-scale tectonic movements of continents. Wegener noted the similarities in continental coasts, but what really drew him to the idea that the continents had once been a single land mass was the similarity of fossils (Hallam, 1972: 88). Between 1920 and 1930 Wegener's theory, known as continental drift, drew great criticism, and by the late 1930s it had been discarded. However, interest in continental drift began to revive in the 1950s and 1960s with evidence of sea-floor spreading. The spreading of the sea floor has been linked with Wegener's continental drift theory into a new unifying tectonic theory called PLATE TECTONICS (Alexander, 1975). Alfred Wegener and researchers like him knew nothing about the continents resting on movable plates, but they did see the similarities in continental coastlines and developed a tectonic theory that has made revolutionary changes in the earth sciences.

The fact that tectonic activity or deformation leaves an imprint on the earth's surface has recently led to detailed and sophisticated studies of surface features and has led to the development of a special branch of physical geography known as tectonic GEOMORPHOLOGY. Most studies have, according to Dale Ritter, "been directed at understanding the geomorphic response to vertical and/or horizontal displacement along faults and the tilting or warping associated with broad uplifts" (1986: 43).

Recent studies in tectonic geomorphology have used sophisticated techniques of REMOTE SENSING and computer analysis. These studies have covered a wide variety of tectonic processes "including differences in tectonic activity between the earth and planets" (Arvidson and Guinness, 1982: 88), analysis of single landforms like ALLUVIAL FANS (Keller et al., 1982), environmental planning and tectonic activity (Schowengerot and Glass, 1983), landform stability (Bull and McFadden, 1977), and prediction of fault-movement periodicity (Colman and Watson, 1983).

References

Alexander, T. 1975. "A Revolution Called Plate Tectonics Has Given Us a Whole New Earth." *Smithsonian* 3: 31–36. An easy-to-read article on plate tectonics.
Arvidson, R. E., and Guinness, E. A. 1982. "Clues to Tectonic Styles in the Global Topography of Earth, Venus, Mars." *Journal of Geologic Education* 30: 86–92. An application of tectonic principles to other planets.
Beloussov, V. V. 1980. *Geotectonics*. Berlin: Springer-Verlag.
Bolt, Bruce A. 1978. *Earthquakes*. San Francisco: W. H. Freeman and Co. An easy-to-read book on earthquakes.
Bull, W. B., and McFadden, L. D. 1977. "Tectonic Geomorphology North and South of the Garlock Fault, California." In D. O. Doehring, ed., *Geomorphology in Arid Regions*. Proceedings of the Eighth Annual Geomorphology Symposium. Binghamton, N.Y.: State University of New York, pp. 115–139.
Colman, S. M., and Watson, K. 1983. "Ages Estimated from a Diffusion Model for Scarp Degradation." *Science* 221: 263–265.
Hallam, A. 1972. "Continental Drift and Fossil Record." *Scientific American* 228: 1, 87–95.

Keller, E. A.; Bonkowski, M. S.; Korsch, R. J.; and Shlemon, R. J. 1982. "Tectonic Geomorphology of the San Andreas Fault Zone in the Southern Indio Hills, Coachella Valley, California." *Geological Society of America Bulletin* 93: 46–56.

Ritter, Dale F. 1986. *Process Geomorphology*. 2d ed. Dubuque, Ia.: W. C. Brown. A recent geomorphology textbook with a good discussion of tectonic landforms.

Schowengerot, R. A., and Glass, C. E. 1983. "Digitally Processed Topographic Data for Regional Tectonic Evaluation," *Geologic Society of America Bulletin* 94: 549–556.

Tikhomirov, V. V. 1969. "The Development of the Geological Sciences in the U.S.S.R. from Ancient Times to the Middle of the Nineteenth Century." In Cecil J. Schneer, ed., *Toward a History of Geology*. Cambridge, Mass.: The M.I.T. Press.

Sources of Additional Information

For more detailed information about faulting and earthquakes see Peter Ballance and Harold Reading, *Faults* (Boston: Blackwell Scientific Publications, 1980); Ray Kerr, "Prospects for Short-Term Earthquake Prediction," *Science* 227 (1984): 42; C. Simon, "Beneath California's Hills, Another Earthquake Hazard," *Science News* 124 (December 17, 1983): 388; Hill L. Mason, "San Andreas Fault: History of Concepts," *Geological Society of America Bulletin* 92 (1981): 112–131; and John Zieler, "Tertiary Faulting beneath Wadial-Batin (Kuwait)," *Geological Society of America Bulletin* 91 (1980): 610–618. For a discussion of folding of rocks see Timothy Whitten, *Structural Geology of Folded Rocks* (Chicago: Rand-McNally and Co., 1966); and Arthur N. Strahler, *Physical Geology* (New York: Harper and Row, Publishers, 1981). For a discussion of various aspects of tectonics see Jean Goguel, *Tectonics* (San Francisco: W. H. Freeman and Co., 1962); A. Mitchell and M. S. Garson, *Mineral Deposits and Global Tectonic Settings* (New York: Academic Press, 1981); and Kees DeJong and Robert Scholten, *Gravity and Tectonics* (New York: John Wiley and Sons, 1973). Much recent information is available on tectonic geomorphology, for example, M. Morisawa and J. T. Hack, eds., *Tectonic Geomorphology* (Boston: Allen and Unwin, 1985); C. D. Ollier, *Tectonics and Landforms* (London: Longmans, Green and Co., 1981); and C. A. Cotton, "Tectonic Landscapes," in Rhodes Fairbridge, ed., *The Encyclopedia of Geomorphology* (Stroudsburg, Pa.: Dowden, Hutchinson, and Ross, 1968), pp. 1109–1116.

THORNTHWAITE SYSTEM. See EVAPOTRANSPIRATION.

THUNDERSTORM. An intense convectional storm associated with a cumulonimbus cloud.

Thunder is the noise that accompanies the lightning generated within the storm. According to James Murray et al. (1933), thunder and lightning have been used together to describe this phenomenon since about A.D. 725, although its scientific study came much later. Generation of the electrical charges in the cumulonimbus clouds is thought to be related to the freezing process, as suggested by B. J. Mason (1962a) and J. Latham (1966). R. Gunn (1965) studied electrification in thunderstorms and its relationship to raindrop collisions. B.F.J. Schonland (1950) provided a look at most aspects of thunder and lightning in considerable detail.

Thunderstorms develop in cumulonimbus clouds that may reach heights of 40,000 feet or more and within which there are strong vertical wind currents in the form of updrafts and downdrafts. Energy for thunderstorms comes primarily from latent heat that is being generated within the system.

Geographically, according to Mason, "The greatest thunderstorm activity, with giant clouds towering up to 15 kilometers above the ground and whose debris of anvil cirrus may be stretched out in great sheets hundreds of kilometres long, is to be found in the equatorial regions of the East Indies, the Congo and Brazil" (1962b: 8). It is within this equatorial zone that convection plays a major role in the ATMOSPHERIC CIRCULATION as well, and the warm water available in abundance over the equatorial oceans combines with convection to generate many thunderstorms every day.

Thunderstorms and their associated cumulonimbus clouds are excellent breeding grounds for hail also. *Hail* is precipitation that falls in frozen form, and it can be extremely destructive. Murray et al. (1933) traced an interest in hail and hailstones to 1563 and cited an account of a hailstone that measured fourteen inches around that had fallen at Hertfordshire, England, in 1697.

According to Rhodes Fairbridge (1967: 442), the largest hailstone ever found was one that was 14 centimeters in diameter that fell in Potter, Nebraska. An example of the possible economic consequences of hail, suggested by Fairbridge, is that "in the vineyard regions of the south of France such cumulonimbus tends to build up over the hilly country about the grape harvest, and it has long been customary there to discharge cannons to nucleate precipitation before the convection reaches freezing levels" (1967: 442). S. D. Flora (1956) chronicled hailstorms in the United States and provided a perspective on their frequency, distribution, and destructiveness.

References

Battan, L. J. 1964. *The Thunderstorm*. New York: Signet Science Library. A gem of a book that has withstood the test of time. It is still an excellent introduction to thunderstorms.

Deland, Raymond. 1967. "Thunderstorm." In Rhodes W. Fairbridge, ed., *The Encyclopedia of Atmospheric Sciences and Astrogeology*. Encyclopedia of Earth Sciences Series. Vol. 2. New York: Reinhold Publishing Co., pp. 998–999. This is a concise account of thunderstorm basics.

Fairbridge, Rhodes W. 1967. "Hail." In Rhodes W. Fairbridge, ed., *The Encyclopedia of Atmospheric Sciences and Astrogeology*. Encyclopedia of Earth Sciences Series. Vol. 2. New York: Reinhold Publishing Co., p. 442. This is a brief but informative discussion.

Flora, S. D. 1956. *Hailstorms of the United States*. Norman: University of Oklahoma Press. This is probably the most extensive study of hail phenomena in the United States, and it is an invaluable source for those interested in hailstorms.

Gunn, R. 1965. "Thunderstorm Electrification and Raindrop Collisions and Disjection in an Electric Field." *Science* 150: 888–889.

Latham, J. 1966. "Some Electrical Processes in the Atmosphere." *Weather* 21: 120–127. This is an excellent look at lightning and related atmospheric phenomena.

Mason, B. J. 1962a. "Charge Generation in Thunderstorms." *Endeavour* 21: 156–163.
———. 1962b. *Clouds, Rain, and Rainmaking*. Cambridge: Cambridge University Press. Here is the standard reference on the subjects of clouds and rain, including thunderstorms.
Murray, James A.H., et al., eds. 1933. *The Oxford English Dictionary*. Oxford: Clarendon Press.
Schonland, B.F.J. 1950. *The Flight of Thunderbolts*. London: Oxford University Press. Although not current, this book still contains some very useful information.

Sources of Additional Information

Thunderstorms and hail usually receive only brief attention in introductory physical geography texts, as is apparent in Tom L. McKnight, *Physical Geography: A Landscape Appreciation* (Englewood Cliffs, N.J.: Prentice-Hall, 1984), and E. Willard Miller, *Physical Geography: Earth Systems and Human Interactions* (Columbus, Ohio: Charles E. Merrill Publishing Co., 1985). More about the nature of thunderstorms can be found in Joe R. Eagleman, *Meteorology: The Atmosphere in Action* (New York: D. Van Nostrand Co., 1980), and Paul E. Lydolph, *Weather and Climate* (Totowa, N.J.: Rowman and Allanheld, Publishers, 1985). The geographic distribution of thunderstorms is discussed in Glenn T. Trewartha and Lyle H. Horn, *Introduction to Climate*, 5th ed. (New York: McGraw-Hill Book Co., 1980). Details about various aspects of thunderstorms can be found in the following: H. R. Byers and R. Braham, *The Thunderstorm* (Washington, D.C.: U.S. Weather Bureau, 1949); C. W. Newton, "Severe Convective Storms," in *Advances in Geophysics*, vol. 12 (New York: Academic Press, 1967), pp. 238–308; and Sven Petterssen, *Introduction to Meteorology*, 3d ed. (New York: McGraw-Hill Book Co., 1966).

TIDE. The regular and predictable rise and fall of the water level of the earth's oceans caused by the alignment of sun, moon, and earth.

The regular ebb and flow of tides has been recorded for many centuries. It was Pliney who first wrote about the overall cause of tidal motion: "Much has been said about the nature of waters; but the most wonderful circumstance is the alternate flowing and ebbing of the tides, which exist, indeed, under various forms, but is caused by the sun and the moon" (Clancy, 1968: 3).

Although there is a regular interval for high and low tides (12 hours and 25 minutes), the various alignments of sun, moon, and earth cause considerable variation in tide magnitude. When all three bodies are aligned, whether with the sun and moon on one side of the earth or on opposite sides, we experience the highest tides called *spring tides*. This occurs during the new- and full-moon phases. When the moon, earth, and sun are at right angles so that the moon is at first or last quarter, we experience the smallest tides called *neap tides*. In addition to these alignments, there are many other variables involved in tidal magnitude prediction. The variables are admirably summarized by G. Godin:

> Any attempt at a prediction must be preceded by a scrutiny of the observational material, which consists of measurements on the elevation of the water level, the deflection of the vertical, the value of the gravitational acceleration, and so on at

given time intervals and provides a set of numbers associated with a given locality on the earth. The problem of [tide] prediction becomes one of extending this set of numbers from the past into the future. (1972: v)

It was not until Sir Isaac Newton published his *Principia* in 1687 (Wylie, 1979: 20–21) that the quantitative aspects of tidal motion were initially studied. He established the equilibrium theory of tidal magnitude, which predicted tidal heights. He did not explain, however, the manner in which water moves on the earth's surface to create the tides. This dynamic approach would not be proposed for nearly 100 more years. P. S. LaPlace (1775, 1776) is credited with the solution of the mechanism for movement of the tides. He explained tides as a system of dynamic, rhythmic waves of water that are constantly in motion. He showed that the prediction of tides is a problem of fluid motion not static equilibrium.

For almost two centuries after LaPlace, various methods of tidal analysis have concentrated on the determination of amplitudes, the predominant phase harmonics, and the "noise" of tidal variations (Darwin, 1892; Doodson, 1928; Lecolazet, 1956; Horn, 1960). Recent efforts have also been made to study variables considered nontidal to help predict tidal potential (Munk and Cartwright, 1966). The basic methods for determining tidal potential were initially developed by Doodson (1928) and are still the standard today.

Much research on tides during the past two to three decades has concentrated in three areas: the generation of power using tides, the effects of tides on storm surges, and the geomorphic effects of tides and ocean level changes. Power generation using tidal motion began several centuries ago. There were mills using tides in North America as early as 1607 (Wylie, 1979: 220). Large-scale exploitation of tides has not been widespread and is in its infancy even today. The La Rance tidal power plant in France was completed in 1966 and remains the largest such facility in the world. The largest proposed tidal plant, near Passamaquoddy Bay, which separates Maine from New Brunswick, is in the planning stages and may or may not be developed by the mid-1990s. Initial planning for this giant project started in 1920 and continued into the mid-1980s. Technical, financial, and political problems beset this project, and it may never be built, at least in the original planned form (Clancy, 1968: 147–157; Wylie, 1979: 236–239). There are other smaller scale efforts in Europe and the Soviet Union (Wylie, 1979: 233) and a significant study area in the United Kingdom using the Severn estuary (Shaw, 1980). Most of these efforts are also only in the planning stages.

Storm surges are usually associated with hurricanes (see TROPICAL CYCLONE) but may be present in other less violent, yet large-scale, storms. As a storm approaches land, the water level along the shoreline rises in response. This is the storm surge. When the surge is coincident with the high spring tide, the effects can be devastating (Nummedal, 1983: 114). Extreme flooding, property damage, injury, and death can all accompany such a magnified storm surge. An example of the research in this area is the extensive study of Hurricanes Allen and Frederic. These studies enumerate the effects of the enhanced storm surge (U.S. Army Corps of Engineers, 1981a, 1981b).

The third area of current research combines tidal effects on the geomorphology of coastlines with research on global sea-level changes. The geomorphology of coastlines has been an area of study at least since the early 1900's. Beach EROSION and deposition have been the central themes with tides being a significant contributor to erosional and depositional processes (e.g., Thompson and Thompson, 1919; La Fond, 1939; Otvos, 1965; Schwartz, 1967). More recent work looks at sea-level changes with the concurrent tide-level changes and how they affect shoreline morphology (e.g., Wanless, 1982; Bryant, 1983). Expect continued research in these areas with increasing links to coastal GEOMORPHOLOGY, sea-state changes, and tides.

References

Bryant, E. 1983. "Regional Sea Level, Southern Oscillation, and Beach Change, New South Wales, Australia." *Nature* 305: 213–216. An example of recent research looking at geomorphology and sea-level changes.

Clancy, E. P. 1968. *The Tides.* New York: Doubleday and Co. A very good, easy-to-read synopsis of tides.

Darwin, G. 1892. "On an Apparatus for Facilitating the Reduction of Tidal Observations." In *Proceedings of the Royal Society*, Series A, 52: 345–376. A must reference for tidal predictions.

Doodson, A. T. 1928. "The Analysis of Tidal Observations." In *Proceedings of the Royal Society*, Series A, 227: 223–279. Still the standard for calculating tidal potential.

Godin, G. 1972. *The Analysis of Tides.* Toronto: University of Toronto Press. A detailed, mathematical summation of work done in tidal analysis.

Horn, W. 1960. "Some Recent Approaches to Tidal Problems." *International Hydrographic Review* 37: 65–88.

LaFond, E. C. 1939. "Variations of Sea Level on the Pacific Coast of the United States." *Journal of Marine Research* 2: 17–29.

LaPlace, P. S. 1775. "Recherches sur Plusieurs Points du Système du Monde." *Mémoires de l'Academie Royale des Sciences* 88: 75–182. This and the next reference establashed the dynamic approach to tidal explanation.

———. 1776. "Recherches sur Plusieurs Points du Système du Monde." *Memoires de l'Academie Royale des Sciences* 89: 177–267.

Lecolazet, R. 1956. "Application a l'analyse des Observations de la Marée Gravimétrique de la Méthode de H. et Y. Labrouste dite par Combinaisons Lineaires de'ordonees." *Annales de Geophysique* 12: 59–71.

Munk, W., and Cartwright, D. 1966. "Tidal Spectroscopy and Prediction." *Transactions of the Royal Society*, Series A, 259: 533–581. Looked at nontidal influences on tidal prediction.

Nummedal, D. 1983. "Barrier Islands." In P. D. Komar, ed., *Handbook of Coastal Processes and Erosion* (Boca Raton, Fla.: CRC Press, pp. 77–121.

Otvos, E. G. 1965. "Sedimentation in Erosion Cycles of Single Tidal Periods on Long Island Sound Beaches." *Journal of Sedimentary Petrology* 35: 604–609. Looked at erosional and depositional processes of tides.

Schwartz, M. L. 1967. "Littoral Zone Tidal Cycle Sedimentation." *Journal of Sedimentary Petrology* 37: 677–683.
Shaw, T. L., ed. 1980. *An Environmental Appraisal of Tidal Power Stations—With Practical Reference to the Severn Barrage.* London: Pitman Publications. An indepth analysis of the environmental consequences of tidal power-station production.
Thompson, W. F., and Thompson, J. B. 1919. "The Spawning of the Grunion." *California State Game Commission Fish Bulletin*, no.3. Sacramento: California State Game Commission. Early work on tidal deposition and fish habitat.
U.S. Army Corps. of Engineers. 1981a. *Hurricane Frederic, Past Disaster Report, 30 August–14 September, 1979.* Washington, D.C.: U.S. Army Corps of Engineers. This and the next reference look at storm surges and the influence of tides.
———. 1981b. *Report on Hurricane Allen, 3–10 August 1980.* Washington, D.C.: U.S. Army Corps of Engineers.
Wanless, H. R. 1982. "Editorial: Sea Level is Rising—So What?" *Journal of Sedimentary Petrology* 52: 1051–1054.
Wylie, F. E. 1979. *Tides and the Pull of the Moon.* Brattleboro, Vt.: Stephen Greene Press. An excellent work for lay people concerning all aspects of tides.

Sources of Additional Information

Some texts that are useful for lengthy discussions of tides include F.G.W. Smith, *The Seas in Motion* (New York: Thomas Y. Crowell, 1973); M. G. Gross, *Oceanography, A View of the Earth* (Englewood Cliffs, N.J.: Prentice-Hall, 1972); R. L. Carson, *The Sea Around Us* (London: Oxford University Press, 1951); and E. C. Abendanon, *The Problem of the Tides* (The Hague: C. Blommendaal, 1960). For texts relating to modeling and geomorphology of tides see P. D. Komar, *Beach Processes and Sedimentation* (Englewood Cliffs, N.J.: Prentice-Hall, 1976); D. H. Peregrine, ed., *Floods Due to High Winds and Tides* (New York: Academic Press, 1981); and P.P.G. Dyke, A. O. Moscardini, and E. H. Robson, *Offshore and Coastal Modelling* (Berlin: Springer-Verlag, 1985). For shorter discussions and definitions of tides see W. F. Bynum, ed., *Dictionary of the History of Science* (Princeton, N.J.: Princeton University Press, 1981); *The Cambridge Encyclopedia of Earth Sciences* (Cambridge: Cambridge University Press, 1981); and *Van Nostrand's Scientific Encyclopedia*, vol. 2 (New York: Van Nostrand Reinhold Co., 1983).

TIME. 1. A concept concerned with the order and duration of events and the interval between those events. 2. The time at any given place determined by the diurnal motion of the sun.

"I know what time is," said Augustine, "but if someone asks me, I cannot tell him." These remarks by Augustine many centuries ago still carry much truth today, for although the concept of time seems to be simple at first glance, it becomes a very complex issue after prolonged analysis.

The origins of our concept of time are shrouded in mystery, although we do know that ancient peoples were aware that a sense of rhythm was a vital factor in understanding time. Nature seemed to be regulated by these natural rhythms or stages as well as the temporal journey of humans through life.

Although no one knows exactly what time is and it is very difficult to define and explain, human cultures have spent much energy figuring out ways to measure it. The first universal measure for time was the cycles of the moon (Boorstin, 1983: 4). Although it was easy to measure time using moon cycles, it wasn't very useful, particularly to hunters and farmers who needed a calendar of seasons.

The Egyptians were the first to discover the length of the solar year. Early Egyptian attempts to measure time involved the rise and fall of water in the Nile River. They developed what Boorstin called a *nilometer*, a simple vertical scale that measured the flood level of the river. After realizing that their nilometer did not keep in step with the moon's phases, they determined that 12 months of 30 days per month with 5 days added at the end would provide a useful measure to determine and predict seasonal changes.

Although the Egyptian system of months was more reliable than a lunar time scale, it was not a perfect system. For hundreds of years after the Egyptians, people tried to fit days and months evenly into a year.

The Christianization of Western Europe marks an important change in the perception of time. Precise dates of religious events assumed paramount importance, and as a result Pope Gregory XIII appointed a committee to review the matter. The result was the development of a precise astronomical calendar introduced in March 1582, now called the Gregorian calendar (Whitrow, 1972).

Along with the development of the calendar, methods to measure time more precisely became increasingly important (Landes, 1983). Crude attempts to develop clocks using the sun, water, or sand in an hourglass worked very well for societies that were not interested in very precise measurements. But increasing demands for more precision led to the development of the "equal hour." According to Daniel Boorstin, "There are few greater revolutions in human experience than this movement from the seasonal or 'temporary' hour to the equal hour. Here was man's declaration of independence from the sun, new proof of his mastery over himself and his surroundings" (1983: 39).

Much pressure for the development of accurate clocks came from astronomers and navigators (Landis, 1983: 105). Early explorers found it necessary to be able to locate themselves both on latitude north or south of the equator and longitude east or west of some agreed point (see LATITUDE/LONGITUDE). Determining latitude was a relatively easy process using celestial bodies and astronomical tables. Determining longitude, however, was a more complex process. To determine a longitudinal or east-west position on the earth it is necessary to measure the time differences when the sun is at noon in different places.

To understand the noonday position of the sun and its relationship to longitude we must "imagine that a meridian (north south line) sweeps around the globe at just the right speed to be located always where the sun's rays strike the earth's surface at the highest possible angle" (Strahler and Strahler, 1983: 27–28).

If this noon meridian sweeps around the globe of 360 degrees of longitude every 24 hours, it therefore must cover 1 degree of longitude every four minutes, or 15 degrees of longitude every hour. One hour of time is therefore equivalent to 15 degrees of longitude. Knowing this, it is possible, if we have accurate clocks, to determine our exact longitudinal position (Gould, 1923). For example, if the noon meridian reaches one place two hours after another, we know those places are separated by 30 degrees of longitude.

Since every place that is east or west of another place has noon at a different time, these individual places determined their "local time" based upon the noonday sun when it was directly over their meridian. But with the improvement in transportation and communication facilities in the nineteenth century, the use of local time created many problems. If every community used a different time, it would create chaos for travelers. To avoid these problems, "standard time," based on a standard meridian, was developed. Standard meridians 15 degrees apart were selected; thus adjacent zones have one hour time differences, and all places within a zone have the same time. The current system of standard time was made operational in the United States on November 18, 1883, and in the following year an international congress held in Washington, D.C., adopted standard meridians for the entire world based on the prime meridian at Greenwich, England (Strahler and Strahler, 1983: 29–33).

An adjustment to this idea of standard time came with the development of daylight saving time. The idea of daylight saving time was first proposed by Benjamin Franklin and first used by the European countries during World War I to cut electric power consumption. It has since been adopted in the United States (Bartky and Harrison, 1979).

References

Bartky, I. R., and Harrison, E. 1979. "Standard and Daylight-Saving Time." *Scientific American* 240: 46–53. A thorough discussion of these two time concepts.

Boorstin, Daniel J. 1983. *The Discoverers*. New York: Random House. An excellent book on the history of ideas with much information on the development of the concept of time.

Gould, Rupert T. 1923. *The Marine Chronometer*. London: J. D. Potter. A detailed discussion of the relationship of time measurement to longitude.

Landes, David S. 1983. *Revolution in Time*. Cambridge: Harvard University Press. A thorough analysis of the role of clocks on culture.

Strahler, Arthur N., and Strahler, Alan H. 1983. *Modern Physical Geography*. 2d ed. New York: John Wiley and Son. A good discussion of time and longitude.

Whitrow, G. J. 1972. *The Nature of Time*. New York: Holt, Rinehart, and Winston.

Sources of Additional Information

Short discussions and definitions of time can be found in Sir Dudley Stamp and Audrey M. Clark, eds. *A Glossary of Geographical Terms* (London: Longmans, Green and Co., 1979), p. 30; Douglas M. Considine, ed., *Van Nostrand's Scientific Encyclopedia*, 5th ed., vol. 2, (New York: Van Nostrand-Reinhold Co., 1983), p. 2827; and W. F. Bynum, E. J. Browne, and Ray Porter, eds., *Dictionary of the History of Science* (Princeton, N.J.: Princeton University Press, 1981), pp. 49–50, 252–253, 348–352. For detailed discussions of measuring time see Eric Bruton, *The History of Clocks and Watches* (New York: Rizzoli, 1979); Carlo Cipolla, *Clocks and Culture, 1300–1700* (New York:

Walker, 1967); Daniel W. Hering; *The Lure of The Clock* (New York: New York University Press, 1932); F.A.B. Ward, *Time Measurement: Historical Review*, 4th ed. (London: Science Museum, 1970); and J. T. Fraser, F. C. Haber, and G. H. Muller, *The Study of Time* (New York: Springer-Verlag, 1972).

TOPOGRAPHIC MAP. 1. A graphic representation of the earth's surface that depicts relief features and selected man-made or cultural features. 2. A type of map that portrays the shape of the terrain, using contour lines or hachures, and usually of a large scale.

Attempts to use symbols on maps to portray terrain or relief features began almost six thousand years ago. From Mesopotamia we have several clay tablets and stones with landscape relief depicted with a fish-scale design to represent valleys and mountains (Hodgkiss, 1981: 39). The Egyptians also used crude forms of topographical maps, the most famous being a map of the area around Wadi Hammamat, an area on the Nile between the royal capital at Thebes and the Red Sea part of Kosseir. The hills on the map are shown by lines of peaks, and houses are portrayed as rectangles with a little square at one side to represent a door. According to P.D.A. Harvey, "The map has been variously interpreted in the past, as showing the site of the tomb of the Pharaoh Seti I or as a map of some of the mines in Nubia that were the Egyptian kingdom's principal source of gold" (1980: 51). This map probably dates from about 1200 B.C.

There is little evidence from ancient Greece relative to the development of topographic maps. About all we have are two coin designs, one dating from about 500 B.C. and the other about 335 B.C., in which hills are shown in simple relief. Although we do not have any actual maps, it is probably safe to assume that they really did exist, since to choose a map as a symbol on a coin probably means they were familiar with the map concept (Harvey, 1980: 53).

Although elaborate scale maps based on measured surveys were being made throughout the Roman Empire from the first to the third century A.D., very few of the maps survive today. The picture maps that do survive usually depict mountains with undulating lines of peaks and forests by rows of trees (Harvey, 1980: 53). Illustrations to the treatises on surveying were also important sources for Roman topographic maps.

Relief portrayal as developed by the Greeks and Romans continued for many centuries. Usually highlands and lowlands were marked by color, or hills and mountains were shown in profile (Larsgaard, 1984: 14); in some cases the two methods were combined (Imhof, 1982). More refined methods of relief portrayal came about with the development of copper-plate engraving, which permitted more refined elaboration of hill profiles. By the late fifteenth century great cartographers of the Dutch and Flemish schools produced maps that gave "a vivid, if rough, interpretation of the relief of the land" (Taylor, 1928: 474).

A major breakthrough in depicting relief on maps came with the development of the isoline or contour line at the end of the sixteenth century. Contour lines, or lines connecting points of equal elevation, were first developed in 1584 by the Dutch surveyor Pieter Bruinsz. Bruinsz mapped the River Spaarne and used *isobaths* (lines on a map that connect points of equal depth below the surface of the water body) to portray the changing nature of the stream bed (Robinson, 1982: 210).

By the early seventeenth century the topographic map had begun to take on a form somewhat distinct from the more common geographic or planimetric map (Larned, 1907: 18). The practice of drawing relief features as if they were being looked down upon from directly above became an important new development. Shading of slopes to give the impression of relief was used effectively on Swiss cantonical maps by Hans Konrad Gyger. His map of the canton of Zurich produced in 1667 is considered by many to be the first true relief map (Larsgaard, 1984: 15). Another important innovation at the end of the seventeenth century was the introduction of the hachure. A *hachure* is a short line or scratch that, through a shading technique, makes objects stand out in relief. The earliest map to use the technique of hachuring was a masterpiece by David du Vivier produced in 1674 and published under the title *Carte particuliere des environs de Paris* (Brown, 1979: 247).

Although the isoline, both as a contour and isobath, was first used in the late 1500s, it was not fully accepted until the 1730s, when it was developed by marine cartographers (Jervis, 1938; Kish, 1976). Spot heights also began to be used around this time to measure altitude positions above a prescribed datum. They were generally adopted and used by the French in the 1720s and the 1780s (Hodgkiss, 1981: 43). About the same time, the end of the eighteenth century, an officer in the French corps of engineers, Milet de Mureau, was the first to use contour lines to indicate land elevation (Crone, 1968: 127). Hachures were still being used at the end of the eighteenth century and often in conjunction with contour lines (Thrower, 1972: 78). In spite of all of the progress in methods of relief depiction most maps, however, did not show landforms at all, and if they did, they probably used hills and mountains in profile.

During the nineteenth century many methods of shading to show relief were used but none could clearly and accurately depict the four primary facets of relief required on a topographic map: size, slope, elevation, and shape (Greenhood, 1964: 77). In the early nineteenth century color was used with varying intensity to indicate altitude differences. Contour lines began to be more widely used when the French government survey started to use them in 1816 followed by the United States in 1822. It was not until the 1840s, however, that the use of hachures was displaced by contouring as the most common method of relief depiction. Denmark was the first to adopt them followed by the British in 1850 and the United States in 1878 (Hodgkiss, 1981: 43).

In the twentieth century the use of contours has been the most widespread method for showing relief on most national survey maps. The twentieth century has seen many changes in the technology of topographic map compilation, and the future most certainly will see many more changes. Maps of the future will have a diversity of content much greater than that of the present. New tools for map making such as the high-speed electronic computer, digital applications, REMOTE SENSING technology, and space-science applications will all make important impacts on map compilation and production. Maps will be more complete. According to Morris Thompson, "The image map of the future will offer a new standard for completeness. Whereas the line map or data bank will supply only the data that has been selected, the image map will supply everything that the camera or sensor can record" (1982: 227).

References

Brown, Lloyd A. 1979. *The Story of Maps*. New York: Dover Publications. An excellent book on the history of map making.
Crone, Gerald R. 1968. *Maps and Their Makers, An Introduction to the History of Cartography*. 4th ed. London: Hutchinson University Library.
Greenhood, David. 1964. *Mapping*. Rev. ed. Chicago: University of Chicago Press.
Harvey, P.D.A. 1980. *The History of Topographical Maps: Symbols, Pictures, and Surveys*. New York: Thames and Hudson. An excellent source on the history of the development of topographic maps.
Hodgkiss, A. G. 1981. *Understanding Maps: A Systematic History of Their Use and Development*. Folkestone, Eng.: Dawson.
Imhof, Edvard. 1982. *Cartographic Relief Presentation*. Berlin, N.Y.: DeGruyter. An excellent discussion of the early history of topographic relief methods.
Jervis, Walter W. 1938. *The World in Maps: A Study in Map Evolution*. 2d ed. London: G. Philip.
Kish, George. 1976 "Early Thematic Mapping: The Work of Philippe Buache." *Imago Mundi* 28: 129–136.
Larned, C. W. 1907. "History of Mapmaking and Topography." *Scientific American Supplement* 64: 116–118, 132–134.
Larsgaard, Mary L. 1984. *Topographic Mapping of the Americas, Australia, and New Zealand*. Littleton, Colo.: Libraries Unlimited. A good discussion of topographic mapping in the Americas and Oceania. An excellent bibliography is included.
Robinson, Arthur H. 1982. *Early Thematic Mapping in the History of Cartography*. Chicago: University of Chicago Press.
Taylor, E.G.R. 1928. "A Regional Map of the Early Sixteenth Century." *Geographical Journal* 71: 474–479.
Thompson, Morris M. 1982. *Maps for America*. 2d ed. Washington, D.C.: U.S. Geological Survey. An excellent analysis of the development of topographic maps in America.
Thrower, Norman J. 1972. *Maps and Man, An Examination of Cartography in Relation to Culture and Civilization*. Englewood Cliffs, N.J.: Prentice-Hall.

Sources of Additional Information

For short definitions of the concept see F. J. Monkhouse, *A Dictionary of Geography* (London: E. Arnold, 1965), p. 311; Defense Mapping Agency, *Glossary of Mapping, Charting, and Geodetic Terms* (Washington, D.C.: U.S. Government Printing Office, 1973); Dorothy Sylvester, *Map and Landscape* (London: G. Philip, 1952), p. 3; and U.S. Department of the Army, *Cartographic Aerial Photography*, Technical Manual, no. TM5-243 (Washington, D.C.: U.S. Department of the Army, 1970). For an excellent, easy-to-read discussion of the history of mapping see John Noble Wilford, *The Mapmakers* (New York: Vintage Books, 1981). The sport of orienteering, that combines topographical map reading and navigation is thoroughly discussed in Bjorn Kjellstrom, *Be Expert with Map and Compass* (New York: Scribner's, 1984).

TORNADO. A small, violent rotating storm or vortex of low pressure associated with cumulonimbus clouds that are usually formed along or near a very well-developed cold front.

The tornado receives its name from the Spanish *tronada*, which means thunderstorm. The name is appropriate because virtually all tornadoes form in combination with intense cumulonimbus cloud development, which results in thunderstorms.

Tornadoes are small, but their intensity is greater than that for any other type of storm. Winds within a tornado may reach or even exceed 250 miles per hour (400 kilometers per hour); yet the storm may be only a few hundred feet wide at its base.

Writing about the genesis of a tornado, Arthur Strahler and Alan Strahler noted that "it appears as a dark *funnel cloud* hanging from a cumulonimbus cloud'' (1987: 134). The lower end of this funnel cloud twists violently as the storm moves along, and it alternates between touching down on the earth's surface and rising above the surface. Where it touches down, it is extremely destructive. Joe Eagleman, V. U. Muirhead, and Nicholas Willems (1975) provided an excellent account of the destructive capabilities of tornadoes, along with a thorough look at the nature of tornadoes and thunderstorms.

Because of their nature, tornadoes are extremely difficult to study. The earliest theoretical views of tornado development were summarized by E. M. Brooks (1951). S. D. Flora (1953) provided the first book-length treatment of tornadoes, describing these storms and offering extensive statistics about their properties and geographic distribution. B. Vonnegut (1960) argued in favor of an electrical theory of tornado formation. Louis Battan's (1961) book is a very readable discussion of all types of severe storms, including tornadoes. Despite earlier theories about tornadoes and their formation, Battan wrote that "although meteorologists still do not know how and why tornado funnels form where they do, they can specify the conditions usually associated with their development'' (1961: 88). During the 1960s evidence failed to support an electrical theory of tornado formation, but little was substituted in its place. At the beginning of the next decade R. G. Barry and R. J. Chorley wrote that "the exact tornado mechanism is still not understood.... one idea is that convergence beneath the base of cumulonimbus clouds, aided by the interaction between cold precipitation downdraughts and neighboring updraughts, may initiate rotation'' (1970: 137). They suggested other possibilities also, some of which are related to the picture that we have today.

The exact mechanism that generates tornadoes is still not completely understood, but the conditions associated with tornado development are well known.

A well-developed cold front in the interior of the United States seems to be the perfect breeding ground for tornadoes. Ahead of the advancing cold front moist, unstable air needs to be flowing into it beneath an area of upper air divergence that is occurring with the jet stream overhead. Tornadoes develop where turbulence is at a maximum and intense cumulonimbus formation is occurring.

The vast majority of the world's tornadoes are found in the United States, although they do occur elsewhere. Within the United States tornadoes occur most frequently in Oklahoma and at high frequencies in bordering states, primarily Texas and Kansas.

A phenomenon that resembles the tornado in many respects is the waterspout, which forms at sea under cumulonimbus clouds along a cold front. Typically, the waterspout is smaller and less intense than the tornado.

References

Barry, R. G., and Chorley, R. J. 1970. *Atmosphere, Weather, and Climate*. New York: Holt, Rinehart and Winston. Although somewhat dated, this book still offers much to anyone interested in meteorology and climatology.

Battan, Louis J. 1961. *The Nature of Violent Storms*. Garden City, N.Y.: Anchor Books, Doubleday and Company. Even though it is dated in some ways, this is still a very readable and informative book for nontechnical readers.

Brooks, E. M. 1951. "Tornadoes and Related Phenomena." In T. F. Malone, ed., *Compendium of Meteorology*. Boston: American Meteorological Society, pp. 673–680. An excellent summary of theories about the formation of tornadoes as of 1950.

Eagleman, Joe R.; Muirhead, V. U.; and Willems, N. 1975. *Thunderstorms, Tornadoes, and Building Damage*. Lexington, Mass.: Lexington Books, D. C. Heath and Co. An excellent look at tornadoes and thunderstorms and the havoc that violent storms can wreak on a landscape.

Flora, S. D. 1953. *Tornadoes of the United States*. Norman: University of Oklahoma Press. Contains extensive materials on tornado formation and geographic distribution.

Strahler, Arthur N., and Strahler, Alan H. 1987. *Modern Physical Geography*. New York: John Wiley and Sons. Probably the best physical geography text that is currently available. An excellent reference for anyone interested in the subject.

Vonnegut, B. 1960. "The Role of Electrical Phenomena Associated with Tornadoes." *Journal of Geophysical Research* 65: 203–212. An argument in favor of an electrical theory of formation development.

Wilkins, E. M. 1967. "Tornadoes." In Rhodes W. Fairbridge, ed., *The Encyclopedia of Atmospheric Sciences and Astrogeology*. Encyclopedia of Earth Sciences Series. Vol. 2. New York: Reinhold Publishing Co., pp. 1003–1005.

Sources of Additional Information

Tornadoes are typically discussed in introductory physical geography texts, and very good discussions can be found in William M. Marsh, *Earthscape: A Physical Geography* (New York: John Wiley and Sons, 1987), and Tom L. McKnight, *Physical Geography: A Landscape Appreciation* (Englewood Cliffs, N.J.: Prentice-Hall, 1984). More informative discussions of tornadoes and associated weather conditions can be found in meteorology texts such as Joe R. Eagleman, *Meteorology: The Atmosphere in Action* (New York: D. Van Nostrand Co., 1980), or in climatology texts such as Paul E. Lydolph, *The Climate of the Earth* (Totowa, N.J.: Rowman and Allanheld, Publishers, 1985), and John E. Oliver and John J. Hidore, *Climatology: An Introduction* (Columbus, Ohio: Charles E. Merrill Publishing Co., 1984).

TROPIC OF CANCER. See EQUINOX.

TROPIC OF CAPRICORN. See EQUINOX

TROPICAL CYCLONE. An intense low-latitude storm that typically brings with it both high winds and torrential rains.

Known as typhoons in the western Pacific, hurricanes in the Caribbean, and willy-willies in Australia, tropical cyclones are powerful storms that originate in the low latitudes, typically between 8 and 15 degrees north or south of the equator (Strahler and Strahler, 1987). They develop over warm ocean surfaces and derive their energy from the latent heat drawn up through evaporation and released during condensation.

The tropical cyclone is an almost circular storm with closed isobars and winds in excess of 120 kilometers per hour. Average barometric pressure within the tropical cyclone is around 950 millibars or less. These storms are seasonal, and they occur most often in late summer or fall. As Arthur Strahler and Alan Strahler commented, "The general rule is that tropical cyclones of the northern hemisphere occur in the season during which the ITC has moved north; those of the southern hemisphere occur when it has moved south" (1987: 138).

Within the center of the tropical storm is a central, cloud-free vortex, the eye of the storm. Within this eye air is descending and warming; thus there is no cloud formation.

The tropical cyclone typically moves on a westward path within the tropical easterlies or trade winds, but it may ultimately move poleward and get swept into the westerlies before it dies out.

Knowledge of tropical cyclones developed slowly during the first part of the twentieth century, and Isaac Cline (1926) was the first to attempt to incorporate into one book what had been discovered up to that date. Early attempts to unravel the mystery of the tropical cyclone and its development suggested that convection cells generate a sudden and extensive release of energy in the form of latent heat (Barry and Chorley, 1970). However, subsequent research showed a more complicated picture. According to R. G. Barry and R. J. Chorley, "Energy is apparently transferred from the cumulus-scale to the large-scale circulation of the storm through the organization of the clouds into spiral bands, although the nature of this process is still being investigated" (1970: 227).

After World War II research on tropical cyclones increased considerably. Erik H. Palmen (1948) presented a detailed look at the formation and structure of tropical cyclones, and T. Bergeron (1954) followed with a lengthy study of the problem of such storms.

G. E. Dunn and B. I. Miller (1960) were able to draw together enough information to provide an excellent look at tropical cyclones in the Atlantic. Subsequently, a number of important works on tropical cyclones appeared, including that of W. M. Gray (1968), who provided a more global view of these phenomena; Herbert Riehl (1963), who talked about actually modifying hurricane behavior; and B. I. Miller (1967), who studied internal characteristics of tropical cyclones.

Tropical cyclones received considerable attention in the work of Riehl (1979) and were thoroughly analyzed in the work of R. A. Anthes (1982). These two studies are excellent summaries of where current research is with respect to tropical cyclones. An excellent recent study of the impact of tropical cyclones is found in R. H. Simpson and Riehl (1981).

Not all tropical disturbances turn into tropical cyclones, although tropical cyclones do develop out of smaller disturbances. Development of a tropical cyclone from a lesser disturbance seems to depend on the presence of an upper troposphere anticyclone, which may be essential for high-level outflows that are sufficient to develop the very low pressure that is characteristic of the tropical cyclone. According to Joe Eagleman (1980), a sequence of terms is used to describe the development of a true tropical cyclone. The sequence begins with the tropical disturbance, which has neither strong winds nor closed isobars; proceeds to the tropical depression, which has at least one closed isobar; moves to the tropical storm, which has several closed isobars and distinct rotation; and finally becomes the true tropical cyclone.

References

Anthes, R. A. 1982. *Tropical Cyclones: Their Evolution, Structure, and Effects.* Boston: American Meteorological Society. One of the best studies currently available on the subject of tropical cyclones.

Barry, R. G., and Chorley, R. J. 1970. *Atmosphere, Weather, and Climate.* New York: Holt, Rinehart and Winston. Although somewhat dated, this is still an excellent book for many basic meteorological and climatological topics.

Bergeron, T. 1954. "The Problem of Tropical Hurricanes." *Quarterly Journal of the Royal Meteorological Society* 80: 131–164. A classic study for the time.

Cline, Isaac Monroe. 1926. *Tropical Cyclones.* New York: Macmillan. The first book-length study of tropical cyclones.

Dunn, G. E., and Miller, B. I. 1960. *Atlantic Hurricanes.* Baton Rouge: Louisiana State University Press. A classic that is still worth studying today.

Eagleman, Joe R. 1980. *Meteorology: The Atmosphere in Action.* New York: D. Van Nostrand Co. One of several introductory texts on the science of meteorology.

Gentilli, J. 1967. "Tropical Cyclones." In Rhodes W. Fairbridge, ed., *The Encyclopedia of Atmospheric Sciences and Astrogeology.* Encyclopedia of Earth Sciences Series. Vol. 2. New York: Reinhold Publishing Co. A thorough and straightforward discussion of tropical cyclones.

Gray, W. M. 1968. "Global View of the Origin of Tropical Disturbances and Hurricanes." *Monthly Weather Review* 96: 669–700. Brief but of considerable interest at the time.
Miller, B. I. 1967. "Characteristics of Hurricanes." *Science* 157: 1389–1399.
Palmen, Erik H. 1948. "On the Formation and Structure of Tropical Hurricanes." *Geophysica* 3: 26–38. Perhaps the earliest study that tries to put together an integrated look at tropical cyclones.
Riehl, Herbert. 1963. "On the Origin and Possible Modification of Hurricanes." *Science* 141: 1001–1010.
———. 1979. *Climate and Weather in the Tropics*. New York: Academic Press. An excellent survey that includes an excellent discussion of tropical cyclones.
Simpson, R. H., and Riehl, H. 1981. *The Hurricane and Its Impact*. Baton Rouge: Louisiana State University Press.
Strahler, Arthur N., and Strahler, Alan H. 1987. *Modern Physical Geography*. 3d ed. New York: John Wiley and Sons. One of the best of the current texts in physical geography.

Sources of Additional Information

Tropical cyclones are discussed in most introductory physical geography texts, including the following: William M. Marsh, *Earthscape: A Physical Geography* (New York: John Wiley and Sons, 1987); Tom L. McKnight, *Physical Geography: A Landscape Appreciation* (Englewood Cliffs, N.J.: Prentice-Hall, 1984); and E. Willard Miller, *Physical Geography: Earth Systems and Human Interactions* (Columbus, Ohio: Charles E. Merrill Publishing Co., 1985). Books in meteorology and climatology also contain discussion of tropical cyclones, sometimes in more detail than those found in the books above. Two good examples are John E. Oliver and John J. Hidore, *Climatology: An Introduction* (Columbus, Ohio: Charles E. Merrill Publishing Co., 1984), and Paul E. Lydolph, *The Climate of the Earth* (Totowa, N.J.: Rowman and Allanheld, Publishers, 1985). Those interested in pursuing more about hurricanes may also want to look at T. Helm, *Hurricanes: Weather at Its Worst* (New York: Dodd, Mead and Co., 1967).

TROPOPAUSE. See TROPOSPHERE.

TROPOSPHERE. The lowermost of the identifiable layers of the atmosphere.

The troposphere is characterized by the steady decline in temperature that is described by the normal environmental LAPSE RATE. The troposphere extends upward an average of about 14 kilometers (9 miles), although it varies with location and the seasons. Within the troposphere it is possible to find temperature inversions.

The troposphere is important to people because most of the weather that affects life on our planet develops in the troposphere. Water vapor is present in the troposphere, and when conditions are favorable, precipitation can occur. Aside from water vapor and the other gases that comprise the lower atmosphere, dust, smoke, and salt particles are present and act to form condensation nuclei (see CLOUD). The upper boundary of the troposphere is marked by the tropopause, although it is neither a simple nor fixed boundary in its elevation, as noted by J. S. Sawyer (1954). The tropopause is not always a single layer; rather it may be characterized at times by at least two layers, one above the other. Furthermore, transfers between the tropopause and the STRATOSPHERE have been studied by R. E. Newell (1963).

References

Newell, R. E. 1963. "Transfer through the Tropopause and within the Stratosphere." *Quarterly Journal of the Royal Meteorological Society* 89: 167–204. A detailed discussion of such transfers.

Sawyer, J. S. 1954. "Day-to-day Variations in the Tropopause." *Geophysical Memoirs, London* 11: 1–40. An early, but important, study of the tropopause and its complexity.

Sources of Additional Information

The troposphere and tropopause are discussed in most physical geography texts, including the following: Arthur N. Strahler and Alan H. Strahler, *Modern Physical Geography*, 3d ed. (New York: John Wiley and Sons, 1987), and William M. Marsh, *Earthscape: A Physical Geography* (New York: John Wiley and Sons, 1987). Most meteorological texts also discuss the troposphere and tropopause in some detail, including Joe R. Eagleman, *Meteorology: The Atmosphere in Motion* (New York: D. Van Nostrand Co., 1980). The following climatology texts also discuss the troposphere and tropopause: R. G. Barry and R. J. Chorley, *Atmosphere, Weather, and Climate* (New York: Holt, Rinehart and Winston, 1970); Glenn T. Trewartha and Lyle H. Horn, *An Introduction to Climate* (New York: McGraw-Hill Book Co., 1980); and John E. Oliver and John J. Hidore, *Climatology: An Introduction* (Columbus, Ohio: Charles E. Merrill Publishing Co., 1984).

TSUNAMI. A sea wave, with a long wave length, generated by submarine earthquakes or volcanic activity. It is often incorrectly referred to as a tidal wave.

The term *tsunami* comes from the Japanese (*tsu* means harbor and *nami* means wave) and is used to describe what also is called a seismic sea wave. Most commonly these waves are generated by earthquake shocks that occur on the ocean floor, and they can be a real threat in coastal areas when they come ashore.

According to James Gilluly, Aaron Waters, and A. O. Woodford, about 60 percent of tsunami "are caused by dip-slip reverse faults . . . so that shoreward water motion would be expected, in most tsunami the first water movement is actually a withdrawal, followed, after a few minutes, by a great inrush of the sea'' (1975: 94). They cited a number of tsunami that have occurred, including one that had a height of 64 meters as it broke on the shore of the Kamchatka Peninsula in 1737 and one with a height of 28 meters that struck the Japanese city of Miyako in 1896. Because of some of the disasters that have resulted from tsunami in the past, there is now as tsunami warning system that is used whenever seismographs record submarine earthquakes that may have the capacity to generate tsunami.

Although tsunami hardly operate to etch coastlines on a regular basis, they do make a contribution to overall coastal geomorphic processes in some regions, as discussed by P. J. Coleman (1968) in his study of tsunami and geologic processes in coastal areas. Their major effect is on the rearrangement of unconsolidated materials and shore features that are easily eroded.

References

Coleman, P. J. 1968. "Tsunamis as Geological Agents." *Australian Geological Society Journal* 15: 268–273. An excellent discussion of geological effects of tsunami, it is often cited in geomorphology texts.

Gilluly, James; Waters, Aaron C.; and Woodford, A. O. 1975. *Principles of Geology*. 4th ed. San Francisco: W. H. Freeman and Co.

Sources of Additional Information

In introductory physical geography texts, the term *tsunami* is not always mentioned, and when it is, there is no discussion beyond a definition, or perhaps, as in Arthur N. Strahler and Alan H. Strahler, *Elements of Physical Geography*, 3d ed. (New York: John Wiley and Sons, 1984), a brief discussion of possible disastrous effects associated with tsunami that break on inhabited shorelines. More advanced geology texts offer little in addition to what has already been mentioned, although the following are useful references: H. F. Garner, *The Origin of Landscapes: A Synthesis of Geomorphology* (New York: Oxford University Press, 1974), and F. P. Shepard, *Submarine Geology*, 2d ed. (New York: Harper and Row, Publishers, 1969). A more in-depth discussion of the causes and consequences of tsunami can be found in Karl V. Steinbrugge, *Earthquakes, Volcanoes, and Tsunamis* (New York: Skandia America Group, 1982). A study of the effects of tsunami in Hawaii is H. G. Loomis, *Tsunami Wave Runup Heights in Hawaii* (Honolulu: Hawaii Institute of Geophysics).

U

ULTRAVIOLET. The energy in the electromagnetic spectrum with wavelengths of approximately 0.01–0.40 micrometers, where 1 micrometer = 10^{-6} meters (from 100-4,000 angstroms).

Ultraviolet radiation, a part of the electromagnetic spectrum (EMS) (see ELECTROMAGNETIC RADIATION), was first discovered by Johann Wilhelm Ritter in 1801. He found that silver nitrate plates reacted not only to visible light but also to energy wavelengths shorter than visible light: ultraviolet radiation (UV) (Fischer et al., 1975: 39). UV energy has been found to have several interesting and important characteristics. Very little of the UV energy from the sun ever reaches the surface of the earth because it is absorbed by OZONE and molecular oxygen in the atmosphere. What does get through is the energy source primarily responsible for suntans and burns and for much of certain types of skin cancer. UV radiation can also be used to kill certain bacteria and fungi, and it is a key ingredient in producing vitamin D (Koller, 1965). It can also be used for specialized applications in REMOTE SENSING. The two research areas of most interest for geographers are the UV/ozone-oxygen and nitrogen oxide interactions in the atmosphere and the remote-sensing applications.

The first principal role of UV radiation deals with the atmosphere. UV energy's role in the changing of stratospheric chemistry has been studied for decades, but the critical aspects of the UV/ozone relationship has been realized only since the early to mid-1970s (e.g., Caldwell, 1971; Crutzen, 1974). Ozone absorbs UV solar radiation, particularly in the 0.28 to 0.32 micrometer range, the part of the EMS called UV-B. Evidence that has been building during the past decade suggests that ozone in the STRATOSPHERE is destroyed by several complex chemical reactions involving carbon, oxygen, hydrogen, chlorine, and nitrogen compounds such as chlorofluorocarbons (Committee on Causes and Effects of Changes in Stratospheric Ozone, 1984: 3). This destruction of ozone allows much larger quantities of UV-B to reach the earth's surface, causing a large number of potential problems including the aggravation of dozens of human skin diseases, DNA changes, animal photocarcinogenesis, stunted plant growth, and increased atmospheric heating (Committee on Causes and Effects of Changes in Stratospheric Ozone, 1984: 135–143; Committee on Chemistry and Physics of Ozone Depletion, 1982: 6–12). Because of the vastness and complexity of the atmosphere, research in the coming years in this area will depend a great deal on atmosphere/UV interaction modeling. To quote the Committee on Causes and Effects of Changes in the Stratosphere of the National Academy of Sciences,

> definitive studies of this region [the stratosphere] will require very sophisticated three-dimensional models. Current models are only beginning to be able to provide adequate simulation of the current climate, and several years of effort will be required to develop credible coupled dynamical-photochemical simulation models. However, in the interim more work should be done with simplified models to test the sensitivity of chemical processes to transport. (1968: 64)

The remote-sensing applications of UV energy is the other principal area of research interest. Although humans have known about the UV portion of the EMS for almost two centuries, when compared with almost all other parts of the spectrum, UV is still the least used and explored band for remote-sensing applications. This is true in spite of the fact that UV energy is contiguous with the visible portion of the spectrum (Fischer et al., 1975: 39). The reasons for this involve the same atmospheric attention of UV discussed above. The fact that ozone and molecular oxygen absorb most of the UV energy passing through the atmosphere makes it very difficult to use UV in remote sensing. E. F. Horn and R. N. Hubbard (1967), for example, noted that high aerosol concentration and atmospheric backscatter reduce UV reflections from a scene and create undesirable background interference.

Some work has been undertaken using the UV in spite of these drawbacks. In fact, UV radiation has been used in photography since Ritter's discovery that silver halide crystals react with shortwave radiation. Laboratory work using UV spectra indicates that whereas many materials have low reflectance in UV, some materials give high reflectance values in the UV. They include carbonate rocks, evaporites, snow, asphalt, and some metals (Mangold, 1966; Watts, 1966; Watts and Goldman, 1967; Cronin et al., 1968). There has also been a specialized use of UV for detecting oil spills in offshore areas. Significant detection capabilities for observing oil slicks in the UV and blue-light portions of the spectrum have been noted in the literature (e.g., Vizy, 1974; Lintz and Simonett, 1976). In recent years, however, little use has been made of the UV for remote-sensing purposes, whereas most other portions of the spectrum have had increased use. The majority of UV sensing that is currently done occurs in specialized projects using laboratory spectrometers.

References

Caldwell, M. M., 1971. "Solar UV Irradiation and the Growth and Development of Higher Plants." In A. C. Giese, ed., *Photophysiology*. Vol. 6. New York: Academic Press, pp. 131–177.

Committee on Causes and Effects of Changes in Stratospheric Ozone. 1984. *Causes and Effects of Changes in Stratospheric Ozone: Update 1983*. Washington, D.C.: National Academy Press. This and the next volume are excellent works describing the sources and effects of UV radiation increases.

Committee on Chemistry and Physics of Ozone Depletion, 1982. *Causes and Effects of Stratospheric Ozone Reduction: An Update*. Washington, D.C.: National Academy Press.

Cronin, J. F.; Rooney, T. P.; Williams, R. S.; Molineaux, C. E.; and Blianystis, E. E. 1968. *Ultraviolet Radiation and the Terrestrial Surface*. AFCRL 68-0572 Special Reports, no. 83. Cambridge, Mass.: Air Force Cambridge Research Laboratories. A good reference for the uses of UV in spectral analysis.

Crutzen, P. J. 1974. "Estimates of Possible Future Ozone Reductions from Continued Use of Chlorofluoromethanes (CF_2CL_2, $CFCL_2$)." *Geophysical Research Letter* 3: 169–172. One of the first reports on the damage to the ozone in the stratosphere by chlorofluorocarbons.

Fischer, W. A.; Badgley, P.; Orr, D. G.; and Zissis, G. J. 1975. "History of Remote Sensing." In R. G. Reeves, ed., *Manual of Remote Sensing*. Vol. 1. Falls Church, Va.: American Society of Photogrammetry, pp. 27–50. A good source for information on the history of the use of UV in remote sensing.

Horn, E. F., and Hubbard, R. N. 1967. *Consideration for the Application of an Ultraviolet Illumination to Night Aerial Photographic Reconnaissance*. Technical Report AFAL-TR-67-252. Wright-Patterson Air Force Base, Ohio: United States Air Force.

Koller, L. R. 1965. *Ultra-Violet Radiation*. New York: John Wiley and Sons. A good text on the physical characteristics of UV radiation.

Lintz, J., and Simonett, D. S., ed. 1976. *Remote Sensing of Environment*. Reading, Mass.: Addison-Wesley Publishing Co.

Mangold, V. L. 1966. *Narrowband Ultraviolet Photography*. Technical Report AFFDL-TR-66140. Wright-Patterson Air Force Base, Ohio: United States Air Force.

Tousey, R. 1985. "Ultraviolet Radiation." In R. M. Besanson, ed., *The Encyclopedia of Physics*. New York: Van Nostrand-Reinhold Co., pp. 1273–1277. An excellent short reference on the history and the physics of UV radiation.

Vizy, K. N. 1974. "Detecting and Monitoring Oil Slicks with Aerial Photos." *Photogrammetric Engineering* 40: 697–698.

Watts, H. V. 1966. "Reflectance of Rocks and Minerals to Visible and Ultraviolet Radiation." *Technical Letter, NASA*, no. 32. Washington, D.C.: U.S. Geological Society.

Watts, H. V., and Goldman, H. J. 1967. *Visible and Ultraviolet Reflectance and Luminescence from Various Saudi Arabian and Indiana Limestone Rocks*. Open File Report. Washington, D.C.: U.S. Geological Society.

Sources of Additional Information

Works dealing with the history and physics of ultraviolet energy include: R. M. Besanson, ed., *The Encyclopedia of Physics* (New York: Van Nostrand-Reinhold Co., 1985); W. F. Bynum, E. J. Browne, and R. Porter, *Dictionary of the History of Science*

(Princeton, N.J.: Princeton University Press, 1981); and R. W. Fairbridge, *The Encyclopedia of Atmospheric Sciences and Astrogeology* (New York: Van Nostrand-Reinhold Co., 1967). For additional references dealing with ultraviolet energy and remote sensing see R. K. Holz, ed., *The Surveillant Science—Remote Sensing of the Environment* (Boston: Houghton Mifflin Co., 1973); R. G. Reeves, *Manual of Remote Sensing*, vols. 1 and 2, 2d ed. (Falls Church, Va.: American Society of Photogrammetry, 1983). For additional work on climate and ultraviolet radiation see J. Jager, *Climate and Energy Systems* (New York: John Wiley and Sons, 1983). For a good review article on lasers and shortwave radiation see R. W. Waynant and R. C. Elton, "Review of Short Wavelength Laser Research," in *Proceedings, Institute of Electrical and Electronics Engineers* 64 (1976): 1059–1092.

V

VALLEY GLACIER. 1. A glacier that flows down a valley usually bounded by exposed bedrock. 2. A glacier that occupies a preexisting valley and is made up of coalescing cirque glaciers (Alpine type). 3. A glacier formed in a preexisting valley at the edge of an ice cap or ice sheet (outlet glacier).

Although the most extensive glaciation on earth has been continental and ice-cap glaciation, the glacial theory had its birth in the Alpine valleys of Europe. The Alps are a predominantly valley-glacier environment, so it is understandable that most of the early work concerning glaciers dealt with valley glaciers. The ease of recognition and the difficulty of early exploration were well stated by Clifford Embleton and Cuchlaine A.M. King: "The complex of forms associated with mountain glaciation, the U-shaped valleys, rock-basin lakes, cirques and other features, are so well known that they are among the most easily recognized of landforms. They are, however, by no means easy to explain in detail, and many less striking but equally important features of glaciation may even be entirely overlooked" (1975: 1).

J. J. Scheuchzer (1723) proposed a theory of glacial movement that involved the entering of water in crevasses and freezing, thus providing a movement mechanism for glaciers. This was one of the earliest research works on glacial theory. The list of other early researchers is an impressive collection of prominent scientists of the late eighteenth and early nineteenth centuries. James Hutton (1795: 218), John Playfair (1802), R. Bernhardi (1832), and H. T. de la Beche (1832) all espoused the glacial theory in some form. However, it was not until Louis Agassiz turned his attention to the glacial theory that the scientific community began to take seriously the idea of widespread glaciation. Louis Agassiz contributed little new to the theory but added his name and scientific stature to the idea. In 1837 he delivered his famous address on the study of glaciers, the "Discourse of Neuchatel," and his *Etudes sur les Glaciers* was published in 1840 (Embleton and King, 1975: 1–2). The glacial theory was not popular in Britain until A. C. Ramsey (1860) wrote about the glaciers of Switzerland and Wales.

A helpful focus of glacial research during the twentieth century was the derivation of several glacial classification schemes. An important approach to glacier classification has been suggested by H. W. Ahlmann (1948). He devised three classification schemes: thermal, dynamic, and morphologic. Each of these systems either implicitly or explicitly defines valley glaciers. However, in these schemes there is seldom a sharp boundary between glacial types. More precisely, there are transitions between glacier types, with considerable overlap between categories.

The *thermal classification scheme* entails a twofold system: temperate glaciers and cold or polar glaciers. *Temperate glaciers* are approximately at their pressure melting points throughout the glacier and have melt water at their bases, whereas *cold glaciers* exhibit colder temperatures throughout the glacier and are frozen throughout the basal area. This scheme is useful for the concept of valley glaciers because present-day valley glaciers are almost all of the temperate type; their flow characteristics exhibit sliding patterns over bedrock, and thus they move with greater speed than cold glaciers. This in turn leads to erosional differences between the two glacier types.

The *dynamic classification scheme* is also useful. Embleton and King stated the case for this classification:

> The activity of glaciers is influenced to a certain extent by their thermal characteristics: a cold-based glacier requires a larger shear stress to induce movement than a temperate glacier. However, the dynamic activity of a glacier is also closely associated with its mass balance, . . . A classification based on the glacier's dynamic activity consists of three main types: 1. active, 2. passive or inactive, 3. dead glaciers. (1975: 93)

Most valley glaciers are fed by cirques (Alpine type) or plateaus (outlet type) and are active glaciers. Usually, this entails a positive mass balance and a large total budget. Temperate (e.g., valley) glaciers generally have large total budgets.

The most generally useful classification system is the *morphologic scheme*. A modification of H. W. Ahlmann's morphologic system is the ten-class system as listed in Embleton and King (1975: 95):

1. Niche, wall-sided or cliff glacier
2. Cirque glacier
3. Valley glacier—Alpine type
4. Valley glacier—outlet type
5. Transection glacier
6. Piedmont glacier
7. Floating glacier tongues and ice shelves

8. Mountain ice caps
9. Glacier cap or ice cap
10. Continental ice sheet

Although the specific classes of Alpine and outlet glaciers are the main classes of concern, classes 1 (niche) through 8 (mountain ice cap) are all related to valley glaciers. The niche and cirque glaciers are small, isolated mountain glaciers that, when allowed to expand, become valley glaciers. Transection and mountain ice-cap glaciers can be the feeding grounds for radiating valley glaciers. Piedmont and floating glaciers can be fed, in turn, by valley glaciers upgradient from them. Examples of research conducted on these various glacier types include that of J. D. Ives and King (1955), R. P. Sharp (1958), G. E. Groom (1959), W. V. Lewis (1960), F. Ahnert (1963), and G. deQ. Robin (1972).

In 1963 D. L. Linton (1963) established a system of glacial trough classification including the Alpine, Icelandic, composite, and intrusive types. Each of them is categorized by how the ice reacts with its preexisting terrain and helps to explain the present-day postglacial landscape. However, "probably none of [the valleys] are primarily the work of glaciers; rather they are pre-existing valleys, mostly stream valleys, occupied for a time and remodeled with various degrees of thoroughness by glaciers" (Flint, 1971: 127).

Much of the research on valley glaciers in the past two to three decades has dealt with the quantitative analysis of the valley shapes after glacial retreat. The characteristic parabolic cross-section shape and size have been analyzed in the work of H. Svensson (1959), D. J. Drewry (1972), and V. M. Haynes (1972). The study of the longitudinal valley parameters have been even more prolific and comprehensive. The specific features of rock steps (riegels) and lake basins have had particular attention. For some of the classic works in this area of study see J. F. Nye (1952) and W. V. Lewis (1954).

As seen from the above discussion, most work throughout the 1960s dealt with glacial terrain morphology, with additional work on the dating of glacial retreats during and after the Pleistocene. A revolution of sorts began in 1970s when researchers such as G. S. Boulton (1972, 1974) used ice physics to theorize on ice thermal regions and landform production. Glacial physics and renewed interest in the reconstruction of pleistocene events have been the main thrusts of research during the past decade. For examples of this work see D. E. Sugden (1977, 1978), G. S. Boulton (1982), J. O. Hagen and associates (1983), and Robin (1983).

Glacial chronology and the pleistocene are covered in greater detail in CONTINENTAL GLACIER.

References

Agassiz, Louis, 1840. *Etudes sur les Glaciers*. Neuchatel, Switz. This is probably the most important written work promoting the initial acceptance of the glacial theory.
Ahlmann, H. W. 1948. "Glaciological Research on the North Atlantic Coasts." *Royal Geographical Society Research Serial*, no.1: 83pp.

Ahnert, F. 1963. "The Terminal Disintegration of Sleensby Gletscher, North Greenland." *Journal of Glaciology* 4: 537–545.

Beche, H. T. de la. 1832. *Geological Manual*. London: Treuttel and Wurtz, Treuttel Jun. and Richter. This is an early text that prepared the glacial theory for analysis.

Bernhardi, R. 1832. "An Hypothesis of Extensive Glaciation in Prehistoric Times." *Jahrbuch für Minerologie, Geognosie und Petrefahtenkunde* 111: 257–267.

Boulton, G. S. 1972. *The Role of Thermal Regions in Glacial Sedimentation*. Institute of British Geographers Special Publication, no. 4. London: Institute of British Geographers, pp. 1-19.

———. 1974. "Processes and Patterns of Glacial Erosion." In Coates, D. R., ed., *Glacial Geomorphology*. Binghamton: State University of New York, pp. 41–87.

———. 1982. "Subglacial Processes and the Development of Glacial Bedforms." In R. Davidson-Arnott, W. Nickling, and B. D. Fahey, eds., *Research in Glacial, Glaciofluvial and Glaciolacustrine Systems*, Norwich, Conn.: GeoBooks, pp. 1–31.

Drewry, D. J. 1972. "The Contribution of Radio Echo Sounding to the Investigation of Cenozoic Tectonics and Glaciation in Antarctica." *Institute of British Geographers Special Publication* 4: 43–57.

Embleton, C., and C.A.M. King. 1975. *Glacial Geomorphology*. New York: John Wiley and Sons. This is an excellent reference for any area of glacial geomorphology.

Flint, R. F. 1971. *Glacial and Quaternary Geology*. New York: John Wiley and Sons. Although now dated to some degree, this is an invaluable asset for anyone studying glacial history.

Groom, G. E. 1959. "Niche Glaciers in Bunsow-land Vestspitsbergen." *Journal of Glaciology* 3: 369–376.

Hagen, J. O.; Wold, B.; Liestol, O.; Ostrem, G.; and Sollid, J. L. 1983. "Subglacial Processes at Bondhusbreen, Norway: Preliminary Results." *Annals of Glaciology* 4: 91–98.

Haynes, V. M. 1972. "The Relationship between the Drainage Areas and Sizes of Outlet Troughs of the Sukkertoppen Ice Cap, West Greenland." *Geografiska Annaler* 54A: 66–75.

Hutton, J. 1795. *Theory of the Earth*. Vol. 2. Edinburgh: William Creech. The seminal work in modern geology including a section on glaciation.

Ives, J. D., and King, C.A.M. 1955. "Glaciological Observations at Morsarjokull, S. W. Vatnajokull, Iceland." *Journal of Glaciology* 2: 477–482.

Lewis, W. V. 1954. "Pressure Release and Glacial Erosion." *Journal of Glaciology* 2: 417–422.

———. 1960. 'Norwegian Cirque Glaciers." *Royal Geographical Society Research Serial* 4: 104p.

Linton, D. L. 1963. "The Forms of Glacial Erosion." *Transactions, Institute of British Geographers* 33: 1–28.

Nye, J. F. 1952. "The Mechanics of Glacier Flow." *Journal of Glaciology* 2: 82–93.

Playfair, J. 1802. *Illustrations of the Huttonian Theory of the Earth*. Edinburgh: William Creech. The classical follow-up of Hutton's work.

Ramsey, A. C. 1860. *The Old Glaciers of Switzerland and North Wales*. In Ball, J., ed., *Peaks, Passes and Glaciers*. London.

Robin, G. deQ. 1972. "Polar Ice Sheets: A Review." *Polar Record* 16: 5–22.

———. 1983. *The Climatic Record in Polar Ice Sheets.* Cambridge: Cambridge University Press. An excellent resource for climate records in ice caps.
Scheuchzer, J. J. 1723. *Itinera per Helvetiae Regiones Alpinas.* Leyden.
Sharp, R. P. 1958. "Malaspina Glacier Alaska." *Geological Society of America Bulletin* 69: 617–646.
Sugden, D. E. 1977. "Reconstruction of the Morphology, Dynamics, and Thermal Characteristics of the Laurentide Ice Sheet at its Maximum." *Arctic and Alpine Research* 9: 21–47.
———. 1978. "Glacial Erosion by the Laurentide Ice Sheet." *Journal of Glaciology* 20: 367–391.
Svensson, H. 1959. "Is the Cross-Section of a Glacial Valley a Parabola?" *Journal of Glaciology* 3: 362–363.

Sources of Additional Information

For examples of work on the surging of valley glaciers see R. A. Bindschadler, "A Numerical Model of Temperate Glacier Flow Applied to the Quiescent Phase of a Surge-Type Glacier," *Journal of Glaciology* 28 (1982): 239–266; C. M. Clapperton, "The Debris Content of Surging Glaciers in Svalbard and Iceland," *Journal of Glaciology* 14 (1975): 395–406; D. N. Collins, "Quantitative Determination of the Subglacial Hydrology of Two Alpine Glaciers," *Journal of Glaciology* 23 (1979): 347–362; and J. S. Walder, "Stability of Sheet Flow of Water beneath Temperate Glaciers and Implications for Glacier Surging," *Journal of Glaciology* 28 (1982): 273–294. As examples of the current work of the physics of glacial ice see K. Hutter, "Dynamics of Glaciers and Large Ice Masses," *Annual Review of Fluid Mechanics* 14 (1982): 87–130; W.S.B. Paterson, *The Physics of Glaciers,* 2d ed. (Oxford: Pergamon, 1981); and J. Weertman and G. E. Birchfield, "Basal Water Film, Basal Water Pressure and Velocity of Traveling Waves on Glaciers,"*Journal of Glaciology* 29 (1983): 20–27. For examples of the work being done on the hydrology of glaciers see R. Bindschadler, "The Importance of Pressurized Subglacial Water in Separation and Sliding at the Glacier Bed," *Journal of Glaciology* 29 (1983): 3–19; D. Hantz and L. Llibourty, "Waterways, Ice Permeability at Depth, and Water Pressures at Glacier d'Argentiere, French Alps," *Journal of Glaciology* 29 (1983): 227–239; and P. Holmlund and R. leB. Hooke, "High Water-Pressure Events in Moulins, Storglaciaren, Sweden," *Geografiska Annaler* 65A (1983): 19–25. Good references for glaciers and glacial geomorphology include D. R. Coates, ed., *Glacial Geomorphology* (Binghamton: State University of New York, 1974); R. Davidson-Arnott, W. Nickling, and B. D. Fahey, eds., *Research in Glacial, Glaciogluvial, and Glaciolacustrine Systems* (Norwich, Conn.: GeoBooks, 1982); H. E. Wright and D. G. Frey, eds., *The Quaternary of the United States* (Princeton, N.J.: Princeton University Press, 1965); and the entire set of *Journal of Glaciology.*

VEGETATION. The plant life of the earth, usually limited in geographic studies to natural vegetation.

The study of natural vegetation has a long tradition in geography and occurs under the subdiscipline of biogeography. The existence of plants on the earth's surface, according to S. R. Eyre (1968), began some 400 million to 500 million years ago. The term *vegetation* dates at least to the sixteenth century, according to James Murray and associates (1933).

The scientific study of natural vegetation, though anticipated earlier, awaited the development of a useful classification scheme. As Rexford Daubenmire noted, "When Linnaeus made his epic contribution to standardized nomenclature in the eighteenth century, taxonomic concepts were greatly improved and distribution patterns could be better defined" (1978: 1). The scientific study of natural vegetation began. The next major influence on the study of natural vegetation came a century later with the work of Charles Darwin (1859).

Plant geography, or biogeography, traces its roots to the late eighteenth and early nineteenth centuries and the work of Alexander von Humboldt (1793, 1808). The first activities of plant geographers involved finding, naming and mapping plants in different regions of the world, although by the end of the nineteenth century they became more interested in explaining plant distributions and the relationship between plants and their environments. The latter field became known as plant ecology in the twentieth century, although it is also referred to by some as ecological plant geography.

The major work in biogeography has been done during the current century and has focused on the classification and distribution of natural vegetation. Many of the earlier contributions by biogeographers are summarized in Hugh Raup (1942).

Within geography, much of the early work in climatology, especially climate classification, was closely related to studies of the distribution of natural vegetation, as is apparent in the work of Vladimer Köppen (1931).

Today there are two schools of thought in biogeography with respect to the study of vegetation, floristic plant geography and ecologic plant geography. According to Daubenmire, *floristic plant geography* is "primarily a study of evolutionary divergence, migration, and decline of taxa, as influenced by past events of the earth's history," whereas *ecologic plant geography* "takes plant communities as units having ranges to be interpreted" (1978: 2).

Arthur Strahler and Alan Strahler (1987) suggested that geographers can approach the study of the distribution of the earth's vegetation through a combination of both structural and floristic criteria. With respect to the structure of vegetation, they consider the following: (1) life form, (2) size and stratification, (3) coverage, (4) periodicity, (5) leaf shape and size, and (6) leaf texture. They then consider the broadest ecosystems, which they classify first into five biomes—forest, savanna, grassland, desert, and tundra—and next, when necessary, into formation classes. The geographic distribution of these biomes and formation classes can then be related to the geographic distribution of climate.

One important idea in the study of natural vegetation is the concept of ecological succession. According to Strahler and Strahler, "Succession leads to formation of the most complex community of organisms possible in an area, given its physical controlling factors of climate, soil, and water" (1987: 437). The end point of succession, in which there is an equilibrium between the environment and the assemblage of plants within it, is called the *climax*, or *climax vegetation*.

Geographers are always aware of the influence that people have over their environment, and so some biogeographers have looked at natural and other vegetation complexes as they have been altered by people. John Curtis (1965) studied how people had modified the mid-latitude grasslands, and Thomas Detwyler (1972) studied the vegetation in cities, including gardens and parks.

References

Curtis, John. 1965. "The Modification of Mid-Latitude Grasslands and Forests by Man." In William L. Thomas, Jr., ed., *Man's Role in Changing the Face of the Earth*. Chicago: University of Chicago Press, pp. 721–736. One of many excellent essays in this widely acclaimed book.

Darwin, Charles R. 1859. *On the Origin of Species*. London: J. Murray. Classic study of evolution.

Daubenmire, Rexford. 1978. *Plant Geography, with Special Reference to North America*. An excellent introduction to floristic plant geography, with a focus on vegetation in the United States.

Detwyler, Thomas R. 1972. "Vegetation of the City." In Thomas R. Detwyler and Melvin G. Marcus, eds., *Urbanization and Environment: The Physical Geography of the City*. Belmont, Calif.: Duxbury Press, pp. 229–259. An excellent though somewhat dated attempt to focus on the physical characteristics of cities, including vegetation.

Eyre, S. R. 1968. *Vegetation and Soils: A World Picture*. 2d ed. Chicago: Aldine Publishing Co. A useful geographic perspective on plants and soils.

Köppen, Wladimer. 1931. *Grundriss der Klimakunde*. Berlin: De Gruyter. Contains a discussion of Köppen's climatic classification system and its relationship to vegetation distributions.

Murray, James A. H., et al. eds. 1933. *The Oxford English Dictionary*. Oxford: Clarendon Press.

Raup, Hugh M. 1942. "Trends in the Development of Geographic Botany." *Annals of the Association of American Geographers* 32: 319–354. An excellent survey of biogeography up to 1941.

Strahler, Arthur N., and Strahler, Alan H. 1987. *Modern Physical Geography*. New York: John Wiley and Sons. One of the best physical geography texts currently on the market and an excellent source for anyone wanting an introduction to biogeography.

von Humboldt, Alexander. 1793. *Florae Fribergensis Subterraneas Exhibens*. Berlin: H. A. Rottman. Among the first major works by this outstanding German geographer.

———. 1808. *Ansichten der Natur, mit wissenschaftlichten Erlauterung*. Stuttgart: Cotta. Another major work by this prolific writer.

Sources of Additional Information

Good introductions to the study of biogeography can be found in the following introductory physical geography texts: William M. Marsh, *Earthscape: A Physical Geography* (New York: John Wiley and Sons, 1987); Tom L. McKnight, *Physical Geography: A Landscape Appreciation* (Englewood Cliffs, N.J.: Prentice-Hall, 1984); and E. Willard Miller, *Physical Geography: Earth Systems and Human Interactions* (Columbus,

Ohio: Charles E. Merrill Publishing Co., 1985). Excellent studies of plant ecology and biogeography include the following: Martin Kellman, *Plant Geography*, 2d ed. (London: Methuen & Co., 1980); W. D. Billings, *Plants, Man, and the Ecosystem*, 2d ed. (Belmont, Calif.: Wadsworth Publishing Co., 1970); Heinrich Walter, *Vegetation of the Earth and Ecological Systems of the Geobiosphere* (New York: Springer-Verlag, 1979); A. W. Kuchler, *Vegetation Mapping* (New York: Ronald Press, 1967); and Pierre Dansereau, *Biogeography: An Ecological Perspective* (New York: Ronald Press, 1957). Methodological considerations in the study of natural vegetation are thoroughly explored in P. D. Moore and S. B. Chapman, *Methods in Plant Ecology*, 2d ed. (Oxford: Blackwell Scientific Publications, 1986). The relationship between vegetation and the atmosphere is documented in J. L. Monteith, ed., *Vegetation and the Atmosphere*, vol. 2, Case Studies (New York: Academic Press, 1976). Many aspects of the relationship between vegetation and microclimates are considered in Hamlyn G. Jones, *Plants and Microclimate: A Quantitative Approach to Environmental Plant Physiology* (Cambridge: Cambridge University Press, 1983). Specific vegetation complexes and the vegetation of particular geographic regions are considered in many books, including the following: R. T. Coupland, ed., *Grassland Ecosystems of the World* (Cambridge: Cambridge University Press, 1979); David J. de Laubenfels, *Deserts and Grasslands: The World's Open Spaces* (Garden City, N.Y.: Doubleday, 1976); Eric A. G. Duffey, *The Forest World* (London: Orbis, 1980); C.R.W. Speeding, *Grassland Ecology* (Oxford: Oxford University Press, 1976); John L. Vankat, *The Natural Vegetation of North America: An Introduction* (New York: John Wiley and Sons, 1979).

VOLCANO. A cone-shaped mountain created by the flow of magma up through a central vent.

According to James Murray et al. (1933), use of the word *volcano* dates at least to the early seventeenth century. Much earlier, however, volcanoes had represented one of the great mysteries for people to ponder. Superstition and fear were natural responses to volcanic eruptions, and only considerably later were these responses to be replaced by scientific study and an improved nature of volcanoes and volcanic activity.

Early Roman interest in volcanoes manifests itself in the form of Vulcan, the god of fire, who was believed to be a blacksmith of the gods. The volcano was his workshop, and the smoke that periodically belched forth was evidence that he was indeed at work. The Polynesians reckoned with volcanoes also, and one of their myths tells of Maui, who stole fire from a volcano and used it to light fire to the local vegetation, thus giving people fire for their use. The Hawaiian created Pele, goddess of the Hawaiian volcanoes and resident of Kilauea, Hawaii's most active volcano. They believed that Pele made an appearance before each eruption of the volcano, a story that is still believed by many.

The Roman god Vulcan gave his name to the modern scientific study of volcanoes, which is called vulcanology. However, long before modern scientists began their studies of volcanoes and volcanism, the early Greeks and Romans took their turn. In the fourth century B.C. Aristotle suggested that volcanoes were "pent up" winds that had been trapped underground, finally ignited and erupting when they struck sulphur or coal under the earth's surface. Some 350 years later Strabo suggested that volcanoes represented a combination of imprisoned fire and wind, a combination that periodically overcame its confinement and burst to the surface. The most impressive and useful early study of volcanoes, however, was provided by Pliny the Younger, who wrote a detailed account of the eruption of Mt. Vesuvius in A.D. 79.

Scientific interest in volcanoes and volcanic eruptions waned over the centuries but emerged again in the eighteenth century, with studies of Mt. Vesuvius, Mt. Etna, and other Italian volcanoes, as well as the discovery that basalt apparently was a rock that had crystallized from molten magma, as noted by Lazarro Spallanzani (1798). Further advances in volcanology were slowed, however, by arguments that were developing between two schools of thought about volcanic features and their origins. The Neptunists argued that all rocks originated in the ocean; the Plutonists argued that some rocks were formed by cooling from molten magma, either underground or at the surface. Arguments also developed about whether volcanic cones were a result of doming of the earth's surface or actual ejection of molten material from below. Some of these early ideas turned out to be correct, at least to some degree, but the arguments slowed progress in volcanology.

Paulett Scope (1825) was the first to classify volcanic eruptions by labeling them as (1) permanent, (2) moderate, and (3) paroxysmal. *Permanent eruptions* were those that were of a quiet and regular nature, *paroxysmal eruptions* were those that were violent and irregular, and *moderate eruptions* were those that fell somewhere in between the other two categories. Eruptions were later classified analogously into explosive, intermediate, and quiet, with virtually the same descriptions as those suggested by Scope. Other classifications of volcanoes appeared also in the nineteenth century.

Alfred Lacroix (1904) developed a classification that included the following types of volcanic eruptions: (1) *Hawaiian*—a non-explosive eruption that is characterized by quiet ejection of magma onto the surface without sending it far into the air; (2) *Strombolian*—continuous mild discharges of viscous lava ejected in recurring explosions, typically accompanied by the formation of luminous clouds that result from the ejection of incandescent materials; (3) *Vulcanian*—more explosive type, created when magma accumulates below a plug that prevents its escape until an explosive eruption blows materials out of the way and into the air; and (4) *Peleean*—the most violent of all eruptions, resulting in the emission of hot, gas-charged fragments of lava and superheated steam, accompanied by an incandescent cloud of toxic, burning gases.

Subsequently, two additional types have been added to Lacroix's initial four categories. One is the *Icelandic* type, similar to the Hawaiian type but creating a plateau rather than a shield volcano. The other is the *Solfataric* stage, a late stage in the development of a volcano that is characterized by magmatic movement beneath the surface, which then allows gases to escape.

Volcanic activity is reflected in landform types, and major volcanic features are directly related to the types of activities that resulted in their creation. Major volcanic landforms include the following: (1) cinder cone, (2) shield volcano, (3) composite volcano, (4) volcanic plateau, and (5) caldera. Each of them is discussed below.

The *cinder cone* is a relatively small volcanic feature, seldom more than a few hundred meters in height, which forms as the result of a violent explosive eruption that sends volcanic materials high into the air. As these materials descend back to the ground they literally pile up to form a neat, conical hill of unconsolidated volcanic cinder and ash. Typically, they are found clustered within an area of considerable volcanic activity.

Shield volcanoes are normally constructed on the ocean floor and may, as in the case of the Hawaiian Islands, ultimately reach heights that bring them above sea level. Gentle slopes and broad tops, as well as multiple vents and fissures, are characteristic of shield volcanoes, as are quiet, gentle eruptions. Shield volcanoes are sometimes also called domes, although they should not be confused with structural domes.

Composite volcanoes are quintessential examples of volcanic activity. They are tall, conical, steep-sided, and often magnificent. Examples include Fujiyama in Japan, Mt. Mayone in the Philippine Islands, Mt. Hood in Oregon, Mt. Baker in Washington, Mt. Shasta in California, and Mt. Shishaldin in Alaska's Aleutian Islands. They represent a composite of the previous two volcanic types, shields and cinder cones. Explosive eruptions help to give them their steep-sided structure and conical shape, whereas flows of felsic lava stabilize the unconsolidated fragments and ash, giving them their permanence. Thus they are typically formed over a long period as the pattern of volcanic activity switches back and forth between the periods of explosive eruptions and the generation of cinders and ash and the quieter eruptions of lava that then flow down the sides and cover the layers of unconsolidated materials. Composite volcanoes are sometimes called stratovolcanoes because of the alternating layers of ash and lava.

The *volcanic plateau* is most similar to the Hawaiian shield volcano. Volcanic flows, referred to as flood basalts, flow out through cracks and fissures onto the surface, where they often cover hundreds or even thousands of square miles with volcanic rock. The Columbia Plateau in the northwestern United States, which covers parts of Oregon, Washington, and Idaho, is an excellent example. According to H. F. Garner, "The Columbia Plateau has an estimated average lava thickness of 0.5 mile and an area of about 200,000 square miles" (1974: 201).

The *caldera* is a great depression left in the top of a volcano. It forms as a result of a violent explosion that blows away the entire central part of the volcano, followed by subsidence of much of what is left. Crater Lake, in Oregon, is a typical example of a caldera. Some 6,600 years ago a mountain known as Mt. Mazama exploded and then subsided, leaving what is now Crater Lake at an elevation estimated to be some 4,000 feet lower than the height of old Mt. Mazama. According to Arthur Strahler and Alan Strahler, "Valleys previously cut by streams and glaciers into the flanks of Mount Mazama were beheaded by the explosive subsidence of the central portion and now form distinctive notches in the rim'' (1987: 246). Wizard Island is a cinder cone that has formed within Crater Lake subsequent to the development of the caldera.

Although volcanoes had been studied scientifically since at least the eighteenth century, it was not until the 1960s that scientists began to gain a better understanding of their formation and distribution. With the new ideas of plate tectonics, continental drift, and sea-floor spreading, it was easier to see broad patterns of volcanic activity as the result of other features, primarily plate boundaries and subduction zones.

For example, the distribution of numerous volcanoes around the rim of the Pacific Ocean, known as the "Circum-Pacific Ring of Fire," had been recognized for a long time. However, an explanation of its location awaited the plate-tectonic view, in which the volcanic activity around the Pacific rim is related directly to subduction zones. These subduction zones are areas of considerable volcanism; thus they are areas in which magma is present in large amounts under the surface. "Reaching the earth's surface," wrote Strahler and Strahler, "quantities of this magma build volcanoes, which tend to form a volcano chain lying about parallel with the deep oceanic trench that marks the line of descent of the oceanic plate" (1987: 225). From the Andes in South America to the Cascades in North America and on to the volcanic ranges of Japan and the Philippine Islands, we see mountains that have been formed in association with subduction along the Pacific Rim.

References
Bullard, Fred M. 1962. *Volcanoes: In History, in Theory, in Eruption*. Austin: University of Texas Press. Remains an excellent general reference on volcanoes, although it is in some ways outdated.
Francis, Peter G. 1976. *Volcanoes*. Middlesex, Eng.: Penguin Books, Harmondsworth. Still a worthwhile volume for anyone interested in pursuing the study of volcanoes.
Garner, H. F. 1974. *The Origin of Landscapes: A Synthesis of Geomorphology*. New York: Oxford University Press. A thorough work on geomorphology.
Lacroix, Alfred. 1904. *La Montagne Pelee et ses Eruptions*. Paris: Masson. Classic study of volcanism around the turn of the century.
Murray, James A. H. et al. eds. 1933. *The Oxford English Dictionary*. Oxford: Clarendon Press.
Scope, Paulett G. 1825. *Volcanoes, the Character of their Phenomena, Their Share in the Structure and Composition of the Surface of the Globe, and Their Relation to Its Internal Forces with a Description Catalogue of All Known Volcanoes and Volcanic Formations*. London: W. Phillips. One of the earliest studies of volcanoes and volcanic activity.
Sheets, Payson D., and Grayson, Donald K. 1979. *Volcanic Activity and Human Ecology*. New York: Academic Press. An excellent account of human problems associated with volcanic activity.
Spallanzani, Lazaro. 1798. *Travels in the Two Sicilies, and Some Parts of the Apennines*. 4 Vols. London: G. G. and J. Robinson. Careful observations of the eruptions of Italian volcanoes and their effects.

Strahler, Arthur N., and Strahler, Alan H. 1987. *Modern Physical Geography*. New York: John Wiley and Sons. The best of current physical geography texts and a solid reference for most topics in physical geography, including volcanoes.

Tilling, Robert J. 1979. *Volcanoes*. Washington, D.C.: U.S. Government Printing Office. A useful general reference on volcanoes and volcanic landforms.

Sources of Additional Information

The following introductory texts in physical geography and physical geology provide good introductions to the world of volcanoes and volcanic activity: Tom L. McKnight, *Physical Geography: A Landscape Appreciation* (Englewood Cliffs, N.J.: Prentice-Hall, 1984); William M. Marsh, *Earthscape: A Physical Geography* (New York: John Wiley and Sons, 1987); E. Willard Miller, *Physical Geography: Earth Systems and Human Interactions* (Columbus, Ohio: Charles E. Merrill Publishing Co., 1985); and Edgar W. Spencer, *Physical Geology* (Reading, Mass.: Addison-Wesley Publishing Co. Two useful books that discuss volcanic activity in relation to plate tectonics are Tjeerd H. Van Andel, *New Views on an Old Planet: Continental Drift and the History of Earth* (Cambridge: Cambridge University Press, 1985), and S. Uyeda, *The New View of the Earth* (San Francisco: W. H. Freeman and Co., 1978). More detailed discussions of volcanoes and related phenomena can be found in the following: H. Williams and A. R. McBirney, *Volcanology* (San Francisco: W. H. Freeman and Co., 1979; C. Ollier, *Volcanoes* (Cambridge: Cambridge University Press, 1969); and M. Fred Bullard, *Volcanoes of the Earth* (Austin, Tex.: University of Texas Press, 1976). A discussion of most prominent volcanic landscapes in the United States can be found in William D. Thornbury, *Regional Geomorphology of the United States* (New York: John Wiley and Sons, 1965). A look at the eruption of Mt. St. Helens can be found in Bruce L. Foxworthy and Mary Hill, *Volcanic Eruptions of 1980 at Mt. St. Helens: The First 100 Days* (Washington, D.C.: U.S. Government Printing Office, 1982). A detailed discussion of volcanic activity in Hawaii can be found in Robert Y. Koyanagi et al., *Seismicity of the Lower East Rift Zone of Kilauea Volcano, Hawaii, 1960 to 1980* (Washington, D.C.: U.S. Government Printing Office, 1983). A readable study of Cascade volcanoes is provided by Stephen L. Harris, *Fire and Ice: The Cascade Volcanoes* (Seattle: The Mountaineers, Pacific Search Books, 1976). A detailed discussion of problems associated with forecasting volcanic activity is presented by Haroun Tazieff and Jean-Christophe Sabroux, eds., *Forecasting Volcanic Events*, Developments in Volcanology 1 (Amsterdam: Elsevier, 1983).

W

WATER TABLE. See GROUNDWATER.

WATER VAPOR. See PRECIPITATION.

WEATHERING. 1. The breakdown of rock material at or near the earth's surface due to atmospheric exposure. 2. The chemical decomposition or physical disintegration of earth materials because of exposure to the atmosphere.

There are two general types of weathering: *chemical*, in which ROCKS decompose or break up due to chemical processes, and *mechanical*, or *physical*, in which large rock masses disintegrate into smaller particles by physical forces. The processes involved in weathering are complicated, and according to E. D. Ollier, "At first sight weathering seems a hopelessly complicated subject, with a multitude of processes operating on an endless range of rocks and minerals [see MINERALS] under a great variety of climatic and hydrological conditions" (1975: 2). Although both physical and chemical weathering occur at any particular place simultaneously, physical weathering is more predominant in cold and dry climates (see CLIMATOLOGY), whereas chemical weathering is more predominant in warm humid climates where chemical and biochemical changes occur at much faster rates (Keller, 1978: 852).

Attempts to understand the weathering processes can be traced to early Greek times and the observations of Thales (Fenton and Fenton, 1952: 3). Thales was interested in how rocks were broken into pieces. It was not until the eighteenth century, however, that earth scientists tried in a systematic way to understand the mechanisms associated with weathering. Of particular importance at this time were the studies by Emile Gurttard, Nicolas Desmarest, and John Playfair (Fenton and Fenton, 1952: 30–60; Rappaport, 1969: 272–288).

The nineteenth century saw a great increase in the amount of scientific investigations relative to weathering processes. An impetus to this type of research was the recognition that buildings and monuments made of stone were starting to decay. A wide range of weathering mechanisms relative to this decay were investigated (Henry, 1856; Ansted, 1860; Merrill, 1886) with most of the attention focused on the effects of frost action (Luquer, 1895).

Investigations of salt crystallization damage were also numerous at the end of the nineteenth century. Despite early recognition of this problem (Geikie, 1880; Richmond, 1910) it is really only during the past two decades that recent research has focused on the problem (Price, 1975).

The examination of weathering through exposure trials is another area of research, primarily conducted since the early 1900s. The first well-documented tests were conducted by J. S. Owens (1912) on limestone in and around London. Later work by C. H. Scholer (1928) and J. F. McMahon and C. R. Amberg (1947) used visual inspection and weight loss of materials as measures of the rate of weathering decay.

In 1904 G. P. Merrill wrote a book called *A Treatise on Rocks, Rock-Weathering, and Soils*. This book included both physical and chemical weathering processes and laid the framework for a variety of field studies. Included in these later field studies were investigations into the nature of exfoliation (Blackwelder, 1925), the role of insolation in rock weathering (Blackwelder, 1933), the decomposition of granite (Barton, 1916) and basalt (Palmer, 1927), and the nature of chemical weathering at low temperatures (Williams, 1949).

The literature on weathering has grown rapidly in the past two decades. Much of this research has been of great importance to geographers investigating landscape features as well as those interested in reconstructing past environments (Whalley and McGreevy, 1985). Field studies continue to play an important role (Gardner, 1983; Caine, 1983; Pye and Paine, 1984), particularly in regard to physical weathering processes. New research in chemical weathering has been especially important relative to the role it plays in the formation of SOIL (Bateman and Catt, 1985) and the determination of ages of deposits (Wayne, 1984).

References

Ansted, D. T. 1860. "On the Decay and Preservation of Building Materials." *Journal of the Franklin Institute* 40: 155–163, 217–223. An early study of the effects of weathering on building materials.

Barton D. C. 1916. "Notes on the Disintegration of Granite in Egypt." *Journal of Geology* 24: 382–393.

Bateman, R. M., and Catt, J. 1985. "Modification of Heavy Mineral Assemblages in English Coversands by Acid Pedochemical Weathering." *Catena* 12: 1–21.

Blackwelder, E. 1925. "Exfoliation as a Phase of Rock Weathering." *Journal of Geology* 33: 793–806.

———. 1933. "The Insolation Hypothesis of Rock Weathering." *American Journal of Science* 26: 97–113.

Caine, N. 1983. *The Mountains of Northeastern Tasmania: A Study of Alpine Geomorphology*. Rotterdam: A. A. Balkema.
Fenton, C. L. and Fenton, M. A. 1952. *Giants of Geology*. Garden City, N.Y.: Doubleday and Co.
Gardner, J. S. 1983. "Rockfall Frequency and Distribution in the Highwood Pass Area, Canadian Rocky Mountains." *Zeitschrift fur Geomorphologie* 27: 311–324.
Geikie, A. 1880. "Rock Weathering as Illustrated in Edinburgh Churchyards." In *Proceedings of the Royal Society of Edinburgh* 10: 518–532.
Henry, J. 1856. "On the Mode of Testing Building Materials and an Account of the Marble Used in the Extension of the United States Capitol." *American Journal of Science* 72: 30–38.
Keller, Walter. 1978. "Weathering Processes." In Daniel N. Lapedes, ed., *McGraw-Hill Encyclopedia of the Geological Sciences*. New York: McGraw-Hill Book Co. A good general discussion of the types of weathering processes.
Luquer, L. McI. 1895. "The Relative Effects of Frost and the Sulphate of Soda Efflorescence Tests on Building Stones." *Transactions of the American Society of Civil Engineers* 33: 235–256.
McMahon, J. F., and Amberg, C. R. 1947. "Disintegrating Effect of Repeated Freezing and Thawing on Building Brick." *Journal of the American Ceramic Society* 30: 81–89.
Merrill, G. P. 1886. *The Weathering of Building Stones*. Annual Report of the Smithsonian Institution, Pt. II. Washington, D.C.: Smithsonian Institution, pp. 331–345.
———. 1904. *A Treatise on Rocks, Rock-Weathering and Soils*. New York: Macmillan.
Ollier, C. D. 1975. *Weathering*. London: Longmans, Green and Co. An excellent book on many aspects of weathering, particularly as it relates to landform development.
Owens, J. S. 1912. "Experiments on the Weathering of Portland Stone." *The Surveyor and Municipal and County Engineer*, September 20, pp. 380–382.
Palmer, H. S. 1927. "Lapies in Hawaiian Basalts." *U.S. Geological Survey* 17: 627–631.
Price, C. A. 1975. "The Decay and Preservation of Natural Building Stone." *Chemistry in Britain* 11: 350–353.
Pye, K., and Paine, A.D.M. 1984. "Nature and Source of Aeolian Deposits near the Summit of Ben Arkle, Northwest Scotland." *Geologie en Mijnbouw* 63: 13–18.
Rappaport, Rhoda. 1969. "The Geological Atlas of Guettard, Lavoisier and Monnet." In Cecil J. Schneer, ed., *Toward a History of Geology*. Cambridge: The M.I.T. Press.
Richmond, W. 1910. "Short Life of Stone in London." *Chemical Abstracts* 4: 97.
Scholer, C. H. 1928. "Some Accelerated Freezing and Thawing Tests on Concrete." In *Proceedings of the American Society for Testing and Materials* 28: 472–486.
Wayne, W. J. 1984. "Relative Dating Techniques to Distinguish Late Pleistocene-Holocene Continental Sediments." *Chemical Geology* 44: 337–348.
Whalley, W. B., and McGreevy, J. P. 1985. "Weathering." *Progress in Physical Geography* 9: 559–582. An excellent review of the latest research on weathering.
Williams, J. E. 1949. "Chemical Weathering at Low Temperatures." *Geographical Review* 39: 129–135.

Sources of Additional Information

For short discussions of weathering processes see Stella Stiegeler, ed., *Dictionary of Earth Sciences* (New York: Pica Press, 1976); Robert W. Durrenberger, *Dictionary of the Environmental Sciences* (Palo Alto, Calif.: National Press Books, 1973); and Sir Dudley Stamp, *Longmans Dictionary of Geography* (London: Longmans, Green and Co., 1966). For an excellent book that relates weathering and soil formation to geomorphology see Peter W. Birkeland, *Pedology, Weathering, and Geomorphological Research* (New York: Oxford University Press, 1974). For an easy-to-read book that analyzes slopes and how weathering is related to their development see R. J. Small and M. J. Clark, *Slopes and Weathering* (New York: Cambridge University Press, 1982). For a thorough discussion of research on weathering and building research disciplines see J. P. McGreevy and W. B. Whalley, "Weathering," *Progress in Physical Geography* 8 (1984): 543–569.

WIND. See ATMOSPHERIC.

Z

ZOOGEOGRAPHY. See Biogeography.

Appendix: Outline of Concepts

This appendix is designed to classify the concepts in this book into related groups. Twelve major topics have been chosen (I-XII), with the remainder of the concepts outlined below these major topics. There is considerable overlap when considering where concepts should fit. Some of the concepts are listed in more than one place; however, only the major areas of overlap are handled in this manner.

Physical Geography

I. Biogeography
   A. Biosphere
      1. Biocycle
      2. Biomes
      3. Ecosystem
      4. Life Zone
      5. Vegetation
   B. Phytogeography
   C. Plant Succession/Climax
   D. Zoogeography

*II. Cartography
   A. Latitude/Longitude
      1. Graticule
      2. Great Circle
      3. Meridians
      4. Parallels
   B. Map Projection
      Great Circle
   C. Topographic Map

* Denotes a concept that is not explicitly defined in the text.

III. Climatology
　　A. Evapotranspiration
　　　　1. Actual Evapotranspiration
　　　　2. Potential Evapotranspiration
　　B. Hydrologic Cycle
　　C. Köppen System
　*D. Physical Climatology
　　　　1. Adiabatic Heating and Cooling
　　　　2. Lapse Rate
　　　　3. Latent Heat
　　E. Thornthwaite System

*IV. Earth/Sun Relationships
　　A. Equinox
　*B. Solstice
　　C. Tropic of Cancer
　　D. Tropic of Capricorn

V. Geology
　　A. Geologic Time Scale
　　　　Pleistocene
　　B. Groundwater
　　　　1. Aquifer
　　　　2. Water Table
　　C. Law of Superposition
　　D. Mineral
　　E. Rock
　　　　Petrology
　　F. Tectonics
　　　　1. Earthquake
　　　　　　a. Seismology
　　　　　　b. Tsunami
　　　　2. Faulting
　　　　3. Folding
　　　　4. Geosyncline
　　　　5. Mantle
　　　　6. Plate Tectonics
　　　　　　a. Asthenosphere
　　　　　　b. Continental Drift
　　　　　　c. Lithosphere
　　G. Volcano
　　　　Magma

VI. Geomorphology
　　A. Dune
　　B. Erosion
　　　　1. Erosion Cycle
　　　　2. Geographic Cycle
　*C. Fluvial Geomorphology

OUTLINE OF CONCEPTS

        1. Alluvium
           Alluvial Fan
        2. Base Level
        3. Delta
        4. Drainage Basin
        5. Flood Plain
        6. Horton Analysis
        7. Stream
           River
  *D. Glacial Geomorphology
        1. Continental Glacier
        2. Pleistocene
        3. Valley Glacier
           Cirque
  E. Karst
  F. Landform
  G. Mass Wasting
        1. Landslide
        2. Solifluction
  H. Periglacial
        1. Permafrost
        2. Solifluction
  I. Weathering
        1. Chemical Weathering
        2. Physical Weathering

*VII. Human/Nature Interaction
  A. Environmental Perception
  B. Natural Hazard
  C. Natural Resource
  D. Pollution
      Heat Island

*VIII. Meteorology
  A. Air Mass
      Frontal System
  *B. Atmosphere
        1. Ionosphere
        2. Ozone
        3. Stratopause
        4. Stratosphere
        5. Tropopause
        6. Troposphere
  C. Atmospheric Circulation
        1. Adiabatic Heating and Cooling
        2. Anticyclone
        3. Convection
        4. Hadley Cell
        5. Coriolis Effect

6. Cyclone
   a. Tornado
   b. Tropical Cyclone
      Hurricane/Typhoon
7. Foehn
   Chinook
8. Horse Latitudes
9. Intertropical Convergence Zone
10. Jet Stream
11. Land and Sea Breezes
    Sea Breeze
12. Wind

D. Atmospheric Pressure
  1. Adiabatic Heating and Cooling
  2. Anticyclone
  3. Barometer
  4. Convection
     Hadley Cell
  5. Coriolis Effect
  6. Cyclone
     a. Tornado
     b. Tropical Cyclone
        Hurricane/Typhoon
  7. Intertropical Convergence Zone
  8. Isobar
  9. Millibar

E. Evapotranspiration
  1. Actual Evapotranspiration
  2. Evaporation
  3. Potential Evapotranspiration

F. Heat Island

G. Water Vapor
  1. Cloud
     Fog
  2. Condensation
     a. Condensation Nuclei
     b. Hygroscopic
  3. Evaporation
  4. Humidity
     Dew Point
  5. Lapse Rate
  6. Precipitation
     a. Hail
     b. Orographic Precipitation
     c. Rain Shadow
     d. Thunderstorm

IX. Ocean

OUTLINE OF CONCEPTS

    A. Beach
    B. Tide

X. Remote Sensing
    A. Electromagnetic Radiation
        Electromagnetic Spectrum
            a. Infrared
            b. Ultraviolet
    B. LANDSAT

XI. Soil
    A. Humus
    B. Leaching
    C. Pedology
    D. Soil Classification
    E. Soil Profile
        Soil Horizon

XII. Time
        International Date Line

Index

Actual evapotranspiration, 3
Adiabatic cooling, 3–4, 44, 147
Adiabatic heating, 3–4, 44, 147
Adiabatic lapse rate, 86, 147; dry, 147; wet, 147
Africa, 86
Agassiz, L., 208, 261–62
Air mass, 4–7, 22, 86; classification, 5–6; equatorial, 5; frontal system, 88; polar, 5; regional, 4; tropical, 5–6
Air temperature, 19
Alaska, 193–94; Prudhoe Bay, 194
Alluvial fan, 7–9, 10, 97, 238
Alluvium, 7, 9–11, 93
Alps, 86, 106
Antarctica, 47
Anticyclone, 11, 19
Anvil cloud, 240
Aquifer, 11–14
Arabs, 101
Arctic, 75
Asthenosphere, 14, 157
Atlantic Ocean. *See* Ocean
Atmosphere, 147; ionosphere, 128; stratosphere, 131, 185, 232–33, 255, 257; troposphere, 131, 147, 185, 232, 254–55
Atmospheric circulation, 14–18, 22, 50, 75, 240; hydrologic cycle, 118; intertropical convergence zone, 127; latent heat, 150
Atmospheric pressure, 18–20, 21, 214; land and sea breezes, 143
Australia, 76

Bagnold, R., 63
Barograph, 22
Barometer, 18, 21–23; aneroid, 22; siphon, 22; three-liquid, 22; two-liquid, 22
Base level, 23–25, 76, 102; local, 24; temporary, 24; ultimate, 24
Beach, 25–28
Beach drift, 26
Bergeron, T., 5
Bergschrund theory, 38
Berkeley School of Cultural Geography, 70
Biblical flood, 97
Biocycles, 28, 34
Biogeography, 28–33, 70, 181, 266
Biomes, 33, 34
Biosphere, 33–35, 213
Bjerknes, J., 88
Bjerknes, V., 88
Boyle, R., 21
Bronze Age, 170
Budyko, M. I., 114

Cape Horn, 75
Carbon/nitrogen ratio, 117
Carson, R., 211
Cartography, 217
Cation-exchange capacity, 117
Chapman ozone cycle, 186
Chemical weathering, 37. See also Weathering
China, 10
Chinook, 37. See also Foehn
Chlorofluorocarbons, 186, 257–58. See also Ozone
Circum-Pacific Ring of Fire, 271
Cirque, 37–40. See also Valley glacier
Civil War, 217
Clements, F. E., 199–200
Climate classification, 41, 80; Köppen, 139–40
Climatic zones, 69
Climatology, 29, 33, 40–44, 79, 181, 198, 273; applied, 42; dynamic, 42; heat island, 114; latent heat, 150; precipitation, 215; synoptic, 42
Climax-pattern theory, 201
Clock, 153
Cloud, 44–46, 254; types, 45, 50
Coalescence theory, 215
Coastal zone, 26
Cold-cloud theory, 215
Cold glacier, 262
Colorado River, 102
Condensation, 46, 148
Condensation nuclei, 45, 46, 254
Conduction, 50
Continental drift, 27, 31, 158, 203–5, 238
Continental glacier, 37, 46–49, 102, 208, 263
Contours, 247
Convection, 49–51. See also Orographic precipitation, Thunderstorm
Cook, Captain J., 182
Coriolis effect, 22, 51–53
Cosmic rays, 72–73
Crystallography, 171
Cumulonimbus, 239
Current, 181

Cyclone, 19, 53. See also Atmospheric pressure, Tropical cyclone

Dam, 27
Dana, J. D., 171
Dana System, 171
Darcy, H., 12, 109
Darwin, C., 30, 94, 98, 182, 266
Daubenmire, R. F., 199–200
Davis, W. M., 8, 24, 59, 76, 145, 198; geographical cycle, 102; karst, 136; periglacial, 190; stream, 234
DDT, 212
Delta, 10, 55–58
Deposition, 76, 83, 97; flood plain, 84; water, 55–58; wind, 63
Descartes, R., 21, 22
Desert, 62
Dew point, 50, 59, 148
Diluvial theory, 208
Dokuchaiev, V. V., 226
Doldrums, 127; See also Intertropical convergence zone
Dolomite, 135
Double-sweep process, 215
Drainage basin, 59–61, 76, 83; classification, 59
Dune, 61–66; coastal, 62; classification, 62–63; desert, 62
Dynamic classification scheme (glacier), 262
Dynamic equilibrium, 59

Earth: core, 67; interior, 67
Earth crust, 157
Earthquake, 67–69, 97, 196, 205, 237; classification, 68; lithosphere, 157; prediction, 68; tsunami, 255; wave, 68
Ecology, 64. See also Ecosystem
Ecosphere, 33. See also Biosphere
Ecosystem, 30, 64, 69–72, 266; pollution, 211; urban, 70
Edaphic concept, 225–26
Egypt, 108, 247
Electromagnetic radiation, 72–74, 123, 185, 257
Electromagnetic spectrum, 74, 217, 257
Electron, 73

Engineering concept (soil), 226
Environmental perception, 74
Environmental Protection Agency (EPA), 109
Environmental systems, 70
Epeirogenesis, 237
Equator, 74
Equatorial zone, 15. *See also* Intertropical convergence zone
Equinox, 74–75; autumnal, 74; vernal, 74
Erosion, 34, 75–79, 97, 101, 196; accelerated, 77; beach, 243; ice, 38; periglacial, 190; stream, 23; wind, 14
Erosion cycle, 76, 79, 234. *See also* Geographic cycle
Europe, 62
Evaporation, 79, 148. *See also* Evapotranspiration
Evapotranspiration, 42, 79–82; actual, 79–80; potential, 79–80, 214; potential evapotranspiration calculated, 80
Eye of the storm, 252

Facilitation, 200
Faulting, 83, 255
Faunal region, 29
Flint, R. F., 47
Flood: Rapid City, 84. *See also* Flood plain
Flood plain, 10, 83–85; aggradation, 84; classification, 84; degradation, 84; hazard, 84
Fluvial process, 97
Foehn, 22, 86–88; psychological and physical maladies, 87; thermodynamic theory, 86
Fog, 88; advection, 45; radiation, 45
Folding, 88
Fossil, 222
France, 10, 12, 62, 76, 88
Frequency, 73
Frontal system, 5, 22, 88–91
Frontogenesis, 89
Funnel Cloud, 250. *See also* Tornado

Galileo, 67
Gamma rays, 72–73
Geiger, R., 42, 140

General systems theory, 70
Geochemical process, 34
Geographic concept (soil), 226
Geographic cycle, 79, 93, 102
Geologic time scale, 93–96, 99; epoch, 95; era, 94–95; period, 95
Geology, 76, 94, 96–100, 101, 145; geosyncline, 105; mineral, 170; remote sensing, 217; soil, 226, 229
Geomorphic process, 14; karst, 136–37; mass wasting, 167–68; periglacial, 189–90
Geomorphology, 24, 48, 57, 76, 100–105, 145, 198; coastal, 243; karst, 136; process oriented, 101–2; tectonics, 238
Geosyncline, 105–7
Germany, 76
Gilbert, G. K., 8, 55, 59, 62, 76, 83; geomorphology, 102; stream, 234
Glacial classification, 262–63
Glacier, 37. *See also* Continental glacier, Valley glacier
Global circulation, 51
Gradational analysis, 201
Graded stream, 24
Graticule, 107, 165
Great Britain, 76
Great circle, 107, 152
Greeks, 40, 196; delta, 55; earthquake, 67; geology, 97, 161, 170, 221, 273; groundwater, 11, 108; hydrologic cycle, 79; landforms, 101, 135, 144–45; latitude/longitude, 152; ocean, 181; stream, 233; topographical map, 247; water, 118
Greenland, 47
Greenwich mean time, 246
Greenwich meridian, 152
Gregorian calender, 245
Groundwater, 11, 107–11, 119; karst, 135; soil leaching, 155
Gulf Stream, 182

Hachure, 248
Hadley, G., 14–16; cell, 15, 113, 132; Coriolis, 41, 51
Hail, 113, 240

Heat, 123
Heat balance, 50
Heat Capacity Mapping Mission, 218
Heat island, 113–16
Henry Mountains, 76, 102
Henry the Navigator, 152, 182, 197
Hooke, R., 22, 67
Horse latitudes, 16, 116
Horton, R. E., 59, 234; analysis, 116
Human ecology, 70
von Humboldt, A., 29, 41, 98, 162, 203, 266; Kosmos, 197; rock, 221
Humidity, 116, 214–15; absolute, 214; relative, 214; specific, 214
Humus, 116–18, 228
Hurricane, 118. See also Tropical cyclone
Hutton, J., 76, 98, 102, 221, 261
Hydraulics, 12
Hydrologic cycle, 12, 79, 108, 118–21
Hygroscopic, 121

Ice: cold, 38; isothermic, 38
Ice Age. See Pleistocene
Ice sheet, 46
India, 10
Infrared, 64, 72–73, 123–25, 217; thermal, 217
International Date Line, 126
International Meridian Congress (1884), 126
Intertropical convergence zone (ITCZ), 127–28, 252
Ionosphere, 128
Iron Age, 170
Isobar, 19, 129
Isobath, 248
Isoline, 248
Isostacy, 76
Isotherm, 19

Jet stream, 131–33; polar-front jet, 132; subtropical jet, 132; tornado, 251

Kanat, 108
Kant, E., 197
Karst, 76, 135–39; Jamaica, 136
Kelvin temperature, 123

King, C. A. M., 192–93
Köppen, W., 30, 42, 266; Köppen system, 139–42
Krakatoa, 30

Lake Bonneville, 55
Land and sea breeze, 14, 143–44
Landform, 14, 37, 46, 47, 69, 77, 83, 144–47; classification, 145; geomorphology, 101; glacial, 47, 261; karst, 136; mass wasting, 168; maturity, 76; old age, 76; volcanic, 269–70; youth, 76
LANDSAT, 124, 147, 166, 218
Landscape development, 59
Landslide, 147
Lapse rate, 4, 44, 86, 147–49, 184; dew point, 148; negative, 148; superadiabatic, 148; troposphere, 147
Laser, 124
Latent heat, 127, 149–52; fusion, 149; sublimation, 149; vaporization, 149–50
Latitude/longitude, 40, 74, 139, 152–54, 190; map projection, 165; time, 245
Law of superposition, 93, 98, 154
Leaching, 154–56
Leeward, 184
Lee wave, 87
Leopold, L. B., 83, 234
Life zone, 33, 157
Light: speed of, 75; visible, 72–73. See also Electromagnetic radiation
Limestone, 135
Linnaeus, C., 29, 33
Lithosphere, 157–59, 226
Longshore drift, 26
Lyell, C., 26, 94, 98, 102, 145; karst, 135; Pleistocene, 208

Magma, 161–64, 269
Magnetic anomalies, 205
Magnetic field, 158
Mantle, 157, 164
Map projection, 164–67, 196
Marbut, C. F., 226
Marsh, G. P., 69
Mass wasting, 167–69, 189; classification, 168

Maxwell, J. C., 72
Meltwater theory, 38
Mercator, G., 165
Meridian, 126, 169. *See also* Latitude/longitude
Merriam, C. H., 69
Microwave, 72–73. *See also* Electromagnetic radiation
Mid-ocean ridge, 205
Millibar, 19, 132, 169
Mineral, 94, 97, 157, 169–73, 220; soil, 116, 154; weathering, 273
Mineralogy, 97
Mississippi River, delta, 56, 57
Model, 106; drainage basin, 59–60; evapotranspiration, 80; floristic composition, 200; frontal system, 89; groundwater, 109; heat island, 114; quantitative, 12; tolerance, 200; wind, 52
Monsoon, 132
Morisawa, M., 233
Morphologic scheme (glacier), 262
Moslem, 197
Mountain building, geosyncline, 105. *See also* Tectonics
Mount Mazama (Crater Lake), 270–71
Mount St. Helens, 163
Multispectral scanner (MSS), 124, 128

NASA, 124
National Flood Insurance Program, 84
National Science Foundation, 176
Natural Hazard, 68, 84, 175–77, 209, 255–56
Natural history, 29, 33
Natural resource, 62, 157, 176, 178–80, 209; pollution, 212
Neptunists, 98, 161, 221, 269
Newton, Sir I., 242
New Zealand, 76
Nile River, 55
Nivation, 189

Ocean, 27, 29, 172, 181–83, 193; Atlantic, 62; floor, 161, 203; Pacific, 62
Ocean front, 89
Oceanography, 182

Oil slick, 258
Organic matter. *See* Humus
Orogenic movement, 237
Orographic, 44
Orographic precipitation, 183–85
Oscillating movement, 237
Ozone, 185–88, 232, 257–58

Pacific Ocean. *See* Ocean
Parallel, 189. *See also* Latitude/longitude
Pediplain, 77
Pedologic concept, 226
Pedology, 189
Peltier, L. C., 77, 190
Penck, A., 77, 136
Penck, W., 77, 103
Peneplain, 24, 76
Periglacial, 189–92, 193
Permafrost, 190–91, 192–95
Permeability, 11
Persia, 108
Petrology, 196, 221–22
Péwé, T. L., 190
Photochemical reaction, 185
Photography, 217
Physical geography, 59, 76, 101, 181, 196–99
Physiographic analysis. *See* Geomorphology
Phytogeography, 29, 199
Planck, M., 73
Plant geography, 266
Plant succession/climax, 30, 69, 199–202, 266
Plate tectonics, 27, 31, 68, 99, 181, 202–7, 238; geosyncline, 106; lithosphere, 157; magma, 162; rock, 223
Playfair, J., 59, 102, 145, 234, 261, 273
Pleistocene, 37, 46, 47, 48, 189, 207–10, 263
Plutonists, 98, 162, 221, 269
Polar easterlies, 16
Polar front Zone, 15, 16
Pollution, 12, 119, 176, 210–14; ozone, 185–86
Porosity, 11
Powell, J. W., 23, 62, 234
Precipitation, 40, 44, 47, 50, 79, 190, 214–16; foehn, 86; groundwater, 108;

hydrologic cycle, 119; latent heat, 150; orographic, 183–84; troposphere, 254
Pressure gradient, 14
Prime meridian. *See* Greenwich meridian
Ptolemy, C., 152, 165, 196

Quantitative methods in geomorphology, 103
Quantum behavior, 73

Radiation, 50. *See also* Electromagnetic radiation
Radioactivity, 95
Radiometer, 124. *See also* Remote sensing
Radiowave, 72–73, 128. *See also* Remote sensing
Railroads, 62
Rain shadow, 217
Regolith, 226
Remote sensing, 73, 103, 198, 217–20, 238; dunes, 64; infrared, 123; maps, 249; ultraviolet, 257–58
Renewable resource, 178. *See also* Natural resource
Return beam vidicon, 218
Rhumb line, 165
Richter, C. F., 68
River. *See* Stream
Rock, 11, 93, 97, 101, 157, 168, 220–24; igneous, 161–62, 220; magma, 161, 220; metamorphic, 162, 220; sedimentary, 220; soil, 226; weathering, 273
Romans, 11–12, 101, 161, 196; minerals, 170; ocean, 181, pollution, 211; rock, 221; topographical map, 247; volcano, 268
Rossby, C. G., 15; wave, 132
Rotational slip, 38

Sand, 61–63
Satellite imagery, 64. *See also* Multispectral scanner, SPOT, Thematic mapper
Sauer, C., 70, 178
Scanner, 124
Schumm, S. A., 60, 235

Sea breeze, 225. *See also* Land, Sea breeze
Sea-floor spreading, 27, 205. *See also* Plate tectonics
SEASAT, 218
Sedimentation, 10, 77, 106
Seismic waves, 158. *See also* Earthquake
Seismology, 67, 225
Seventh approximation, 229
Sewer, 211
Siberia, 75
Side looking airborne radar, 218; synthetic aperture radar, 218
Skin cancer, 257–58
Skylab, 218
Smithsonian Institute, 41
Soil, 116, 168, 200, 225–28: alluvial, 10; weathering, 274
Soil classification, 226, 228–31
Soil-forming factors, 154
Soil Horizon, 154, 229, 231
Soil morphology, 229
Soil profile, 229, 231
Soil structure, 117
Solifluction, 189, 231
Southern oscillation, 6
Spatial analysis, 70
Spectrometer, 124, 258
Spherical coordinates, 164
SPOT, 219
Spring, 12
Standard time zones, 246
Storm surge, 242
Strahler, A. N., 42, 60, 74, 145, 234
Stratopause, 232
Stratosphere. *See* Atmosphere
Stream, 10, 12, 56, 59, 60, 83, 97, 196, 233–36; erosion, 76; geomorphology, 101; hydraulic geometry, 60; hydrologic cycle, 118–19; landform, 145; pollution, 211–12
Stream channel, 59
Stream grade, 76
Subglacial, 189
Sublimation, 79. *See also* Latent heat
Subsidence, 105–6
Subtropics, 15, 16

Sun-spot theory, 209
Supersaturated air, 45

Tectonics, 237–39
Temperate glacier, 262
Thematic Mapper (TM), 124, 218
Thermal classification scheme (glacier), 262
Thermometer: Galileo, 40
Thornthwaite, C. W., 42, 80; Köppen, 140; Thornthwaite system, 239
Thunderstorm, 45, 50, 239–41, 250; latent heat, 150
Tidal power, 242
Tide, 241–44; spring, 241; neap, 241
Till, fabric, 47
Time, 93, 244–47
Topographic map, 56, 247–50
Topography, 200
Tornado, 250–52
Trade wind, 16, 252
Transpiration, 79. *See also* Evapotranspiration
Tropical, 5; latitudes, 75
Tropical cyclone, 19, 150, 242, 252–54
Tropic of Cancer, 74–75, 252
Tropic of Capricorn, 74–75, 252
Tropopause, 231, 254
Troposphere. *See* Atmosphere
Tsunami, 255–56
Typhoon. *See* Tropical cyclone

Ultraviolet, 72–73, 185, 257–60
Uniformitarian concept, 98
United Arab Republic, 10, 11
United soil classification system, 229
United Soviet Socialist Republic, 193–94
United States, 10, 19
Uplift, 76. *See also* Tectonics

U.S. Army Corps of Engineers, 84
U.S. Geological Survey, 63, 99, 119
U.S. Weather Service, 22; Bureau, 80

Valley glacier, 37, 46, 102, 208, 261–65
Vegetation, 29, 33, 63, 69, 200, 265–68
Volcano, 97, 157, 161, 205, 237, 268–72; classification, 269–70

Warm-cloud theory, 215
Washburn, A. L., 189–90, 194
Washington, D.C., 114
Water: connate, 108; juvenile, 108; meteoric, 108; vapor, 214. *See also* Precipitation
Water balance, 15. *See also* Evapotranspiration
Water table, 108, 273
Water vapor, 273. *See also* Latent heat, Evapotranspiration
Wave, 181, 196
Wave incidence, 26
Wavelength, 73
Weathering, 76, 155, 199, 225, 273–76
Wegener, A., 158, 204, 237
Westerlies, 16
White, G., 175
Willy-willy. *See* Tropical cyclone
Wind, 14–18, 41, 52, 276. *See also* Atmospheric circulation
Windward, 184
Wischmeier, W. H., 77
Wolman, M. G., 83, 234

X-ray, 72–73, 99

Yugoslavia, 136

Zoogeography, 29, 277. *See also* Biogeography

About the Authors

THOMAS P. HUBER is a member of the Department of Geography at the University of Colorado, Colorado Springs.

ROBERT P. LARKIN is Professor of Geography and Environmental Studies at the University of Colorado, Colorado Springs.

GARY L. PETERS is Professor of Geography at California State University, Long Beach.